# Lecture Notes on Data Engineering and Communications Technologies

**199**

Series Editor

Fatos Xhafa, *Technical University of Catalonia, Barcelona, Spain*

The aim of the book series is to present cutting edge engineering approaches to data technologies and communications. It will publish latest advances on the engineering task of building and deploying distributed, scalable and reliable data infrastructures and communication systems.

The series will have a prominent applied focus on data technologies and communications with aim to promote the bridging from fundamental research on data science and networking to data engineering and communications that lead to industry products, business knowledge and standardisation.

Indexed by SCOPUS, INSPEC, EI Compendex.

All books published in the series are submitted for consideration in Web of Science.

Leonard Barolli
Editor

# Advanced Information Networking and Applications

Proceedings of the 38th International Conference on Advanced Information Networking and Applications (AINA-2024), Volume 1

 Springer

*Editor*
Leonard Barolli
Department of Information and Communication
Engineering
Fukuoka Institute of Technology
Fukuoka, Japan

ISSN 2367-4512 ISSN 2367-4520 (electronic)
Lecture Notes on Data Engineering and Communications Technologies
ISBN 978-3-031-57839-7 ISBN 978-3-031-57840-3 (eBook)
https://doi.org/10.1007/978-3-031-57840-3

This Springer imprint is published by the registered company Springer Nature Switzerland AG
The registered company address is: Gewerbestrasse 11, 6330 Cham, Switzerland

Paper in this product is recyclable.

# Welcome Message from AINA-2024 Organizers

Welcome to the 38th International Conference on Advanced Information Networking and Applications (AINA-2024). On behalf of AINA-2024 Organizing Committee, we would like to express to all participants our cordial welcome and high respect.

AINA is an International Forum, where scientists and researchers from academia and industry working in various scientific and technical areas of networking and distributed computing systems can demonstrate new ideas and solutions in distributed computing systems. AINA is a very open society and is always welcoming international volunteers from any country and any area in the world.

AINA International Conference is a forum for sharing ideas and research work in the emerging areas of information networking and their applications. The area of advanced networking has grown very rapidly and the applications have experienced an explosive growth, especially in the area of pervasive and mobile applications, wireless sensor and ad-hoc networks, vehicular networks, multimedia computing, social networking, semantic collaborative systems, as well as IoT, big data, cloud computing, artificial intelligence, and machine learning. This advanced networking revolution is transforming the way people live, work, and interact with each other and is impacting the way business, education, entertainment, and health care are operating. The papers included in the proceedings cover theory, design and application of computer networks, distributed computing, and information systems.

Each year AINA receives a lot of paper submissions from all around the world. It has maintained high-quality accepted papers and is aspiring to be one of the main international conferences on the information networking in the world.

We are very proud and honored to have two distinguished keynote talks by Prof. Fatos Xhafa, Technical University of Catalonia, Spain, and Dr. Juggapong Natwichai, Chiang Mai University, Thailand, who will present their recent work and will give new insights and ideas to the conference participants.

An international conference of this size requires the support and help of many people. A lot of people have helped and worked hard to produce a successful AINA-2024 technical program and conference proceedings. First, we would like to thank all authors for submitting their papers. We are indebted to Program Track Co-chairs, Program Committee Members and Reviewers, who carried out the most difficult work of carefully evaluating the submitted papers.

We would like to thank AINA-2024 General Co-chairs, PC Co-chairs, Workshops Organizers for their great efforts to make AINA-2024 a very successful event. We have special thanks to the Finance Chair and Web Administrator Co-chairs.

We do hope that you will enjoy the conference proceedings and readings.

# AINA-2024 Organizing Committee

## Honorary Chair

Makoto Takizawa — Hosei University, Japan

## General Co-chairs

Minoru Uehara — Toyo University, Japan
Euripides G. M. Petrakis — Technical University of Crete (TUC), Greece
Isaac Woungang — Toronto Metropolitan University, Canada

## Program Committee Co-chairs

Tomoya Enokido — Rissho University, Japan
Mario A. R. Dantas — Federal University of Juiz de Fora, Brazil
Leonardo Mostarda — University of Perugia, Italy

## International Journals Special Issues Co-chairs

Fatos Xhafa — Technical University of Catalonia, Spain
David Taniar — Monash University, Australia
Farookh Hussain — University of Technology Sydney, Australia

## Award Co-chairs

Arjan Durresi — Indiana University Purdue University in Indianapolis (IUPUI), USA
Fang-Yie Leu — Tunghai University, Taiwan
Marek Ogiela — AGH University of Science and Technology, Poland
Kin Fun Li — University of Victoria, Canada

## Publicity Co-chairs

| | |
|---|---|
| Markus Aleksy | ABB Corporate Research Center, Germany |
| Flora Amato | University of Naples "Federico II", Italy |
| Lidia Ogiela | AGH University of Science and Technology, Poland |
| Hsing-Chung Chen | Asia University, Taiwan |

## International Liaison Co-chairs

| | |
|---|---|
| Wenny Rahayu | La Trobe University, Australia |
| Nadeem Javaid | COMSATS University Islamabad, Pakistan |
| Beniamino Di Martino | University of Campania "Luigi Vanvitelli", Italy |

## Local Arrangement Co-chairs

| | |
|---|---|
| Keita Matsuo | Fukuoka Institute of Technology, Japan |
| Tomoyuki Ishida | Fukuoka Institute of Technology, Japan |

## Finance Chair

| | |
|---|---|
| Makoto Ikeda | Fukuoka Institute of Technology, Japan |

## Web Co-chairs

| | |
|---|---|
| Phudit Ampririt | Fukuoka Institute of Technology, Japan |
| Ermioni Qafzezi | Fukuoka Institute of Technology, Japan |
| Shunya Higashi | Fukuoka Institute of Technology, Japan |

## Steering Committee Chair

| | |
|---|---|
| Leonard Barolli | Fukuoka Institute of Technology, Japan |

# Tracks Co-chairs and Program Committee Members

## 1. Network Architectures, Protocols and Algorithms

### Track Co-chairs

| | |
|---|---|
| Spyropoulos Thrasyvoulos | Technical University of Crete (TUC), Greece |
| Shigetomo Kimura | University of Tsukuba, Japan |
| Darshika Perera | University of Colorado at Colorado Springs, USA |

### TPC Members

| | |
|---|---|
| Thomas Dreibholz | Simula Metropolitan Center for Digital Engineering, Norway |
| Angelos Antonopoulos | Nearby Computing SL, Spain |
| Hatim Chergui | i2CAT Foundation, Spain |
| Bhed Bahadur Bista | Iwate Prefectural University, Japan |
| Chotipat Pornavalai | King Mongkut's Institute of Technology Ladkrabang, Thailand |
| Kenichi Matsui | NTT Network Innovation Center, Japan |
| Sho Tsugawa | University of Tsukuba, Japan |
| Satoshi Ohzahata | University of Electro-Communications, Japan |
| Haytham El Miligi | Thompson Rivers University, Canada |
| Watheq El-Kharashi | Ain Shams University, Egypt |
| Ehsan Atoofian | Lakehead University, Canada |
| Fayez Gebali | University of Victoria, Canada |
| Kin Fun Li | University of Victoria, Canada |
| Luis Blanco | CTTC, Spain |

## 2. Next Generation Mobile and Wireless Networks

### Track Co-chairs

| | |
|---|---|
| Purav Shah | School of Science and Technology, Middlesex University, UK |
| Enver Ever | Middle East Technical University, Northern Cyprus |
| Evjola Spaho | Polytechnic University of Tirana, Albania |

## TPC Members

| | |
|---|---|
| Burak Kizilkaya | Glasgow University, UK |
| Muhammad Toaha | Middle East Technical University, Turkey |
| Ramona Trestian | Middlesex University, UK |
| Andrea Marotta | University of L'Aquila, Italy |
| Adnan Yazici | Nazarbayev University, Kazakhstan |
| Orhan Gemikonakli | Final International University, Cyprus |
| Hrishikesh Venkataraman | Indian Institute of Information Technology, Sri City, India |
| Zhengjia Xu | Cranfield University, UK |
| Mohsen Hejazi | University of Kashan, Iran |
| Sabyasachi Mukhopadhyay | IIT Kharagpur, India |
| Ali Khoshkholghi | Middlesex University, UK |
| Admir Barolli | Aleksander Moisiu University of Durres, Albania |
| Makoto Ikeda | Fukuoka Institute of Technology, Japan |
| Yi Liu | Oita National College of Technology, Japan |
| Testuya Oda | Okayama University of Science, Japan |
| Ermioni Qafzezi | Fukuoka Institute of Technology, Japan |

# 3. Multimedia Networking and Applications

### Track Co-chairs

| | |
|---|---|
| Markus Aleksy | ABB Corporate Research Center, Germany |
| Francesco Orciuoli | University of Salerno, Italy |
| Tomoyuki Ishida | Fukuoka Institute of Technology, Japan |

### TPC Members

| | |
|---|---|
| Hadil Abukwaik | ABB Corporate Research Center, Germany |
| Thomas Preuss | Brandenburg University of Applied Sciences, Germany |
| Peter M. Rost | Karlsruhe Institute of Technology (KIT), Germany |
| Lukasz Wisniewski | inIT, Germany |
| Angelo Gaeta | University of Salerno, Italy |
| Angela Peduto | University of Salerno, Italy |
| Antonella Pascuzzo | University of Salerno, Italy |
| Roberto Abbruzzese | University of Salerno, Italy |
| Tetsuro Ogi | Keio University, Japan |

Yasuo Ebara                          Osaka Electro-Communication University, Japan
Hideo Miyachi                        Tokyo City University, Japan
Kaoru Sugita                         Fukuoka Institute of Technology, Japan

## 4. Pervasive and Ubiquitous Computing

**Track Co-chairs**

Vamsi Paruchuri                      University of Central Arkansas, USA
Hsing-Chung Chen                     Asia University, Taiwan
Shinji Sakamoto                      Kanazawa Institute of Technology, Japan

**TPC Members**

Sriram Chellappan                    University of South Florida, USA
Yu Sun                               University of Central Arkansas, USA
Qiang Duan                           Penn State University, USA
Han-Chieh Wei                        Dallas Baptist University, USA
Ahmad Alsharif                       University of Alabama, USA
Vijayasarathi Balasubramanian        Microsoft, USA
Shyi-Shiun Kuo                       Nan Kai University of Technology, Taiwan
Karisma Trinanda Putra               Universitas Muhammadiyah Yogyakarta,
                                       Indonesia
Cahya Damarjati                      Universitas Muhammadiyah Yogyakarta,
                                       Indonesia
Agung Mulyo Widodo                   Universitas Esa Unggul Jakarta, Indonesia
Bambang Irawan                       Universitas Esa Unggul Jakarta, Indonesia
Eko Prasetyo                         Universitas Muhammadiyah Yogyakarta,
                                       Indonesia
Sunardi S. T.                        Universitas Muhammadiyah Yogyakarta,
                                       Indonesia
Andika Wisnujati                     Universitas Muhammadiyah Yogyakarta,
                                       Indonesia
Makoto Ikeda                         Fukuoka Institute of Technology, Japan
Tetsuya Oda                          Okayama University of Science, Japan
Evjola Spaho                         Polytechnic University of Tirana, Albania
Tetsuya Shigeyasu                    Hiroshima Prefectural University, Japan
Keita Matsuo                         Fukuoka Institute of Technology, Japan
Admir Barolli                        Aleksander Moisiu University of Durres, Albania

## 5. Web-Based Systems and Content Distribution

**Track Co-chairs**

| | |
|---|---|
| Chrisa Tsinaraki | Technical University of Crete (TUC), Greece |
| Yusuke Gotoh | Okayama University, Japan |
| Santi Caballe | Open University of Catalonia, Spain |

**TPC Members**

| | |
|---|---|
| Nikos Bikakis | Hellenic Mediterranean University, Greece |
| Ioannis Stavrakantonakis | Ververica GmbH, Germany |
| Sven Schade | European Commission, Joint Research Center, Italy |
| Christos Papatheodorou | National and Kapodistrian University of Athens, Greece |
| Sarantos Kapidakis | University of West Attica, Greece |
| Manato Fujimoto | Osaka Metropolitan University, Japan |
| Kiki Adhinugraha | La Trobe University, Australia |
| Tomoki Yoshihisa | Shiga University, Japan |
| Jordi Conesa | Open University of Catalonia, Spain |
| Thanasis Daradoumis | Open University of Catalonia, Spain |
| Nicola Capuano | University of Basilicata, Italy |
| Victor Ströele | Federal University of Juiz de Fora, Brazil |

## 6. Distributed Ledger Technologies and Distributed-Parallel Computing

**Track Co-chairs**

| | |
|---|---|
| Alfredo Navarra | University of Perugia, Italy |
| Naohiro Hayashibara | Kyoto Sangyo University, Japan |

**TPC Members**

| | |
|---|---|
| Serafino Cicerone | University of L'Aquila, Italy |
| Ralf Klasing | LaBRI Bordeaux, France |
| Giuseppe Prencipe | University of Pisa, Italy |
| Roberto Tonelli | University of Cagliari, Italy |
| Farhan Ullah | Northwestern Polytechnical University, China |

| | |
|---|---|
| Leonardo Mostarda | University of Perugia, Italy |
| Qiong Huang | South China Agricultural University, China |
| Tomoya Enokido | Rissho University, Japan |
| Minoru Uehara | Toyo University, Japan |
| Lucian Prodan | Polytechnic University of Timisoara, Romania |
| Md. Abdur Razzaque | University of Dhaka, Bangladesh |

## 7. Data Mining, Big Data Analytics and Social Networks

### Track Co-chairs

| | |
|---|---|
| Pavel Krömer | Technical University of Ostrava, Czech Republic |
| Alex Thomo | University of Victoria, Canada |
| Eric Pardede | La Trobe University, Australia |

### TPC Members

| | |
|---|---|
| Sebastián Basterrech | Technical University of Denmark, Denmark |
| Tibebe Beshah | University of Addis Ababa, Ethiopia |
| Nashwa El-Bendary | Arab Academy for Science, Egypt |
| Petr Musilek | University of Alberta, Canada |
| Varun Ojha | Newcastle University, UK |
| Alvaro Parres | ITESO, Mexico |
| Nizar Rokbani | ISSAT-University of Sousse, Tunisia |
| Farshid Hajati | Victoria University, Australia |
| Ji Zhang | University of Southern Queensland, Australia |
| Salimur Choudhury | Lakehead University, Canada |
| Carson Leung | University of Manitoba, Canada |
| Syed Mahbub | La Trobe University, Australia |
| Osama Mahdi | Melbourne Institute of Technology, Australia |
| Choiru Zain | La Trobe University, Australia |
| Rajalakshmi Rajasekaran | La Trobe University, Australia |
| Nawfal Ali | Monash University, Australia |

# 8. Internet of Things and Cyber-Physical Systems

**Track Co-chairs**

| | |
|---|---|
| Tomoki Yoshihisa | Shiga University, Japan |
| Winston Seah | Victoria University of Wellington, New Zealand |
| Luciana Pereira Oliveira | Instituto Federal da Paraiba (IFPB), Brazil |

**TPC Members**

| | |
|---|---|
| Akihiro Fujimoto | Wakayama University, Japan |
| Akimitsu Kanzaki | Shimane University, Japan |
| Kazuya Tsukamoto | Kyushu Institute of Technology, Japan |
| Lei Shu | Nanjing Agricultural University, China |
| Naoyuki Morimoto | Mie University, Japan |
| Teruhiro Mizumoto | Chiba Institute of Technology, Japan |
| Tomoya Kawakami | Fukui University, Japan |
| Adrian Pekar | Budapest University of Technology and Economics, Hungary |
| Alvin Valera | Victoria University of Wellington, New Zealand |
| Chidchanok Choksuchat | Prince of Songkla University, Thailand |
| Jyoti Sahni | Victoria University of Wellington, New Zealand |
| Murugaraj Odiathevar | Sungkyunkwan University, South Korea |
| Normalia Samian | Universiti Putra Malaysia, Malaysia |
| Qing Gu | University of Science and Technology Beijing, China |
| Tao Zheng | Beijing Jiaotong University, China |
| Wenbin Pei | Dalian University of Technology, China |
| William Liu | Unitec, New Zealand |
| Wuyungerile Li | Inner Mongolia University, China |
| Peng Huang | Sichuan Agricultural University, PR China |
| Ruan Delgado Gomes | Instituto Federal da Paraiba (IFPB), Brazil |
| Glauco Estacio Goncalves | Universidade Federal do Pará (UFPA), Brazil |
| Eduardo Luzeiro Feitosa | Universidade Federal do Amazonas (UFAM), Brazil |
| Paulo Ribeiro Lins Júnior | Instituto Federal da Paraiba (IFPB), Brazil |

# 9. Intelligent Computing and Machine Learning

**Track Co-chairs**

| | |
|---|---|
| Takahiro Uchiya | Nagoya Institute of Technology, Japan |
| Flavius Frasincar | Erasmus University Rotterdam, The Netherlands |
| Miltos Alamaniotis | University of Texas at San Antonio, USA |

**TPC Members**

| | |
|---|---|
| Kazuto Sasai | Ibaraki University, Japan |
| Shigeru Fujita | Chiba Institute of Technology, Japan |
| Yuki Kaeri | Mejiro University, Japan |
| Jolanta Mizera-Pietraszko | Military University of Land Forces, Poland |
| Ashwin Ittoo | University of Liège, Belgium |
| Marco Brambilla | Politecnico di Milano, Italy |
| Alfredo Cuzzocrea | University of Calabria, Italy |
| Le Minh Nguyen | JAIST, Japan |
| Akiko Aizawa | National Institute of Informatics, Japan |
| Natthawut Kertkeidkachorn | JAIST, Japan |
| Georgios Karagiannis | Durham University, UK |
| Leonidas Akritidis | International Hellenic University, Greece |
| Athanasios Fevgas | University of Thessaly, Greece |
| Yota Tsompanopoulou | University of Thessaly, Greece |
| Yuvaraj Munian | Texas A&M-San Antonio, USA |

# 10. Cloud and Services Computing

**Track Co-chairs**

| | |
|---|---|
| Salvatore Venticinque | University of Campania "Luigi Vanvitelli", Italy |
| Shigenari Nakamura | Tokyo Denki University, Japan |
| Sajal Mukhopadhyay | National Institute of Technology, Durgapur, India |

**TPC Members**

| | |
|---|---|
| Giancarlo Fortino | University of Calabria, Italy |
| Massimiliano Rak | University of Campania "Luigi Vanvitelli", Italy |
| Jason J. Jung | Chung-Ang University, Korea |

| | |
|---|---|
| Dimosthenis Kyriazis | University of Piraeus, Greece |
| Geir Horn | University of Oslo, Norway |
| Dario Branco | University of Campania "Luigi Vanvitelli", Italy |
| Dilawaer Duolikun | Cognizant Technology Solutions, Hungary |
| Naohiro Hayashibara | Kyoto Sangyo University, Japan |
| Tomoya Enokido | Rissho University, Japan |
| Sujoy Saha | NIT Durgapur, India |
| Animesh Dutta | NIT Durgapur, India |
| Pramod Mane | IIM Rohtak, India |
| Nanda Dulal Jana | NIT Durgapur, India |
| Banhi Sanyal | NIT Kurukshetra, India |

# 11. Security, Privacy and Trust Computing

**Track Co-chairs**

| | |
|---|---|
| Ioannidis Sotirios | Technical University of Crete (TUC), Greece |
| Michail Alexiou | Georgia Institute of Technology, USA |
| Hiroaki Kikuchi | Meiji University, Japan |

**TPC Members**

| | |
|---|---|
| George Vasiliadis | Hellenic Mediterranean University, Greece |
| Antreas Dionysiou | University of Cyprus, Cyprus |
| Apostolos Fouranaris | Athena Research Center, Greece |
| Panagiotis Ilia | Technical University of Crete, Greece |
| George Portokalidis | IMDEA, Spain |
| Nikolaos Gkorgkolis | University of Crete, Greece |
| Zeezoo Ryu | Georgia Institute of Technology, USA |
| Muhammad Faraz Karim | Georgia Institute of Technology, USA |
| Yunjie Deng | Georgia Institute of Technology, USA |
| Anna Raymaker | Georgia Institute of Technology, USA |
| Takamichi Saito | Meiji University, Japan |
| Kazumasa Omote | University of Tsukuba, Japan |
| Masakatsu Nishigaki | Shizuoka University, Japan |
| Mamoru Mimura | National Defense Academy of Japan, Japan |
| Chun-I Fan | National Sun Yat-sen University, Taiwan |
| Aida Ben Chehida Douss | National School of Engineers of Tunis, ENIT Tunis, Tunisia |
| Davinder Kaur | IUPUI, USA |

## 12. Software-Defined Networking and Network Virtualization

### Track Co-chairs

| | |
|---|---|
| Flavio de Oliveira Silva | Federal University of Uberlândia, Brazil |
| Ashutosh Bhatia | Birla Institute of Technology and Science, Pilani, India |

### TPC Members

| | |
|---|---|
| Rui Luís Andrade Aguiar | Universidade de Aveiro (UA), Portugal |
| Ivan Vidal | Universidad Carlos III de Madrid, Spain |
| Eduardo Coelho Cerqueira | Federal University of Pará (UFPA), Brazil |
| Christos Tranoris | University of Patras (UoP), Greece |
| Juliano Araújo Wickboldt | Federal University of Rio Grande do Sul (UFRGS), Brazil |
| Haribabu K. | BITS Pilani, India |
| Virendra Shekhavat | BITS Pilani, India |
| Makoto Ikeda | Fukuoka Institute of Technology, Japan |
| Farookh Hussain | University of Technology Sydney, Australia |
| Keita Matsuo | Fukuoka Institute of Technology, Japan |

## AINA-2024 Reviewers

| | |
|---|---|
| Admir Barolli | Burak Kizilkaya |
| Aida ben Chehida Douss | Carson Leung |
| Akimitsu Kanzaki | Chidchanok Choksuchat |
| Alba Amato | Christos Tranoris |
| Alberto Postiglione | Chung-Ming Huang |
| Alex Thomo | Dario Branco |
| Alfredo Navarra | David Taniar |
| Amani Shatnawi | Elinda Mece |
| Anas AlSobeh | Enver Ever |
| Andrea Marotta | Eric Pardede |
| Angela Peduto | Euripides Petrakis |
| Anne Kayem | Evjola Spaho |
| Antreas Dionysiou | Fabrizio Messina |
| Arjan Durresi | Feilong Tang |
| Ashutosh Bhatia | Flavio Silva |
| Beniamino Di Martino | Francesco Orciuoli |
| Bhed Bista | George Portokalidis |

Giancarlo Fortino
Giorgos Vasiliadis
Glauco Gonçalves
Hatim Chergui
Hiroaki Kikuchi
Hiroki Sakaji
Hiroshi Maeda
Hiroyuki Fujioka
Hyunhee Park
Isaac Woungang
Jana Nowaková
Jolanta Mizera-Pietraszko
Junichi Honda
Jyoti Sahni
Kazunori Uchida
Keita Matsuo
Kenichi Matsui
Kiki Adhinugraha
Kin Fun Li
Kiyotaka Fujisaki
Leonard Barolli
Leonardo Mostarda
Leonidas Akritidis
Lidia Ogiela
Lisandro Granville
Lucian Prodan
Luciana Oliveira
Mahmoud Elkhodr
Makoto Ikeda
Mamoru Mimura
Manato Fujimoto
Marco Antonio To
Marek Ogiela
Masaki Kohana
Minoru Uehara
Muhammad Karim
Muhammad Toaha Raza Khan
Murugaraj Odiathevar
Nadeem Javaid
Naohiro Hayashibara
Nobuo Funabiki
Nour El Madhoun
Omar Darwish

Panagiotis Ilia
Petr Musilek
Philip Moore
Purav Shah
R. Madhusudhan
Raffaele Guarasci
Ralf Klasing
Roberto Tonelli
Ronald Petrlic
Sabyasachi Mukhopadhyay
Sajal Mukhopadhyay
Salvatore D'Angelo
Salvatore D'Angelo
Salvatore D'Angelo
Salvatore Venticinque
Santi Caballé
Satoshi Ohzahata
Serafino Cicerone
Shigenari Nakamura
Shinji Sakamoto
Sho Tsugawa
Sriram Chellappan
Stephane Maag
Takayuki Kushida
Tetsuya Oda
Thomas Dreibholz
Tomoki Yoshihisa
Tomoya Enokido
Tomoya Kawakami
Tomoyuki Ishida
Vamsi Paruchuri
Victor Ströele
Vikram Singh
Wei Lu
Wenny Rahayu
Winston Seah
Yong Zheng
Yoshitaka Shibata
Yusuke Gotoh
Yuvaraj Munian
Zeezoo Ryu
Zhengjia Xu

# AINA-2024 Keynote Talks

# Agile Edge: Harnessing the Power of the Intelligent Edge by Agile Optimization

Fatos Xhafa

Technical University of Barcelona, Barcelona, Spain

**Abstract.** The digital cloud ecosystem comprises various degrees of computing granularity from large cloud servers and data centers to IoT devices, leading to the cloud-to-thing continuum computing paradigm. In this context, the intelligent edge aims at placing intelligence to the end devices, at the edges of the Internet. The premise is that collective intelligence from the IoT data deluge can be achieved and used at the edges of the Internet, offloading the computation burden from the cloud systems and leveraging real-time intelligence. This, however, comes with the challenges of processing and analyzing the IoT data streams in real time. In this talk, we will address how agile optimization can be useful for harnessing the power of the intelligent edge. Agile optimization is a powerful and promising solution, which differently from traditional optimization methods, is able to find optimized and scalable solutions under real-time requirements. We will bring real-life problems and case studies from Smart City Open Data Repositories to illustrate the approach. Finally, we will discuss the research challenges and emerging vision on the agile intelligent edge.

# Challenges in Entity Matching in AI Era

Juggapong Natwichai

Chiang Mai University, Chiang Mai, Thailand

**Abstract.** Entity matching (EM) is to identify and link entities originating from various sources that correspond to identical real-world entities, thereby constituting a foundational component within the realm of data integration. For example, in order to counter-fraud detection, the datasets from sellers, financial services providers, or even IT infrastructure service providers might be in need for data integration, and hence, the EM is highly important here. This matching process is also recognized for its pivotal role in data augmenting to improve the precision and dependability of subsequent tasks within the domain of data analytics. Traditionally, the EM procedure composes of two integral phases, namely blocking and matching. The blocking phase associates with the generation of candidate pairs and could affect the size and complexity of the data. Meanwhile, the matching phase will need to trade-off between the accuracy and the efficiency. In this talk, the challenges of both components are thoroughly explored, particularly with the aid of AI techniques. In addition, the preliminary experiment results to explore some important factors which affect the performance will be presented.

# Contents

# Game Theory-Based Efficient Message Forwarding Scheme for Opportunistic Networks

Vinesh Kumar[1], Jagdeep Singh[2(✉)], Sanjay Kumar Dhurandher[3,4], and Isaac Woungang[5]

[1] Department of Computer Science, Bharati College, University of Delhi, New Delhi, India
[2] Department of Computer Science and Engineering, Sant Longowal Institute of Engineering and Technology, Longowal, India
jagdeepknit@gmail.com
[3] Department of Information Technology, Netaji Subhas University of Technology, New Delhi, India
[4] National Institute of Electronics and Information Technology, New Delhi, India
[5] Department of Computer Science, Toronto Metropollitan University, Toronto, ON M5B2K3, Canada
iwoungan@torontomu.ca

**Abstract.** In order to efficiently transmit data packets, this paper proposes a method known as *GT-EMFT* for game-theoretic efficient message forwarding in opportunistic networks. In this protocol, the optimal approach for choosing the next hop depends on a cooperative game between two players which frame the game by taking into account the context information, channel interference, meeting likelihood, and successful delivery of the related node from the destination. Using the Opportunistic Network Environment Simulator, simulation results demonstrate that the proposed protocol *GT-EMFT* outperforms the benchmark protocols *Epidemic* and *GTEER* in terms of average latency, delivery ratio, and average residual energy.

**Keywords:** Game Theory · Routing Protocol · Cooperative Game · ONE Simulator

## 1 Introduction

It is never easy to recover from an emergency situation, especially when there have been many casualties. To increase the effectiveness of rescue teams and save as many lives as possible in these situations, a swift and coordinated response is required. Additionally, since the emergency scenario can last for a while, systems might need to continue functioning for a longer time. There is a genuine need for communication technologies, and recent mass casualty events have prompted the development of new applications [1]. These programs facilitate first responders' tasks by allowing for quicker triage of the victims (ascertaining their medical status), greater coordination, and improved communication in the absence of network infrastructure. The lack of a network to transmit and share information generated during the emergency is the outcome of this. The usage

L. Barolli (Ed.): AINA 2024, LNDECT 199, pp. 1–10, 2024.
https://doi.org/10.1007/978-3-031-57840-3_1

of mobile ad-hoc networks is a common solution to the issue (MANET) [2] . This is the situation with Improvisa, which suggests resolving the issue by dispersing antennas around the disaster region. Despite being conceivable, this might not be practical in major crises. According to some writers, the data generated and gathered in the disaster region should be sent to a coordinating point using a wireless opportunistic network based on mobile devices carried by emergency personnel. Because they lack infrastructure, nodes can store, convey, and forward messages, and the pathways between the nodes are built dynamically, opportunistic networks are a great choice for use in emergencies. Because of this, opportunistic networks can withstand delays and interruptions, and nodes can still communicate with one another even if a route does not exist to link them [3]. This is crucial because, in emergencies, it is crucial to make sure that messages and data generated in the disaster area reach their intended recipients without being lost. After all, they contain crucial information for coordinating the emergency response on a global scale and victim information.

In opportunistic networks [4], numerous routing protocols have been developed in recent years to facilitate end-to-end messaging. Probabilistic routing protocols make up a significant component of these current protocols. When a node carrying a message encounters another node in a probabilistic routing protocol, it calculates the likelihood that the latter node will be able to deliver the message to the intended location. Using this probability, it is decided whether or not to relay the message to the second node. Probabilistic routing is a very useful and efficient method for opportunistic networking, according to the available research. Probabilistic routing protocols may function effectively if every node in opportunistic networks behaves faithfully by the specified protocol. In reality, users have more and more freedom to control their wireless devices, allowing them to allow their nodes to stray from the recommended protocol to maximize their profit. However, when nodes behave selfishly, the adoption of such probabilistic routing methods could result in decreased network performance. Opportunistic networks in particular, like many distributed autonomous systems, experience typical incentive issues including the free-rider dilemma, which occurs when nodes are self-centered. Selfish nodes lack incentives to act peacefully if they are not properly compensated. In other words, we take into account the selfish nodes, who only wish to use other nodes to send their own messages and will not send messages to other nodes if they are not adequately compensated. Therefore, because only a tiny portion of user nodes would prefer to provide their resources unconditionally, the network performance could suffer dramatically. Therefore, having a strong incentive system that encourages selfish nodes to collaborate in probabilistic routing is essential.

While creating routing schemes for wireless networks has advanced significantly, the majority of these schemes are still dependent on contemporaneous end-to-end connections and do not take probabilistic routing into account. Designing efficient plans for probabilistic routing is still a challenge. This paper's goal is to investigate this unsolved issue. We specifically want to know how to make a probabilistic routing protocol compatible if we are given context information. Specifically, how can a probabilistic routing protocol be improved so that selfish nodes will be motivated to cooperate when utilizing it? We proposed a strategy that relies on negotiation to provide an answer to this query. Therefore, nodes will be motivated to engage in packet forwarding if we can create a system where a node with a greater delivery probability can profit by enabling the forwarding of the packet from a node with a lower delivery possibility. To resolve the routing problem, we use concepts from game theory and apply them in this work.

The rest of the paper is organized as follows: Section 2 contains related work and motivation towards this work. The system model and the algorithm/scheme of the proposed work are explained in Sect. 3. The evaluation simulation results are discussed in Sect. 4. Finally, the conclusion of the work is drawn in Sect. 5.

## 2 Related Work

Numerous game theory-based routing and forwarding strategies for opportunistic networks have been put out in the literature. Here, a few recent works are discussed.

In [5], The authors proposed a novel optimized routing protocol using game theoretic modeling. They have designed a non-zero-sum game of two players by using the different parameters namely the context statement, energy, and delivery of the data to the destination. Energy was considered the main parameter in the game. Based on this game theoretic approach, they proposed an algorithm for energy-efficient routing in an opportunistic network. The proposed algorithm had been simulated using one simulator on the haggle-Infocomm-2006 real data trace. From the simulation and result analysis, it is found that the proposed algorithm outperforms the existing algorithm in terms of various performance matrices. They have not considered the channel interference and meeting for this scheme.

In [6], The authors designed and developed a routing scheme based on probability using game theoretic modeling. The game theory was used to encourage cooperation among the selfish nodes in the network. The selfishness of the nodes is divided based on the residual energy and buffer of the node, named self-centric selfishness. They also considered the concept of social selfishness. Using these two selfishness parameters, they introduce a new selfishness metric namely $SM_{i,j}$. Using the metric $SM_{i,j}$, the authors developed a Nash bargaining game. Using this game, the author proposed a routing algorithm named as PRGT algorithm. The simulation and result analysis show that the proposed scheme outperforms the existing scheme such as Epidemic, Prophet, and Spray-And-Wait, and found that the proposed algorithm is very effective. They have not considered the prevention mechanism from the malicious node.

In [7], The authors proposed an optimized routing scheme using the game-theoretic tools to select the best possible next hop. They designed and developed a two-person zero-sum cooperative game using the context information, encounter value, and distance from the destination. The simulation and result analysis proved the efficiency and effectiveness of the algorithm. It is found that the GT-ACR scheme outperforms to Epidemic, Prophet, HBPR, and ProWait routing protocols. They have not considered the energy efficiency and security of the proposed technique.

In [8], A knowledge-based multiplayer collaborative (KBMC) routing scheme has been proposed in this work. The authors considered the information on various factors such as energy consumption, delivery probability, node velocity, and message carry time. Based on these factors, they defined the utility function/payoff function for a game, and this utility function was utilized to design and develop a Nash bargaining game to find the suitable node that can deliver the message. The proposed scheme was simulated and analyzed in terms of overhead, expected latency, delivery probability, and expected hops. Further, it is found that the proposed scheme/technique outperforms the classical approaches.

In [9], The authors adopted the evolutionary game theory to analyze and assist the changes in the network. They also considered the defense mechanism of routing paths and forwarded messages. Using the above information, They designed and developed an evolutionary game theoretic scheme. An algorithm for that was also proposed. The simulation and result analysis clearly reflects that the algorithm proposed in this work successfully handle different type of routing attacks. They have not considered the intelligent/cognitive mechanism.

### 2.1   Motivation

From the above discussion, it is clearly seen that Game theory plays an important role in opportunistic networks. Further, none of the above work [5–9, 11–14, 16, 17] consider the parameters namely channel interference, meeting probability, delivery probability, and contact duration to design a game and define the utility function. This motivates us to propose the game theoretic scheme that considers all the parameters to enhance the efficiency of routing and forwarding.

## 3   System Model

The nodes $N$ in opportunistic networks [17] are created at random, and the links for communication are impressively simulated by sparse dispersal in the surrounding area. The main issue in opportunistic Networks is the routing and forwarding of data to the next node. Let $S$ be the source node that is ready to send data to the next node, and there are three nodes, namely $N_1$, $N_2$, $N_3$ in the neighborhood of $S$.Therefore the best node should be selected. For this, there exist various parameters, namely channel interference, Meeting probability, Delivery probability, and Contact duration. The transmitting nodes with connected links are represented in Figs. 1 and 2. In this work, we have considered the four different parameters namely channel interference, Meeting probability, Delivery probability, and Contact duration, and defined as below:

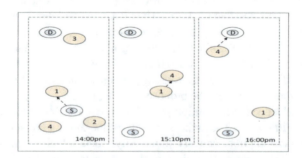

**Fig. 1.** Example of OppNet Routing

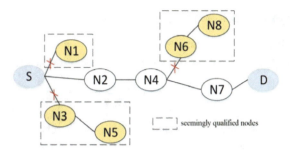

**Fig. 2.** Relay node selection

## 3.1 Channel Interference (CI)

The use of the same channel by some transceivers is obvious. Therefore, it is true that interference [11] will occur when using the same channel, but not otherwise. An interference function $I(S_i, S_j)$ that represents this interference between pairs $i$ and $j$ is defined as

$$I(S_i, S_j) = \begin{cases} 1, & \text{if } S_i = S_j. \\ 0, & S_i \neq S_j. \end{cases} \tag{1}$$

## 3.2 Contact Duration (CD)

It is the duration between the carrier to the destination node and given in Eq. (2).

$$\text{Contact Duration} = \frac{T_{rd}}{(T_r + T_d)/2} \tag{2}$$

Where,

$T_{rd}$ = Total span between carrier (r) and destination node (d).

$$= T_{rd} = \Sigma_{i=1}^{p} t_{rd}(i)$$
$$= \Sigma_{i=1}^{p} \left( t_{rd_{end}}(C_{N_i N_j}) - t_{rd_{start}}(C_{N_i N_j}) \right)$$

$t_{rd_{end}}(C_{N_i N_j})$ = connection's ending duration between the carrier and destination node.

$t_{rd_{start}}(C_{N_i N_j})$ = connection's starting duration between the carrier and destination node.

$T_r$ = Total duration of carrier node.

$T_d$ = Total duration spent in the environment until the destination node is in contact with other nodes.

### 3.3   Delivery Probability Rate (DP)

It is the probability of the total number of data packets successfully received by the destination within a given instant of time and calculated from Eq. (3).

$$DP = \frac{p_{i,j}}{\Sigma_{i=1}^{n} p_{i,j}} \tag{3}$$

Where $p_{i,j}$ represent the delivery probability of a node $i$ in relation to recipient node $j$ and obtained from Prophet and $\Sigma_{i=1}^{n} p_{i,j}$ is the sum of all the node's probability.

### 3.4   Meeting Probability (MP)

The meeting probability is calculated as the Average Number of contacts per frame/Meeting probability. It can be described as

$$\text{Meeting Probability} = \Sigma P_{C_{ij}}(S_n) \tag{4}$$

Based on these parameters, A Game-theoretic model has been designed using the concept of Nash Equilibrium [15] as described below:

In this work, four strategies are used and defined:-

1. The first strategy $S_1$ is defined based on the channel interference described as

$$S_1 = Min(\text{Channel Interference})$$

2. The second strategy $S_2$ is taken based on the Meeting probability of the nodes in the network as

$$S_2 = Max(\text{Meeting probability})$$

3. The Third strategy $S_3$ is considered based on the delivery probability ratio of the nodes and given as

$$S_3 = Max(\text{Delivery Probability})$$

4. The fourth strategy $S_4$ is defined on the contact duration between the nodes as

$$S_4 = Max(\text{Contact Duration})$$

Using the above strategies a utility function $(U)$ is formulated as the product of the strategies:

$$\text{Utility function}(U) = S_1 \times S_2 \times S_3 \times S_4 \tag{5}$$

Let $a$ and $a_i$ be two different values of communication nodes and satisfy the condition below:

$$U(a_i, a_{-i}) \geq U(a_i', a_{-i}')\forall i \tag{6}$$

That clearly shows that the nodes in the networks satisfy the Nash equilibrium condition. Therefore the node $N_1$ forwarded the message to neighboring node $N_r$. The routing and forwarding of the message to the next node are depicted in the algorithm:

---

**Algorithm 1.** GT-EMFT

---

**Input:** $N, N_i, M, a, a' \varepsilon N$, four strategies: $S_1, S_2, S_3$, and $S_4$
**Output:** Message $M$ transmitted successfully to the destination
1: Consider Nodes encounter each other
2: **for** each Node $N_i, i = 0, 1, 2, 3 \dots n$ **do**
3:     Calculate channel interference, meeting probability, delivery probability, and contact duration by using Eqs. (1), (2), (3) and (4).
4:     Calculate the four strategies:

$$S_1 = Min(\text{Channel Interference})$$

$$S_2 = Max(\text{Meeting probability})$$

$$S_3 = Max(\text{Delivery Probability})$$

$$S_4 = Max(\text{Contact Duration})$$

5:     Calculate Utility Function $(U)$

$$\text{Utility function}(U) = S_1 \times S_2 \times S_3 \times S_4$$

6:     **if** The link between the sender node and the neighboring node is available and in contact and $N_r = N_d$ **then**
7:        $N_r$ transmits the message $M$ to $N_d$
8:     **end if**
9:     **if** link availability$>0$; $U(a_i, a_{-i}) \geq U(a'_i, a'_{-i})$ **then**
10:        Node $N_1$ forwarded the message $M$ to Node $N_r$
11:     **else**
12:        Go to step 2 for the next node search.
13:     **end if**
14: **end for**
15: Message delivered successfully.

---

## 4 Evaluation

In this part, the Opportunistic Network Environment Simulator [19] is used to evaluate the performance of the proposed GT-EMFT protocol. The ONE Simulator is a Java-based simulator with different mobility patterns. We have implemented game theory in ONE Simulator v1.6 and simulated the proposed protocol GT-EMFT by considering default parameters. With the aid of the benchmark routing methods Game Theory-Based Energy Efficient Routing in Opportunistic Networks (GTEER) [5] and Epidemic, we have contrasted the performance of the proposed GT-EMFT.

### 4.1 Simulation Results

The delivery probability of the GT-EMFT, GTEER, and Epidemic reduces as the TTL (Time To Live) rises, as shown in Fig. 3's link between the two variables. As a result, when the TTL increases, more messages are retained in the node's buffer, which decreases the likelihood that a message will be sent. With the use of game theory in the network, the likelihood that a message will arrive at its intended location rises. This occurs as a result of the game's application only forwarding messages to nodes with high cooperativeness rather than to any node without first evaluating its related context settings. The delivery probability for GT-EMFT is higher (0.3615) than for GTEER (0.2962) and Epidemic (0.2864). In fact, GT-EMFT outperforms GTEER and Epidemic in terms of delivery probability by 16.15 percent and 18.1 percent, respectively.

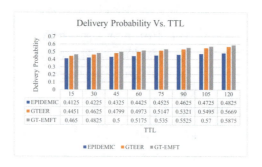

**Fig. 3.** Delivery Probability vs TTL

Variations in the time to live are used to examine the overhead ratio. Figure 4 illustrates how this modification has an impact. After that, the network scenario's node count is altered, and the impact of this change on the probability of delivery, typical delay, and overhead ratio is assessed as shown in Figs. 5 and 6.

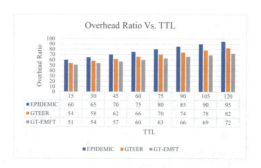

**Fig. 4.** Overhead Ratio vs TTL

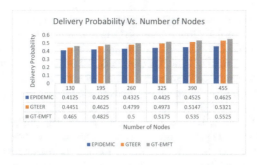

**Fig. 5.** Delivery Probability vs No. of Nodes

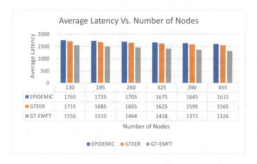

**Fig. 6.** Average Latency vs No. of nodes

## 5 Conclusion

This work proposes a new routing for OppNets GT-EMFT that depends on the recognition of a node's key characteristics. In this technique, a cooperative gaming strategy is used to select the best message forwarder for a specific node. It has been shown that the best response analysis approach may be used to efficiently discover and select the next best message forwarder for a given node from the collection of its possible neighbors. When TTL and buffer size are changed, simulation results using ONE Simulator show that GT-EMFT outperforms GTEER and Epidemic routing protocols in terms of average latency and delivery probability. In our forthcoming work, we are going to look at the security properties of the proposed GT-EMFT system using real trace mobility models.

## References

1. Wei, K., Liang, X., Xu, K.: A survey of social-aware routing protocols in delay tolerant networks: applications, taxonomy and design-related issues. IEEE Commun. Surv. Tutorials **16**(1), 556–578 (2013)
2. Spyropoulos, T., Psounis, K., Raghavendra, C.S.: Spray and wait: an efficient routing scheme for intermittently connected mobile networks. In: Proceedings of the 2005 ACM SIGCOMM Workshop on Delay-Tolerant Networking, pp. 252–259 (2005)

3. Lindgren, A., Doria, A., Schelén, O.: Probabilistic routing in intermittently connected networks. ACM SIGMOBILE Mobile Comput. Commun. Rev. **7**(3), 19–20 (2003)
4. Akbari, Y., Tabatabaei, S.: A new method to find a high reliable route in IoT by using reinforcement learning and fuzzy logic. Wireless Pers. Commun. **112**(2), 967–983 (2020)
5. Singh, J., Dhurandher, S.K., Woungang, I.: Game theory-based energy efficient routing in opportunistic networks. In: International Conference on Advanced Information Networking and Applications, pp. 627–639 (2022)
6. Qin, X., Wang, X., Wang, L., Lin, Y., Wang, X.: An efficient probabilistic routing scheme based on game theory in opportunistic networks. Comput. Netw. **149**, 144–153 (2019)
7. Borah, S.J., Dhurandher, S.K., Woungang, I., Kumar, V.: A game theoretic context-based routing protocol for opportunistic networks in an IoT scenario. Comput. Netw. **129**, 572–584 (2017)
8. Deng, X., Chen, H., Cai, R., Zeng, F., Xu, G., Zhang, H.: A knowledge-based multiplayer collaborative routing in opportunistic networks. In: 2019 IEEE International Conference on Dependable, pp. 16–21. Autonomic and Secure Computing, International Conference on Pervasive Intelligence and Computing, International Conference on Cloud and Big Data Computing, International Conference on Cyber Science and Technology Congress (2019)
9. Guo, H., Wang, X., Cheng, H., Huang, M.: A routing defense mechanism using evolutionary game theory for delay tolerant networks. Appl. Soft Comput. **38**, 469–476 (2016)
10. Deshpande, S.: Cost Efficient Predictive Routing in Disruption Tolerant Networks-Doctoral Dissertation, The Ohio State University (2011)
11. Shrivastav, V., Dhurandher, S.K., Woungang, I., Kumar, V., Rodrigues, J.J.: Game theory-based channel allocation in cognitive radio networks. In: 2016 IEEE Global Communications Conference (GLOBECOM), pp. 1–5 (2016)
12. Dhurandher, S.K., Borah, S.J., Woungang, I., Bansal, A., Gupta, A.: A location prediction-based routing scheme for opportunistic networks in an IoT scenario. J. Parallel Distrib. Comput. **118**, 369–378 (2018)
13. Dhurandher, S.K., Sharma, D.K., Woungang, I., Saini, A.: Efficient routing based on past information to predict the future location for message passing in infrastructure-less opportunistic networks. J. Supercomput. **71**, 1694–1711 (2015)
14. Kumar, V., Dhurandher, S.K., Woungang, I., Gupta, S., Singh, S.: Channel allocation in cognitive radio networks: a game-theoretic approach. In: International Conference on Network-Based Information Systems, pp. 182–192 (2022)
15. Nash, J.: Non-cooperative games. Ann. Math. **54**, 286–295 (1951)
16. Wu, F., Chen, T., Zhong, S., Qiao, C., Chen, G.: A game-theoretic approach to stimulate cooperation for probabilistic routing in opportunistic networks. IEEE Trans. Wireless Commun. **12**(4), 1573–1583 (2013)
17. Dede, J., et al.: Simulating opportunistic networks: survey and future directions. IEEE Commun. Surv. Tutorials **20**(2), 1547–1573 (2017)
18. Singh, J., Dhurandher, S.K., Woungang, I.: Game theory-based energy efficient routing in opportunistic networks. In: International Conference on Advanced Information Networking and Applications, pp. 627–639 (2022)
19. Keränen, A., Ott, J., Kärkkäinen, T.: The ONE simulator for DTN protocol evaluation. In: Proceedings of the 2nd International Conference on Simulation Tools and Techniques, pp. 1–10 (2009)

# Improved UCB MAB-Based Algorithm for Relay Selection in Cooperative Narrowband PLC Communication

Wided Belhaj Sghaier[✉], Hela Gassara, Fatma Rouissi, and Fethi Tlili

Lab. GRES'COM, Ecole Supérieure des Communications de Tunis, University of Carthage, Ariana, Tunisia
{wided.belhajsghaier,hela.gassara,fatma.rouissi,
fethi.tlili}@supcom.rnu.tn

**Abstract.** In this paper, we apply artificial intelligence to find the best relay for a two-hop cooperative narrowband Power Line Communication (NB PLC). There is no channel side information, and the channel is corrupted by the periodic impulsive noise modelled using Middleton Class-A noise. The main idea is to exploit statistical properties of the arms rewards distributions, basing on the realistic physical characteristics of the channel, to enhance the Multi-Armed Bandit (MAB) Upper-Bound Confidence (UCB)-based algorithm decisions. Two new variants of the considered machine learning method are detailed and their performances to quickly and accurately choose the best relay are discussed. Simulation results showed that the proposed algorithms outperform the conventional UCB algorithm in terms of cumulative regret, probability of good selection, and Bit Error Rate (BER).

## 1 Introduction

Various researches are being carried out with the advancement of smart grid applications that rely on narrow-band power line communication (NB-PLC) [1]. Since the PLC channel is a harsh environment for transmission due to the high attenuation and the presence of several types of noise [2], establishing a reliable communication between two network nodes that are separated by a long distance might be challenging.

To address this problem, cooperative communication can be utilized by employing one or more relays, known as intermediate nodes. Multi-hop communication, also known as relaying, has gained popularity as a useful method for enhancing communication reliability [3,4]. Recently, the selection problem in the context of a two-hop cooperative PLC scenario has been studied in the literature [5].

Multi-Armed-Bandit (MAB) based relay selection approach, which is a family of Machine Learning (ML) based algorithms [6,7] was used and their performances was evaluated in order to exploit the optimal relay selection for data transmission [8]. However, these studies do not address the real transmission constraints as those in the NB-PLC channel, especially the multipath attenuation and the impulsive noise.

In this study, our proposal involves applying the MAB technique to resolve the relay selection issue in the NB-PLC channel affected by particularly the impulsive

L. Barolli (Ed.): AINA 2024, LNDECT 199, pp. 11–21, 2024.
https://doi.org/10.1007/978-3-031-57840-3_2

noise. We use the confidence interval of the arms reward distributions statistics, basing on the properties of the channel noise, in order to improve the computation of the padding function in the UCB (Upper-Confidence Bound) algorithm, which leads to two improved variants of the conventional algorithm. Classifying the relays implies an additional improvement using appropriate arm selection in order to more enhance The performance of the solution is evaluated based on their ability to accurately choose the good node, minimize regret, and reduce the Bit Error Rate (BER).

The work is arranged in the following manner. The system model and NB-PLC channel characteristics are described in Sect. 2. In Sect. 3, an overview of MAB-based solutions is presented. Proposed improved versions of the UCB algorithm are detailed in Sect. 4. Simulation results are presented and discussed in Sect. 5.

## 2   System Model Overview

### 2.1   Communication System Design

This paper discusses a two-hop cooperative PLC system where a single intermediate relay is selected for each transmission between the source and destination nodes. Figure 1 demonstrate the cooperative two-hop communication process.

S is the sender and D represents the receiver, $R_i, i \in \{1, ... , N\}$ are $N$ available nodes are those that can receive data from the source (S), decode it, and then send it to the destination [9]. The link's signal-to-noise ratio $(SNR)$ is regarded as a performance indicator that displays the overall transmission performance. The good node is thus chosen depending on the path which is give the highest $SNR$. The total instantaneous $SNR$ via $R_i$ from the transmitter to the receiver is calculated by [5].

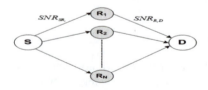

**Fig. 1.** A cooperative NB-PLC that transmits in two hops.

$$SNR_{total,i} = min\{SNR_{SR_i}, SNR_{R_iD}\}, i \in \{1, ..., N\} \qquad (1)$$

where $SNR_{SR_i}$ and $SNR_{R_iD}$ denote the $SNR$ of the link from the sender to the relay and the link from the relay to the receiver, respectively. The communication scheme employed in this paper, which is based on G3-PLC [10], has been presented in the subsequent table (Table 1).

The paper concentrates on analyzing a NB-PLC channel affected by noise, with our tests conducted without taking into account the impact of the channel transfer function.

We intend to solve the following maximizing problem given in expression (2) as we want a strategy of node selection that produces a high $SNR_{total}$ between the sender and the intermediate node:

$$maximize_{R_i \in \{1,..., N\}} SNR_{total,i} \qquad (2)$$

**Table 1.** System Parameters

| Parameters | Values |
|---|---|
| Number of sub-carriers | 256 |
| Number of used sub-carriers | 72 |
| OFDM interval (us) | 213.248 |
| Modulation | DBPSK |
| Sampling rate (MHz) | 1.2 |
| Inter-carrier spacing (KHz) | 4.6875 |
| Frequency range | 154.68 kHz–492.18 kHz |

## 2.2 Noise Model Description in Impulsive NB PLC Environment

Impulsive noise has a significant negative impact on communication quality in NB-PLC environment. The statistical characterization and modeling of this type of noise has been the topic of several research works [11,12], and the Middleton class-A noise has been the most popular impulsive noise model. This model is completely defined by 3 parameters: the density of impulses during one observation period which is represented by the impulsive index (A); the variance of the impulse noise ($\sigma_I^2$), and the ratio $\Gamma = \frac{\sigma_G^2}{\sigma_I^2}$ where $\sigma_G^2$ is the variance of the background noise. The normalized probability density function ($pdf$) of the noise $f_A(x_k)$ is expressed by [11]:

$$f_A(x_k) = \sum_{m=0}^{\infty} \frac{A^m e^{-A}}{m!} N(x_k, 0, \sigma_m^2),\qquad(3)$$

where $N(x_k, 0, \sigma_m^2)$ is the zero-mean Gaussian $pdf$, its variance $\sigma_m^2$ is given by (4)

$$\sigma_m^2 = \sigma_I^2 \frac{m}{A} + \sigma_G^2 = \sigma_G^2 \left( \frac{m}{A\Gamma} + 1 \right)\qquad(4)$$

The idea is to investigate the consequence of the periodic impulsive noise [13] on the relay selection problem. This kind of impulsive noise is observed in the PLC channel, and is generated by the Middleton Class A model with a probability of occurrence A. It is assumed to be a discrete time series with periodic intervals $Tp$, as described in [11]. $P = Tp * A$ is the duration of the impulsive noise over a period $Tp$.

## 3 State of the Art on UCB Multi-Armed Bandit-Based Algorithm

In the field of ML, MAB is a set of decision-making techniques that are developed by agents to maximize their long-term cumulative reward [14,15].
Every round, the agent is given information about the current state. Then, it chooses an action based on this information in addition to the experience gained in previous rounds. More precisely, it receives the reward associated to the selected arm $(a_t)$ at the end of every round. Every time slot, the source selects one relay from a range of available relays, resulting in a particular total $SNR$ link that is regarded as a reward.

This selection is denoted an action at to which corresponds a reward $X_t(a_t)$, calculated depending on the received signal. Then, using a robust transmission mode, this information is sent back to the transmitter. Following this at each trial, the transmitter chooses the arm according to the decision making policy $\pi$. $X_t(k)$ by $U_{k,t}$ is the expected reward. The idea is to detect the good arm every time $t$ denoted by $k_t^*$, with expected reward $U_{k^*,t}$ that is to say

$$U_{k^*,t} = U_t^* = max_{k \in \{1,..,N\}} U_{k,t} \tag{5}$$

The cumulative expected regret of the decision-making policy $\pi$ after $T$ trials is defined by

$$R_\pi = \Xi_\pi \left[ \sum_{i=1}^{T} (U_t^* - U_{k^*,t}) \right] \tag{6}$$

where E[.] reflects the mathematical expectation.

The UCB is a MAB-based solution for stationary bandits problems, which performs as follows: each arm plays once during the exploration phase in order to gather information about all the arms. Then, the basic idea is to select an action $a_k$, $k \in \{1, ..., N\}$ at time $t$ which maximizes the UCB index defined as

$$I_{k,t} = \overline{X_t}(k) + C_t(k) \tag{7}$$

Therefore, the selection policy for UCB can be expressed as

$$a_t = argmax_{1 \leq k \leqslant N} I_{k,t} \tag{8}$$

$\overline{X_t}(k)$ the sum of the rewards earned by executing arm $k$ until time $t$, as determined by (10).

$$\overline{X_t}(k) = \frac{1}{N_t(k)} \sum_{s=1}^{t} X_s 1 \{a_s = k\} \tag{9}$$

with $N_t(k) = \sum_{s=1}^{t} 1 \{a_s = k\}$ is the number of times the arm $k$ has been chosen up to time t, $1 \{a_s = k\}$ returns one if $a_s = k$ and zero otherwise.

$C_t(k)$ is called padding function. It defines the uncertainty factor of the corresponding arm, the less selected arms will have higher values than the most selected arms. A standard choice for the padding function is:

$$C_t(k) = B \sqrt{\frac{\varepsilon log(t)}{N_t(k)}} \tag{10}$$

After multiple simulations, the maximum observed reward is selected as $B$, and the confidence level is adjusted with the tuning parameter $\varepsilon$. Its value is empirical and fixed through many trials of the algorithm. After multiple simulations, the maximum observed reward is selected as $B$, and the confidence level is adjusted with the tuning parameter $\varepsilon$. Its value is empirical and fixed through many trials of the algorithm.

The UCB algorithm is highly effective for stationary bandit problems, demonstrating expected regret that aligns with the lower bound regret of all policies. However, its optimality diminishes in non-stationary PLC environments, as it operates under the assumption of constant reward distributions over time [15]. Additionally, a drawback of this approach lies in the empirical determination of the padding function used to facilitate exploration of all arms, with its fixed value remaining unchanged over time.

## 4    New Version of UCB Algorithm Tailored for Dynamic NB-PLC Environments

To enhance decision outcomes of the UCB algorithm and align it with the evolving nature of NB-PLC channel characteristics, two extra processing are proposed by authors. The first one is based on the statistical study and estimation of the confidence interval of the arms rewards distributions for calculating a more appropriate and time-varying padding function. The second approach is determined by a rectification in the arm selection procedure.

### 4.1    New Padding Function Calculation in UCB Algorithm

Our technique has the same basic concept as UCB, but incorporates prior information about the distribution of an arm's rewards for a powerful exploration. We use the Hoeffding inequality's approach to generate a UCB confidence bound that will adjust the value of the padding function. Thus, we obtain a time varying and arm-dependant padding value according to the characteristics of the periodic impulse noise which affects the arm. Details of the steps to calculate the confidence interval of the arms rewards distributions are given as follows, resulting on a new UCB index value.

1. First, we calculate the sample standard deviation for each arm using expression (11).

$$S_t(k) = \frac{\sum_{i=1}^{t}(X_i - \overline{X}_t(k))^2}{N_t(k) - 1} \tag{11}$$

2. If $N_t(k) < 30$, we use the Student's t-distribution [16] to determine the confidence interval at a desired confidence level $c$ [17], which determines the size of the confidence interval to be added to the empirical observed mean of the arm reward. Thus, the UCB index for the arm $k$ is given by (12)

$$I_{k,t} = \overline{X}_t(k) + t_c \frac{S_t(k)}{\sqrt{N_t(k)}} \tag{12}$$

where the confidence interval is fixed by $t_c \frac{S_t(k)}{\sqrt{N_t(k)}}$. $t_c$ is the critical value for the Student's t-distribution at the confidence level $c$. Its value can be determined from statistics tables or generated by software or statistical calculators.
In general, the most commonly used confidence level $c$ for statistical estimation is between 80% and 95% [17].

3. If $N_t(k) \geqslant 30$, according to the central limit theorem [16], the distribution of the arm rewards can be approximated by a normal one, then the UCB index is fixed according to expression (13)

$$I_{k,t} = \overline{X_t}(k) + z_c \frac{S_t(k)}{\sqrt{N_t(k)}} \tag{13}$$

where the confidence interval is fixed by $z_c \frac{S_t(k)}{\sqrt{N_t(k)}}$. Similarly, $z_c$ is the critical value for the normal distribution at confidence level $c$. Its values can be found in statistics tables or produced by software or statistical calculators.

4. Finally, the selection policy can be expressed as:

$$a_t = argmax_{1 \leq k \leq N} I_{k,t} \tag{14}$$

Algorithm 1 summarizes the proposed UCB algorithm based on the new approach of the padding function calculation.

| Algorithm1:The Enhanced UCB Based on New Padding Function Calculation |
|---|
| 1: for $t = 1, \ldots, N$ |
| 2:        let $a_t = t$ play and observe the reward. |
| 3:        Calculate the empirical mean |
| $\overline{X_t}(k,\tau) = \frac{1}{N_t(k,\tau)} \sum_{s=1}^{t} X_s(k) 1\{a_s = k\}$, $N_t(k) = \sum_{s=1}^{t} 1\{a_s = k\}$ , $S_t(k) = \frac{\sum_{i=1}^{t}(X_i - \overline{X})^2}{N_t(k) - 1}$ |
| 4:        Calculate the padding function as |
| if $N_t(k) < 30$, determine $t_c$ and $C_t(k) = t_c \frac{S_t(k)}{\sqrt{N_t(k)}}$ |
| else determine $z_c$ and $C_t(k) = z_c \frac{S_t(k)}{\sqrt{N_t(k)}}$ |
| 5:        Update the UCB index $I_{k,t} = \overline{X_t}(k) + C_t(k)$. |
| 6: end for |
| 7: for $t = N + 1, \ldots, T$ do |
| 8:        let $a_t = \arg\max_{1 \leq k \leq N} I_{k,t}$, play and observe reward. |
| 9:        Update the UCB index $I_{k,t}$ of the selected arm. |
| 10: end for |

### 4.2 Extra Processing for Arm Selection in UCB Algorithm

The confidence interval is important for improving UCB algorithm performance, because is utilized to adjust the algorithm, ensuring that all arms are considered in the selection process, even those with minimal probability of being chosen. In order to consider the prior knowledge of the channel noise characteristics in the arm selection procedure, we suggest to include an additional test at each instant, which allows to classify the last chosen arm as impulsive or Gaussian channel, depending on its previous reward values. This test is helpful to revise the class of the selected arm. The additional classification process is defined by the following steps.

1. Establish a threshold value, announced as $SNR_{th}$, to distinguish between impulsive and Gaussian noise based on (SNRs). Consequently, if a reward surpasses or equals $SNR_{th}$, it is categorized as Gaussian arm, otherwise, it is classified as impulsive arm. Additionally, for each transmission, the received reward of the $k^{th}$ arm upon selection is recorded in a table denoted $hist_t(k)$. The chosen of arm at the $(t-1)^{th}$ round is denoted as $k^{**}$

2. Choose the best arm according to the following test:
   if $a_t == k^{**}$ and $hist(k^{**}, t-1) \geqslant SNR_{th}$
   $a_t == k^{**}$
   elseif $a_t == k^{**}$ and $hist(k^{**}, t-1) < SNR_{th}$
   $a_t = \arg \max_{1 \leq k \leqslant N} I_{k,t}$ where $a_t \neq k^{**}$
   end

The detailed steps of our method's execution are illustrated in Algorithm 2.

---

**Algorithm2: The improved UCB Incorporating an Additional Classification process**

1: for t=1,...,N
2: let $a_t = t$ Execute a play action and observe the corresponding reward.
3: Calculate the empirical mean
$$\overline{X_t}(k) = \frac{1}{N_t(k)} \sum_{s=1}^{t} X_s(k) 1\{a_s = k\} \quad, N_t(k) = \sum_{s=1}^{t} 1\{a_s = k\},$$
$$S_t(k) = \frac{\sum_{i=1}^{t}(X_i - \overline{X})^2}{N_t(k) - 1}$$
4: Save the observed reward of the chosen arm
$$Hist_t(k) = X_t(k)$$
5: Calculate the padding function as
   if $N_t(k) < 30$ determine $t_c$ and $C_t(k) = t_c \frac{S_t(k)}{\sqrt{N_t(k)}}$
   else determine $z_c$ and $C_t(k) = z_c \frac{S_t(k)}{\sqrt{N_t(k)}}$
6: Update the UCB index $I_{k,t} = \overline{X_t}(k) + C_t(k)$.
7: end for
8 : for t=N+1,...,T do
9 : Define a threshold value, $SNR_{th}$
10: let $a_t = \arg \max_{1 \leq k \leqslant N} I_{k,t}$, choose the best arm according to the following test:
   if $a_t == k^{**}$ and $hist(k^{**}, t-1) \geqslant SNR_{th}$
       $a_t == k^{**}$
   elseif $a_t == k^{**}$ and $hist(k^{**}, t-1) < SNR_{th}$
       $a_t = \arg \max_{1 \leq k \leqslant N} I_{k,t}$ where $a_t \neq k^{**}$
   end
10: Update the UCB index $I_{k,t}$ of the chosen arm.
11: end for

---

## 5 Simulation Results

We examine a two-hop cooperative communication system designed for the NB-PLC channel, elaborated in Sect. 2. The system comprises transmitter, receiver, and relay nodes, all equipped with DBPSK-OFDM-based transceivers adhering to the G3-PLC standard. In this study we use six arms. The $SNR$ values for each link are computed using the following procedure:

We consider a two-hop cooperative communication system for the NB PLC channel as explained in Sect. 2. Source, destination, and relay nodes are DBPSK-OFDM-based transceivers compliant with G3-PLC standard, and the number of relays is $N = 6$. The $SNR$ values for a link are calculated as follows:

$$SNR = \frac{P_s}{\sigma_G^2 + \sigma_I^2} \tag{15}$$

where $P_s$ is the power of the received OFDM symbol, $\sigma_I^2$ is the average impulsive noise power and $\sigma_G^2$ is the average Gaussian noise power, which are calculated over the duration of an OFDM symbol ($T_{OFDM}$). The fixed parameters of impulsive noise which is generated according to Middleton class-A noise, are $\sigma_I^2 = 7.28.10^{-3}$, $\Gamma = 0.01$, and $\sigma_G^2 = 7.28.10^{-5}$. In order to evaluate the performances of the propsed enhanced UCB algorithm, we consider the following scenarios of periodic impulsive noise :

**Scenario 1**: The NB PLC channel is affected by periodic impulsive noise with period $Tp=20*T_{OFDM}$ and $A=0.25$. Thus, the impulsive noise duration over $Tp$ is equal to $P=Tp*A=5*T_{OFDM}$.

**Scenario 2**: the NB PLC channel is affected by periodic impulsive noise with period $Tp=20*T_{OFDM}$ and $A=0.75$. Thus, the impulsive noise duration over $Tp$ is equal to $P=Tp*A=15*T_{OFDM}$.

For confidence interval parameters, when $N_t(k)$ is less than 30, all scenarios get the same value of confidence level $c$, which is set to 80% as the most commonly used value in statistical evaluation. However, when $N_t(k)$ is greater than or equal to 30, the rewards are considered to be normally distributed and the confidence level $c$ is fixed after many trials to 68% and 80% for Scenario 1 and Secnario 2, respectively. The conventional UCB is also applied to the tested system while setting the parameter $\varepsilon$ to 2 and 0.4 for Scenario 1 and Scenario 2, respectively.

Figure 2a and Fig. 2b Exemplify the cumulative regret and average reward curves of various relay selection techniques under scenario 1, characterized by periodic impulsive noise. We notice that (First Proposed Enhancement) and the (Second Proposed Enhancement) Exceed in performance to the conventional UCB, surpass in the context of achieving significantly lower accumulated regret and higher average reward. Figure 3 presents the percentage of correct selection for the various transmission policies.

The proposed algorithms exhibit superior performance compared to the conventional UCB algorithm. Specifically, there's an observed increase of 8.3% in the correct selection percentage incorporating arm classification, compared to the conventional approach.

The resultant accumulated regret and average reward curves for the various tested relay selection policies shown in Fig 4. for the second scenario of impulse noise demon-

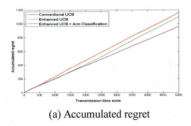

(a) Accumulated regret

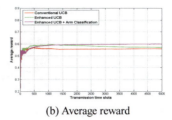

(b) Average reward

**Fig. 2.** (a) Accumulated Regret and (b) Average Reward for both the conventional and new version of UCB algorithms in relation to scenario 1.

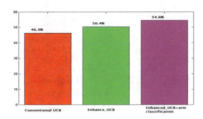

**Fig. 3.** Percentage of accurate selection achieved by various UCB-based algorithms described in scenario 1

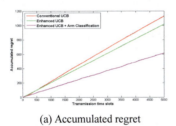

(a) Accumulated regret

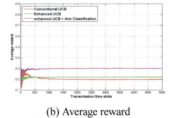

(b) Average reward

**Fig. 4.** (a) Accumulated Regret and (b) Average Reward for the conventional and new version of UCB algorithms in relation to scenario 2.

strate that the two proposed enhanced algorithms always outperform the conventional UCB algorithm.

Figure 5, shows the percentage of correct selection for the different transmission policies. We can notice that the improved version of UCB algorithm increases the good selection of arms with a value of 13% compared to conventional UCB, and with a value of 4.7% when only using the new approach for the padding function computation.

The final evaluation focuses on the system performance, specifically regarding the mean (BER) of the selected node. The BER is computed over every 100 successive transmission time slots using the analytical expression for BER of BPSK-OFDM transceivers.

Figure 6a and Fig. 6b show the cumulative distribution functions (*cdf*) of the average BER corresponding to the various tested relay selection techniques for the two scenarios. We can conclude that the proposed version demonstrate the whole system

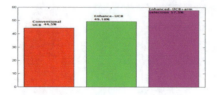

**Fig. 5.** Percentage of good selection of various UCB-based algorithms described by scenario 1.

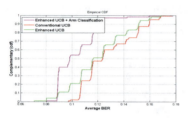

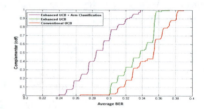

(a) cumulative distribution function (cdf) scenario 1.

(b) cumulative distribution function (cdf) scenario 2.

**Fig. 6.** The average (BER) curves described in the two scenarios.

performances, specifically referring to the probability that the average BER is less than 0.12 is 45% for the ancient UCB, it increases to 80% when using our approach for the first scenario of the noise. Similarly for the second scenario, the probability that the average BER is below 0.316 is expanded from 10% to 75%.

## 6    Conclusion

Artificial intelligence techniques are employed to optimize relay selection in a cooperative NB-PLC channel. The channel's periodic disruptions stem from Middleton Class-A impulsive noise. Our proposed approach involves enhancing the MAB algorithms through statistical estimation. To enhance the likelihood of selecting the most suitable relay, we introduce two innovative variants of the MAB algorithm. Simulation results showcase the superiority of our solution over the traditional UCB algorithm across various performance metrics, including cumulative regret, probability of optimal relay selection, and Bit Error Rate.

## References

1. Sayed, M., Al-Dhahir, N.: Narrowband-PLC/wireless diversity for smart grid communications. In: 2014 IEEE Global Communications Conference, GLOBECOM 2014, pp. 2966–2971 (2015)
2. Gassara, H., Rouissi, F., Ghazel, A.: Top-Down random channel generator for the narrowband power line communication. In: AEU - International Journal of Electronics and Communications, vol. 89 (2018)

3. Lampe, L., Schober, R., Yiu, S.: Distributed space-time coding for multihop transmission in power line communication networks. IEEE J. Sel. Areas Commun. **24**, 1389–1400 (2006)
4. Lampe, L., Vinck, J.: Cooperative multihop power line communications. In: 2012 IEEE International Symposium on Power Line Communications and Its Applications, ISPLC 2012 (2012)
5. Rabie, K., Adebisi, B., Salem, A.: Improving energy efficiency in dual-hop cooperative PLC relaying systems. In: Proceedings of the [Conference Name] (2016)
6. Nomikos, N., Talebi, S., Wichman, R., Charalambous, T.: Bandit-based relay selection in cooperative networks over unknown stationary channels, In: Proceedings of the [Conference Name] (2020). https://doi.org/10.1109/MLSP49062.2020.9231604
7. Agrawal, R.: Sample mean based index policies by O(log n) regret for the multi-armed bandit problem. Adv. Appl. Probab. **27**, 1054–1078 (1995)
8. Nikfar, B., Han Vinck, A.J.: Relay selection in cooperative power line communication: a multi-armed bandit approach. J. Commun. Netw. **19**, 1–9 (2017). https://doi.org/10.1109/JCN.2017.000003
9. Lampe, L., Vinck, J.: On cooperative coding for narrow band PLC networks. AEU-Int. J. Electron. Commun. **65**, 681–687 (2011)
10. Di Bert, L., D'Alessandro, S., Tonello, A.M.: Enhancements of G3-PLC technology for smart-home/building applications. J. Electr. Comput. Eng. 1 (2013)
11. Rouissi, F., Vinck, A.H., Gassara, H., Ghazel, A.: Statistical characterization and modelling of impulse noise on indoor narrowband PLC environment. In: 2017 IEEE International Symposium on Power Line Communications and Its Applications (ISPLC), pp. 1–6 (2017)
12. Shongwey, T., Vinck, A.H., Ferreira, H.C.: On impulse noise and its models. SAIEE Africa Res. J. **106** (2014)
13. Cortés, J., Diez, L., Cañete Corripio, F.J., Lopez, J.: Analysis of the periodic impulsive noise asynchronous with the mains in indoor PLC channels. In: Proceedings of the International Symposium on Power Line Communications and Its Applications (ISPLC), pp. 26–30 (2009). https://doi.org/10.1109/ISPLC.2009.4913398
14. Chen, Y., Su, S., Wei, J.: A policy for optimizing sub-band selection sequences in wideband spectrum sensing. Sensors **19**, 4090 (2019)
15. Bubeck, S.: Regret analysis of stochastic and nonstochastic multi-armed bandit problems (2012). ISBN: 9781601986276
16. Ghasemi, A., Zahediasl, S.: Normality tests for statistical analysis: a guide for non-statisticians. Int. J. Endocrinol. Metab. **10**, 486–489 (2012)
17. Hazra, A.: Using the confidence interval confidently. J. Thoracic Dis. **9**, 4124–4129 (2017)

# Mutual-Visibility in Fibonacci Cubes

Alfredo Navarra[(✉)] and Francesco Piselli

Department of Mathematics and Computer Science, University of Perugia, Perugia, Italy
alfredo.navarra@unipg.it, francesco.piselli@unifi.it

**Abstract.** Hypercubes, Butterfly and other well-structured topologies represent intriguing challenges in computer network architectures, especially for parallel and distributed computations. Within such a context, *mutual-visibility* plays a central role for communication activities. Given a graph $G = (V, E)$ and a subset $X$ of its vertices, $x$ and $y \in V$ are said to be $X$-visible if there exists a shortest $x, y$-path where no internal vertices belong to $X$. The set $X$ is a mutual-visibility set of $G$ if every two vertices of $X$ are $X$-visible. The cardinality of a largest mutual-visibility set is the mutual-visibility number $\mu(G)$ of $G$. In general, computing $\mu(G)$ is NP-complete. In this paper, we study the mutual-visibility in Fibonacci Cube networks, a variant of the Hypercube topology with various interesting properties. In particular, we provide an approximation algorithm for computing $\mu(G)$ in Fibonacci Cubes.

## 1 Introduction

Fibonacci numbers govern many aspects of certain natural growth phenomena as well as applications in mathematical sciences. Their definition has been provided in 1202 by Leonardo Pisano (alias Fibonacci) with the aim to describe *"How Many Pairs of Rabbits Are Created by One Pair in One Year?"*. The sequence has been recursively defined as follows:

$$F_0 = 0; \quad F_1 = 1; \quad F_i = F_{i-1} + F_{i-2}, \tag{1}$$

and it goes like:
0, 1, 1, 2, 3, 5, 8, 13, 21, 34, 55, 89, 144, 233, 377, 610, 987, 1597, 2584, ...

Fibonacci numbers show interesting relationships to other families of well-known sequences, such as the binomial coefficients (see Fig. 1.(1) and [2, 26]):

$$F_i = \sum_{k=0}^{\lceil \frac{i}{2} \rceil} \binom{i-k-1}{k}, \tag{2}$$

or even to the well-known golden ratio (see Fig. 1.(2)):

$$\varphi = \frac{1+\sqrt{5}}{2} = \lim_{i \to \infty} \frac{F_{i+1}}{F_i}. \tag{3}$$

Work funded in part by the Italian National Group for Scientific Computation GNCS-INdAM.

L. Barolli (Ed.): AINA 2024, LNDECT 199, pp. 22–33, 2024.
https://doi.org/10.1007/978-3-031-57840-3_3

The golden ratio can also be used to define Fibonacci numbers according to the Binet's formula (see, e.g., [18]):

$$F_i = \frac{1}{\sqrt{5}}(\varphi^i - (-\varphi)^{-i}).$$  (4)

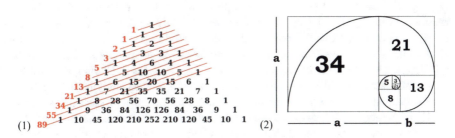

(1)     (2)

**Fig. 1.** (1): Fibonacci numbers calculated from the diagonals of the Tartaglia's triangle, i.e., the binomial coefficient; (2): The relationship between Fibonacci numbers and the golden ratio $\varphi = \frac{a}{b} = \frac{a+b}{a}$.

A natural question in the fields of parallel and distributed computing is whether it is possible to exploit Fibonacci properties when designing computer network architectures. Various proposals exist and we aim at studying reachability/visibility/connectivity capabilities (as in [1, 16, 22, 25]) among the vertices of networks whose topology has relationships with Fibonacci numbers. Investigating on mutual-visibility problems has been a research interest for more than a century because of its strength relationship with routing issues. The famous *no-three-in-line* problem has been posed by Dudeney in 1917 [14], where it is required to find how many points can be placed in a $n \times n$ grid in such a way that there are no three points on the same line. Nowadays, those kind of problems are also interesting in the field of autonomous robots where searching for the mutual-visibility means to look for a configuration where the robots can easily reach each other (see e.g. [12, 27, 29]).

Mutual-visibility in graphs with respect to a set $X$ of vertices has been recently introduced and studied in [13] in the sense of the existence of a shortest path between two vertices $x, y \in X$ not containing a third vertex $z \in X$. The visibility property is then understood as a kind of non-existence of "obstacles" between the two vertices in the mentioned shortest path, which makes them "visible" to each other.

Mutual-visibility in graphs with respect to a subset $X$ of vertices has been recently introduced in [13]. Given a graph, the requirement is to choose a maximal cardinality subset $X$ of vertices so that any two vertices in $X$ admit a shortest path along which no other vertices in $X$ are met. This problem has been called MUTUAL-VISIBILITY problem and has been shown to be NP-complete to solve. Nevertheless, for special graph classes like paths, cycles, blocks, cographs and grids, there exist exact formulae. To formally define this problem, let $G$ be a connected and undirected graph, and $X \subseteq V(G)$ be a subset of the vertices of $G$. Two vertices $x, y \in V(G)$ are $X$-*visible* if there exists a shortest $x, y$-path where no internal vertex belongs to $X$. $X$ is a *mutual-visibility set* if its vertices are pairwise $X$-visible. The cardinality of a largest mutual-visibility set is the *mutual-visibility number* of $G$, and it is denoted by $\mu(G)$. Computing one of such largest sets, referred to as a $\mu$-*set* of $G$, solves the MUTUAL-VISIBILITY problem.

In [10] and [11], exact formulae have been derived for the Cartesian and Strong product of graphs, also showing several interesting connections with other mathematical contexts, like swarm robotics (cf. [5–7]) and the general position problem in graphs (see e.g., [20,24,28]). In [11], it has been proposed a natural extension of the mutual-visibility. Formally, $X \subseteq V(G)$ is a *total mutual-visibility set* of $G$ if every two vertices $x$ and $y$ of $G$ are $X$-visible. A largest total mutual-visibility set of $G$ is a $\mu_t$-set of $G$, and its cardinality is the *total mutual-visibility number* of $G$ denoted as $\mu_t(G)$. Of course, $\mu_t(G) \leq \mu(G)$. In [9], it is shown that also computing $\mu_t(G)$ is a NP-complete problem. This more restrictive setting has proven to be very useful when studying networks with some Cartesian properties in the vertex set, namely, those ones of product-like structures (cf. [11]). In fact, in [21], since the total mutual-visibility number of Cartesian products is bounded, several exact results have been proved. Furthermore, a sufficient and necessary condition to claim when $\mu_t(G) = 0$ is provided. In [23], the authors give several bounds for $\mu_t(G)$ in terms of the diameter, order and/or connected domination number of $G$ and they also determine the exact value of the total mutual-visibility number of lexicographic products. Finally, in [4], the authors studied the mutual-visibility number on Hamming graphs, that is Cartesian products of complete graphs (e.g., hypercubes).

## 1.1   Results Overview

In this work, we study the mutual-visibility in $d$-dimensional *Fibonacci cubes* (denoted as $\Gamma_d$). Such a topology has been introduced by Hsu in the seminal paper [17] in 1993 as a model for interconnection networks. It combines some properties of the Fibonacci numbers with those of the hypercube, which constitutes another fundamental topology in parallel and distributed computing.

We provide an upper and a lower bound with respect to $\mu(\Gamma_d)$, and we propose an approximated solution.

As already pointed out, Fibonacci cubes are close related to hypercubes, which in turn have inspired the definition of the cube-connected cycles.

A *hypercube* of dimension $d$ is denoted as $Q_d$ and consists of $2^d$ vertices, each labelled by a binary string with $d$ bits. Two vertices are adjacent if and only if the two binary strings differ by exactly one bit.

A *cube-connected cycle* of dimension $d$ is denoted as $CCC_d$ and consists of $d \cdot 2^d$ vertices. It models networks that combine the properties of both hypercube graphs and cycle graphs. Informally, $CCC_d$ can be obtained from $Q_d$ by replacing each vertex by a cycle with $d$ vertices. It models networks that maintain some of the properties of the hypercubes with the advantage to decrease to 3 the degree of each vertex.

A *Fibonacci cube* of dimension $d$ is denoted as $\Gamma_d$ and can be defined as the subgraph of $Q_d$ obtained by removing all vertices whose labels contain two consecutive 1's. It consists of $F_{d+2}$ vertices, with $F_i$ being the $i$-th Fibonacci number. It models networks that maintain some of the properties of the hypercubes with the advantage to sensibly decrease the number of vertices. Furthermore, since they have close relationships with Fibonacci numbers, interesting properties arise.

All such graphs usually model interconnection networks, widely used in parallel and distributed computing systems. They offer efficient communication and connectivity patterns that make them suitable for various applications, including parallel processing, supercomputing, and designing efficient interconnection topologies for computer systems. In all applications, the notion of mutual-visibility seems to be naturally relevant. In these systems, two vertices that are $X$-visible can communicate in an efficient way, that is through shortest paths, and their messages can be maintained confidential: the exchanged messages do not pass through vertices in $X$.

For hypercubes and cube-connected cycles, mutual-visibility has been recently investigated in [8], and the obtained results are:

- For hypercubes, first the bounds $\mu(Q_d) \geq \binom{d}{\lfloor \frac{d}{2} \rfloor} + \binom{d}{\lfloor \frac{d}{2} \rfloor + 3}$ and $\mu(Q_d) \leq 2^{d-1}$ have been proved, and then such results have been exploited to get an $O(\sqrt{d})$-approximation algorithm for computing $\mu(Q_d)$. Alternatively, for an $n$-vertex hypercube, the approximation ratio can be expressed as $O(\sqrt{\log n})$.
- For cube-connected cycles, the provided bounds are $\mu(CCC_d) \leq 3 \cdot 2^{d-2}$ and $\mu(CCC_d) \geq 2^{\lceil \frac{d}{2} \rceil - 1}$. They give rise to a $3 \cdot 2^{\lfloor \frac{d}{2} \rfloor - 1}$-approximation algorithm (whose ratio can be expressed as $O(\sqrt{n})$ for an $n$-vertex cube-connected cycle).

In this paper, we investigate on Fibonacci cubes, obtaining optimal solutions for $\mu(\Gamma_d)$ when $d \leq 6$, whereas for larger values of $d$ we prove $\mu(\Gamma_d) \leq F_d + F_{d-2}$ and $\mu(\Gamma_d) \geq \max_{0 \leq k \leq \lceil \frac{d}{2} \rceil} \binom{d-k+1}{k}$. Then, we obtain a $\left( \lceil \frac{d}{2} \rceil + 1 \right)$-approximation algorithm (whose ratio can be expressed as $O(\log n)$ for an $n$-vertex Fibonacci cube).

## 1.2 Outline

Concerning the organization of the paper, in the next section we provide all the necessary notation and preliminary concepts. The subsequent Sect. 3 is specialized for presenting results about mutual-visibility for Fibonacci cubes. In particular, we present an upper and a lower bound to $\mu(\Gamma_d)$ along with the approximation algorithm. Concluding remarks are provided in Sect. 4, suggesting interesting directions for future work.

## 2 Notation and Preliminaries

In this work, we consider undirected and connected graphs. We use standard terminologies from [3,8], some of which are briefly reviewed here.

Given a graph $G$, $V(G)$ and $E(G)$ are used to denote its vertex set and its edge set, respectively, and $n(G)$ is used to represent the size of $V(G)$. If $u, v \in V(G)$ are adjacent, $(u, v) \in E(G)$ represents the corresponding edge. If $X \subseteq V(G)$, then $G[X]$ denotes the subgraph of $G$ *induced* by $X$, that is the maximal subgraph of $G$ with vertex set $X$.

The usual notation for representing special graphs is adopted. $K_n$ is the *complete graph* with $n$ vertices, $n \geq 1$, that is the graph where each pair of distinct vertices are adjacent. $P_n$ represents any *path* $(v_1, v_2, \ldots, v_n)$ with $n \geq 2$ distinct vertices where $v_i$ is adjacent to $v_j$ if $|i - j| = 1$. Vertices $v_2, \ldots, v_{n-1}$ are all *internal* of $P_n$. A *cycle* $C_n$, $n \geq 3$, in $G$ is a path $(v_0, v_1, \ldots, v_{n-1})$ where also $(v_0, v_{n-1}) \in E(G)$. Two vertices $v_i$ and $v_j$ are *consecutive* in $C_n$ if $j = (i+1) \bmod n$ or $i = (j+1) \bmod n$. The *distance* function on a graph $G$ is the usual shortest path distance.

Two graphs $G$ and $H$ are *isomorphic* if there exists a bijection $\phi : V(G) \to V(H)$ such that $(u,v) \in E(G) \Leftrightarrow (\phi(u), \phi(v)) \in E(H)$ for all $u, v \in V(G)$. Such a bijection $\phi$ is called *isomorphism*. Given $G$ and $H$, we consider also the following graph operation: the *Cartesian product* $G \square H$ has the vertex set $V(G) \times V(H)$ and the edge set $E(G \square H) = \{((g,h),(g',h')) : (g,g') \in E(G) \text{ and } h = h', \text{ or, } g = g' \text{ and } (h,h') \in E(H)\}$.

If $H$ is a sugbraph of $G$, $H$ is said to be *convex* if all shortest paths in $G$ between vertices of $H$ actually belong to $H$. Concerning convex subgraphs, we recall the following useful statement:

**Lemma 1** [13]. *Let $H$ be a convex subgraph of any graph $G$, and $X$ be a mutual-visibility set of $G$. Then, $X \cap V(H)$ is a mutual-visibility set of $H$.*

Finally, note that binary strings are used as vertex labels or components of vertex labels for the Fibonacci cubes. For a binary string $x = x_0 x_1 \cdots x_i \cdots x_{d-1}$ with $d$ bits, position 0 corresponds to the leftmost bit, and position $n-1$ to the rightmost bit. Sometimes, we interpret these strings as (binary) numbers. We use $x(i)$ to denote the binary string obtained from $x$ by complementing the bit in position $i$. We will abbreviate "the vertex with label $x$" to "vertex $x$".

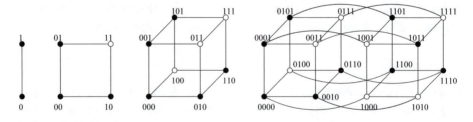

**Fig. 2.** A representation of the hypercube $Q_d$ for $d = 1, 2, 3$ and $4$. In each graph, the black vertices form a $\mu$-set.

## 3   Fibonacci Cubes

In order to define the Fibonacci cubes, it is useful to first revise the definition of the hypercube topology. The $d$-dimensional hypercube $Q_d$ is an undirected graph with vertex set $V(Q_d) = \{0,1\}^d$, and two vertices are adjacent if and only if the two binary strings differ by exactly one bit, that is, the Hamming distance[1] of the two binary strings is 1. It is worth noting that $Q_d$ can also be recursively defined in terms of the Cartesian product of two graphs as follows:

---

[1] The Hamming distance between two strings of equal length is the number of positions at which the corresponding symbols are different.

- $Q_1 = K_2$,
- $Q_d = Q_{d-1} \square K_2$, for $d \geq 2$.

This makes clear that $Q_d$ can also be seen as formed by two subgraphs both isomorphic to $Q_{d-1}$ (cf. Figure 2).

A Fibonacci cube $\Gamma_d$ of dimension $d$ can be defined as the subgraph of a hypercube $Q_d$ obtained by removing from $Q_d$ all the vertices associated with a label containing the substring '11'. This simple definition actually hides why Fibonacci cubes have such a name. To this respect, a more explicit definition may help, see Figs. 3 and 4.

A Fibonacci cube $\Gamma_1$ of dimension $d = 1$ is just a pair of vertices labelled '0' and '1', respectively, connected by an edge.

A Fibonacci cube $\Gamma_2$ of dimension $d = 2$ is obtained from $\Gamma_1$ by prefixing one further bit at '0' to its labels and adding one vertex labelled '10' connected to '00'.

A Fibonacci cube $\Gamma_d$ of dimension $d > 2$ is obtained by considering two Fibonacci cubes, $\Gamma'_{d-1}$ and $\Gamma'_{d-2}$ where the labels of $\Gamma'_{d-2}$ have been modified by prefixing one further bit, set to '0'; then, vertices with the same labels among $\Gamma'_{d-1}$ and $\Gamma'_{d-2}$ are connected; finally, the label of each vertex in $\Gamma_d$ is modified by prefixing the labels of the vertices composing the subgraph $\Gamma'_{d-1}$ (resp. $\Gamma'_{d-2}$) with a bit '0' (resp., '1').

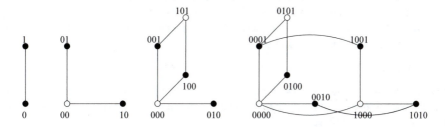

**Fig. 3.** A representation of the Fibonacci cubes $\Gamma_d$ for $d = 1, 2, 3$ and 4. In each graph, the black vertices form a $\mu$-set.

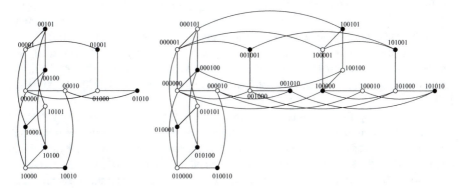

**Fig. 4.** A representation of the Fibonacci cubes $\Gamma_d$ for $d = 5$ and 6. In each graph, the black vertices form a $\mu$-set.

By construction, each label of a Fibonacci cube $\Gamma_d$, $d > 0$, is composed by $d$ bits. As the number of vertices composing $\Gamma_1$ is 2 and that for $\Gamma_2$ is 3, then $|V(\Gamma_d)|$, $d > 2$, is $F_{d+2}$, whereas $|E(\Gamma_d)| = \frac{1}{5}(dF_{d+1} + 2(d+1)F_d)$, see [15].

As shown in [19], Fibonacci cubes preserve the Hamming distance property among vertices as for the hypercubes but significantly reduce the number of vertices to that respect.

## 3.1   An Upper Bound for $\mu(\Gamma_d)$

Here, we first analyze specific optimal solutions for the MUTUAL-VISIBILITY problem in Fibonacci cubes of small dimension $d \leq 6$ and then we provide an upper bound holding for any $d \geq 6$.

First of all, we can claim the following lemma.

**Lemma 2.** *Each subgraph $\Gamma'$ of $\Gamma_d$ that is isomorphic to $\Gamma_{d'}$, $d' < d$, is convex.*

*Proof.* Consider a Fibonacci cube $\Gamma'$ subgraph of $\Gamma_d$ that is isomorphic to $\Gamma_{d'}$ for some $d' < d$. By considering two vertices $x$ and $y$ of $\Gamma'$, we have to show that there not exists a shortest path between $x$ and $y$ passing through a vertex $z \in V(\Gamma_d) \setminus V(\Gamma')$. Since the distance between two vertices in a Fibonacci cube is governed by the Hamming distance, the bits that differ between the labels associated with $x$ and $y$ concern only $\Gamma'$. In fact, there cannot be a shortest path that makes use of a vertex $z \in V(\Gamma_d) \setminus V(\Gamma')$ as, by construction, the corresponding bit leading to (the dimension of) $z$ is different from the one in the labels of both $x$ and $y$. □

**Corollary 1.** $\mu(\Gamma_d) \leq \mu(\Gamma_{d-1}) + \mu(\Gamma_{d-2})$, *for each $d > 2$.*

*Proof.* The proof simply follows by recalling that $\Gamma_d$ can be obtained by suitably combining two Fibonacci cubes of dimensions $d - 1$ and $d - 2$, respectively. Therefore, by Lemma 2 the claim holds. □

We now consider all the dimensions $d \leq 6$, one by one, for which we can provide optimal solutions, see Figs. 3 and 4.

From results provided in [13], we know that $\mu(\Gamma_1) = \mu(K_2) = 2$ and that $\mu(\Gamma_2) = \mu(P_3) = 2$. For $\Gamma_3$ (that contains 5 vertices) there exist $\mu$-sets of size 3. By referring to Fig. 3, the optimal solution is provided, for instance, by the set $\{010, 001, 100\}$.

For $\Gamma_4$ (that contains 8 vertices), Corollary 1 implies that $\mu(\Gamma_4) \leq 5$. Actually we can provide a mutual-visibility set with 5 vertices selected which is then optimal. By referring to Fig. 3, the optimal solution is provided by the set $\{0001, 0001, 0100, 1010, 1001\}$.

Concerning $\Gamma_5$ (that contains 13 vertices), since $\mu(\Gamma_4) = 5$ $\mu(\Gamma_3) = 3$, by Lemma 2 we may obtain a mutual-visibility set with at most 8 vertices. By means of a computer-assisted case-by-case analysis, we have that the optimal solution is of size 8, and it can be provided by the set $\{00010, 00100, 00101, 01000, 01001, 10001, 10010, 10100\}$.

Also for $\Gamma_6$ (that contains 21 vertices) we can obtain an optimal mutual-visibility set. This is composed of 11 vertices: $\{000010, 000100, 001001, 001010, 010000, 010001, 010101, 100001, 100010, 100101, 101000\}$. It is worth noting that $\mu(\Gamma_6)$ turns out to be smaller than $\mu(\Gamma_5) + \mu(\Gamma_4) = 13$.

For $d \geq 7$, we can state the next corollary that provides an upper bound to $\mu(\Gamma_d)$.

**Corollary 2.** $\mu(\Gamma_d) \leq F_d + F_{d-2}$, for each $d \geq 6$.

*Proof.* We prove the statement by induction on $d$. For $d = 6$ we know that $\mu(\Gamma_6) = 11 \leq F_6 + F_4$. Assume the claim holds for any $d' < d$, we show it for $d$. By construction and by Corollary 1, $\mu(\Gamma_d) \leq \mu(\Gamma_{d-1}) + \mu(\Gamma_{d-2})$. By induction hypothesis, $\mu(\Gamma_{d-1}) + \mu(\Gamma_{d-2}) \leq F_{d-1} + F_{d-3} + F_{d-2} + F_{d-4} = F_d + F_{d-2}$, hence the claim holds. $\square$

Summarizing, in Table 1 we report all the obtained results concerning the optimal solutions and the upper bounds for $\mu(\Gamma_d)$.

## 3.2 Lower Bound and an Approximation Algorithm

Here, we first provide a lower bound for $\mu(\Gamma_d)$ and then we derive an approximation algorithm for the mutual-visibility problem in the class of Fibonacci cubes.

**Table 1.** On the size of a mutual-visibility set $X$ for $\Gamma_d$ as $d$ varies. We found $\mu$-sets (i.e., optimal solutions) for $d \leq 6$ and an upper bound for $\mu(\Gamma_d)$ when $d \geq 7$.

| $d$ | $n(\Gamma_d)$ | $\mu(\Gamma_d)$ |
|---|---|---|
| 1 | 2 | 2 |
| 2 | 3 | 2 |
| 3 | 5 | 3 |
| 4 | 8 | 5 |
| 5 | 13 | 8 |
| 6 | 21 | 11 |
| $d \geq 7$ | $F_{d+2}$ | $\mu(\Gamma_d) \leq F_d + F_{d-2}$ |

**Theorem 1.** $\mu(\Gamma_d) \geq \max_{0 \leq k \leq \lceil \frac{d}{2} \rceil} \binom{d-k+1}{k}$, for any $d \geq 1$.

*Proof.* Let $X_k$ be the subset of $V(\Gamma_d)$ containing all the vertices that are at distance $k$ from the vertex labeled with all zeroes. By construction, and by the Hamming distance, the elements of $X_k$ are all the vertices whose labels contain exactly $k$ 1's. Hence, it is easy to find a shortest path between two of the selected vertices, say $x$ and $y$ that are at a distance $j$ from each other. It suffices to detect the $j$ differences among the labels associated with the two vertices. Then, by first replacing (one per step) the 1's present in $x$ but not in $y$ with 0's and then similarly replacing the 0's with 1's, equals to determine a shortest path between $x$ and $y$. Such a path is shortest because at each step makes a change in the direction of the destination, i.e., its length is exactly $j$ (the Hamming distance). Furthermore, along the chosen path, there are no vertices in $X_k$ as the number of 1's is always smaller than $k$ until the destination.

The number of vertices with a fixed number $k$ of 1's in a Fibonacci cube of dimension $d$ is $\binom{d-k+1}{k}$ (see [2]). Since we are looking for a lower bound of $\mu(d)$, we can maximize the obtained binomial coefficient, hence obtaining the claim. $\square$

An immediate consequence of the above theorem is the existence of an approximation algorithm for the mutual-visibility problem in the context of Fibonacci cubes, as stated by the next two corollaries.

From Corollary 2 and Theorem 1, we obtain

**Corollary 3.** *There exists an* $\left(\lceil \frac{d}{2} \rceil + 1\right)$*-approximation algorithm for the* MUTUAL-VISIBILITY *problem on a Fibonacci cube* $\Gamma_d$, *for any* $d \geq 1$.

*Proof.* Concerning $d \leq 6$, we have shown optimal solutions. For $d \geq 7$, we can compare the upper bound provided by Corollary 2 with the lower bound obtained by means of the set $X_k$ provided in the proof of Theorem 1. By exploiting the relationship of Fibonacci numbers with the binomial coefficient as shown in Equation (2), we obtain an approximation algorithm that guarantees a ratio of:

**Table 2.** Numerical comparisons among the actual lower bound (LB), the upper bound (UB) and the approximation ratio of $\left\lceil \frac{d}{2} \right\rceil + 1$ guaranteed for $\mu(\Gamma_d)$, $d \geq 7$.

| $d$ | $n(\Gamma_d) = F_{d+2}$ | LB: Algorithm | UB: $F_d + F_{d-2}$ | UB/LB | $\lceil \frac{d}{2} \rceil + 1$ |
|---|---|---|---|---|---|
| 7 | 34 | 15 | 18 | 1.200 | 5 |
| 8 | 55 | 21 | 29 | 1.380 | 5 |
| 9 | 89 | 35 | 47 | 1.342 | 6 |
| 10 | 144 | 56 | 76 | 1.357 | 6 |
| 11 | 233 | 84 | 123 | 1.464 | 7 |
| 15 | 1597 | 495 | 843 | 1.703 | 9 |
| 20 | 17711 | 5005 | 9349 | 1.867 | 11 |
| 25 | 196418 | 50388 | 103682 | 2.057 | 14 |
| 30 | 2178309 | 497420 | 1149851 | 2.311 | 16 |
| 50 | $3.295 \times 10^{10}$ | $6.107 \times 10^9$ | $1.739 \times 10^{10}$ | 2.848 | 26 |
| 100 | $9.273 \times 10^{20}$ | $1.225 \times 10^{20}$ | $4.895 \times 10^{20}$ | 3.994 | 51 |
| 200 | $7.345 \times 10^{41}$ | $6.856 \times 10^{40}$ | $3.877 \times 10^{41}$ | 5.654 | 101 |
| 500 | $3.650 \times 10^{104}$ | $2.168 \times 10^{103}$ | $1.394 \times 10^{104}$ | 8.883 | 251 |
| 1000 | $1.137 \times 10^{209}$ | $4.792 \times 10^{207}$ | $6.006 \times 10^{208}$ | 12.553 | 501 |

$$\frac{\mu(\Gamma_d)}{|X|} \leq \frac{F_d + F_{d-2}}{\max\limits_{0 \leq k \leq \lceil \frac{d}{2} \rceil} \binom{d-k+1}{k}} < \frac{F_{d+2}}{\max\limits_{0 \leq k \leq \lceil \frac{d}{2} \rceil} \binom{d-k+1}{k}} =$$

$$= \frac{\sum_{k=0}^{\lceil \frac{d}{2} \rceil} \binom{d-k+1}{k}}{\max\limits_{0 \leq k \leq \lceil \frac{d}{2} \rceil} \binom{d-k+1}{k}} = \left\lceil \frac{d}{2} \right\rceil + 1,$$

where the last step comes by the fact that the summation at the numerator is composed by $\lceil \frac{d}{2} \rceil + 1$ elements, each of which admits a ratio with the maximum of at most 1.                                                                                                  □

We remind that in a Fibonacci cube $\Gamma_d$ of $n$ vertices we have $n = F_{d+2}$. By the Binet's formula shown in Eq. 4, we obtain $d = O(\log n)$. Consequently, the approximation provided by the above theorem for an $n$-vertex Fibonacci cube can be expressed as $O(\log n)$.

## 4   Concluding Remarks

In this paper, we have studied mutual-visibility in Fibonacci cubes. In particular, we have provided an upper and a lower bound to $\mu(\Gamma_d)$, hence obtaining a $\left(\lceil \frac{d}{2} \rceil + 1\right)$-approximation algorithm. In Table 2, we report some values for a numerical comparison of lower and upper bounds. This gives rise to posing the question whether a more stringent analysis may lead to tighter bounds. In fact, the growth of the approximation ratio seems rather slower with respect to a linear growth with $d$ guaranteed by the approximation ratio of $\lceil \frac{d}{2} \rceil + 1$. In the last two columns of the table we can compare the actual ratio with that guaranteed by Corollary 3. For instance, when $d = 100$, $d = 200$, $d = 500$ and $d = 1000$, the actual ratio is indeed smaller than 4, 6, 9 and 13, respectively. Furthermore, in practical scenarios is rather unfeasible to have a network structured as a Fibonacci cube of such dimensions with a number of vertices from $10^{20}$ up to $10^{200}$.

An interesting direction for future work is certainly constituted by the study of the total mutual-visibility in Fibonacci cubes.

For Hamming graphs, total mutual-visibility has been solved in [4]. However, in [19] it has been shown that Fibonacci cubes are not Hamming graphs. In particular, it is shown that Fibonacci cubes are prime graphs with respect to the Cartesian product of graphs. In fact, as shown above, while a hypercube $Q_d$ can be defined as the Cartesian product of $d$ copies of $K_2$, Fibonacci cubes do not admit a similar representation, hence results from [4] cannot be exploited.

However, still in [19], it is proven that Fibonacci cubes are *Median* graphs, i.e., undirected graph in which every three vertices $a$, $b$, and $c$ have a unique median: a vertex $m(a,b,c)$ that belongs to shortest paths between each pair of $a$, $b$, and $c$. Perhaps this property might be exploited for the evaluation of the total mutual-visibility in Fibonacci cubes.

Other challenging directions of research for both mutual and total mutual-visibility may concern other topologies, including generalizations of Fibonacci cubes like Fibonacci$(p,r)$-cubes. Let $p, r \le d$, then a Fibonacci $(p,r)$-string of length $d$ is a binary string of length $d$ in which there are at most $r$ consecutive 1's and at least $p$ 0's between two substrings composed of (at most $r$) consecutive 1's. Then the Fibonacci $(p,r)$-cube $\Gamma_d^{(p,r)}$ is the subgraph of $Q_d$ induced by the Fibonacci $(p,r)$-strings (of length $d$). Note that $\Gamma^{(1,d)} = Q_d$ and $\Gamma_d^{(1,1)} = \Gamma_d$.

# References

1. Aloisio, A., Navarra, A.: Constrained connectivity in bounded x-width multi-interface networks. Algorithms **13**(2), 31 (2020). https://doi.org/10.3390/a13020031
2. Azarian, M.K.: Fibonacci identities as binomial sums. Int. J. Contemp. Math. Sci. **7**(38), 1871–1876 (2012)
3. Brandstädt, A., Le, V.B., Spinrad, J.P.: Graph Classes: A Survey. SIAM, Philadelphia (1999)
4. Bujtá, C., Klavžar, S., Tian, J.: Total mutual-visibility in hamming graphs. CoRR **abs/2307.05168** (2023). https://arxiv.org/abs/2307.05168
5. Cicerone, S., Di Fonso, A., Di Stefano, G., Navarra, A.: The geodesic mutual visibility problem for oblivious robots: the case of trees. In: 24th International Conference on Distributed Computing and Networking, ICDCN 2023, pp. 150–159. ACM (2023). https://doi.org/10.1145/3571306.3571401
6. Cicerone, S., Di Fonso, A., Di Stefano, G., Navarra, A.: The geodesic mutual visibility problem: oblivious robots on grids and trees. Pervasive Mob. Comput. **95**, 101842 (2023). https://doi.org/10.1016/J.PMCJ.2023.101842
7. Cicerone, S., Di Fonso, A., Di Stefano, G., Navarra, A.: Time-optimal geodesic mutual visibility of robots on grids within minimum area. In: Dolev, S., Schieber, B. (eds.) Stabilization, Safety, and Security of Distributed Systems. SSS 2023. LNCS, vol. 14310, pp. 385–399. Springer, Cham (2023). https://doi.org/10.1007/978-3-031-44274-2_29
8. Cicerone, S., Di Fonso, A., Di Stefano, G., Navarra, A., Piselli, F.: Mutual visibility in hypercube-like graphs. ArXiv **abs/2308.14443** (2023). https://arxiv.org/abs/2308.14443
9. Cicerone, S., Di Stefano, G., Droždek, L., Hedžet, J., Klavžar, S., Yero, I.G.: Variety of mutual-visibility problems in graphs. Theoret. Comput. Sci. **974**, 114096 (2023). https://doi.org/10.1016/j.tcs.2023.114096
10. Cicerone, S., Di Stefano, G., Klavžar, S.: On the mutual visibility in cartesian products and triangle-free graphs. Appl. Math. Comput. **438**, 127619 (2023). https://doi.org/10.1016/j.amc.2022.127619
11. Cicerone, S., Di Stefano, G., Klavžar, S., Yero, I.G.: Mutual-visibility in strong products of graphs via total mutual-visibility. CoRR **abs/2210.07835** (2022). https://arxiv.org/abs/2210.07835
12. Di Luna, G.A., Flocchini, P., Chaudhuri, S.G., Poloni, F., Santoro, N., Viglietta, G.: Mutual visibility by luminous robots without collisions. Inf. Comput. **254**, 392–418 (2017). https://doi.org/10.1016/j.ic.2016.09.005
13. Di Stefano, G.: Mutual visibility in graphs. Appl. Math. Comput. **419**, 126850 (2022). https://doi.org/10.1016/j.amc.2021.126850
14. Dudeney, H.E.: Amusements in Mathematics. Nelson, Edinburgh (1917)
15. Egecioglu, Ö., Klavžar, S., Mollard, M.: Fibonacci Cubes with Applications and Variations. WorldScientific, Singapore (2023). https://doi.org/10.1142/13228
16. Flammini, M., Moscardelli, L., Navarra, A., Pérennes, S.: Asymptotically optimal solutions for small world graphs. Theory Comput. Syst. **42**(4), 632–650 (2008). https://doi.org/10.1007/s00224-007-9073-y
17. Hsu, W.: Fibonacci cubes-a new interconnection technology. IEEE Trans. Parallel Distrib. Syst. **4**(1), 3–12 (1993). https://doi.org/10.1109/71.205649
18. Kiliç, E.: The Binet formula, sums and representations of generalized fibonacci p-numbers. Eur. J. Comb. **29**(3), 701–711 (2008). https://doi.org/10.1016/j.ejc.2007.03.004
19. Klavžar, S.: Structure of fibonacci cubes: a survey. J. Comb. Optim. **25**(4), 505–522 (2013). https://doi.org/10.1007/s10878-011-9433-z
20. Klavžar, S., Neethu, P., Chandran, S.: The general position achievement game played on graphs. Discret. Appl. Math. **317**, 109–116 (2022). https://doi.org/10.1016/j.dam.2022.04.019

21. Klavžar, S., Tian, J.: Graphs with total mutual-visibility number zero and total mutual-visibility in cartesian products. Discussiones Math. Graph Theory (2023). https://doi.org/10.7151/dmgt.2496

22. Kosowski, A., Navarra, A., Pinotti, M.C.: Exploiting multi-interface networks: connectivity and cheapest paths. Wirel. Netw. **16**(4), 1063–1073 (2010). https://doi.org/10.1007/s11276-009-0188-8

23. Kuziak, D., Rodríguez-Velázquez, J.A.: Total mutual-visibility in graphs with emphasis on lexicographic and cartesian products. CoRR **abs/2306.15818** (2023). https://arxiv.org/abs/2306.15818

24. Manuel, P.D., Klavžar, S.: The graph theory general position problem on some interconnection networks. Fundam. Informaticae **163**(4), 339–350 (2018). https://doi.org/10.3233/FI-2018-1748

25. Mostarda, L., Navarra, A., Nobili, F.: Fast file transfers from IoT devices by using multiple interfaces. Sensors **21**(1), 36 (2021). https://doi.org/10.3390/s21010036

26. Munarini, E., Salvi, N.Z.: Structural and enumerative properties of the fibonacci cubes. Discret. Math. **255**(1–3), 317–324 (2002). https://doi.org/10.1016/S0012-365X(01)00407-1

27. Poudel, P., Aljohani, A., Sharma, G.: Fault-tolerant complete visibility for asynchronous robots with lights under one-axis agreement. Theor. Comput. Sci. **850**, 116–134 (2021). https://doi.org/10.1016/j.tcs.2020.10.033

28. Prabha, R., Devi, S.R., Manuel, P.: General position problem of butterfly networks. ArXiv **/abs/2302.06154** (2023). https://arxiv.org/abs/2302.06154

29. Sharma, G., Vaidyanathan, R., Trahan, J.L.: Optimal randomized complete visibility on a grid for asynchronous robots with lights. Int. J. Netw. Comput. **11**(1), 50–77 (2021)

# Evaluation of the Trust Zone Model with the Information Flow Control

Shigenari Nakamura[1]([✉]) and Makoto Takizawa[2]

[1] Tokyo Denki University, Tokyo, Japan
s.nakamura@mail.dendai.ac.jp
[2] Hosei University, Tokyo, Japan
makoto.takizawa@computer.org

**Abstract.** A ZT model is proposed and discussed these days. Here, since it is assumed that all the accesses are not trustworthy, every access can be allowed is checked. Hence, the ZT model is regarded as the method to protect networks of enterprises from any threats. Since it is critical to make the shift from the present system model to the ZT model smooth, a TZ (Trust Zone) model is proposed. Here, every object is in a trust zone and an authorization decision is made for each trust zone. However, prevention of the illegal information flow is not discussed in the TZ model. In our previous studies, a TZMAC (Trust Zone with Mandatory Access Control) model is proposed. Here, not only objects but also subjects belong to trust zones. Access rules are given based on the relation between the trust zones of the subject and object so that the illegal information flow does not occur. We evaluate the TZMAC model compared with the TZ model in this paper. In both models, operations which do not follow each access rule are rejected to protect the networks from illegal accesses. Evaluation results show that there are operations rejected but illegal information flow occurs in the TZ model. Although the number of operations rejected in the TZMAC model is larger than the TZ model, illegal information flow does not occur.

**Keywords:** Information flow · Zero trust model · Trust zone model · Mandatory access control model

## 1 Introduction

A perimeter model is known as the method to protect networks of enterprises from any threats. In the perimeter model, every access from the inside is regarded as trustworthy. Therefore, it is difficult to protect the networks from any threats these days. In a ZT model [8, 19], since it is assumed that all the accesses are not trustworthy, every access from both the inside and outside can be allowed is checked. Here, many types of information on the access is used for an authorization decision to protect networks from any threats.

However, to change the present system model to the ZT model is not easy. Therefore, a TZ model is proposed [2]. Here, every object is in a trust zone. An authorization decision is made for each trust zone.

© The Author(s), under exclusive license to Springer Nature Switzerland AG 2024
L. Barolli (Ed.): AINA 2024, LNDECT 199, pp. 34–42, 2024.
https://doi.org/10.1007/978-3-031-57840-3_4

A subject may get information of an object via other subjects and objects even if the subject is not allowed to get the information in access control models, i.e. illegal information flow may occur [1]. In the paper [3], a scheduler is proposed where multiple conflicting transactions are synchronized in order to serialize conflicting operations from these transactions and to prevent illegal information flow in the RBAC (Role-Based Access Control) model [21]. In our previous studies [9, 11], types of protocols to prevent illegal information flow in the RBAC and TBAC (Topic-Based Access Control) models are proposed. Furthermore, we consider the IoT (Internet of Things) [18] with the CapBAC (Capability-Based Access Control) model [6, 7]. Here, protocols to prevent illegal information flow are proposed [10, 12, 13] and improved [14, 15]. The electric energy consumption of the protocols is evaluated in the paper [16].

In the TZ model, the prevention of the illegal information flow is not discussed. In this paper, we consider a MAC model [5, 20] where subjects and objects belong to security classes. A relation between a pair of security classes of a subject and an object decides access rules so that illegal information flow does not occur.

In our previous studies, a TZMAC model is proposed [17]. Here, trust zones are used as the security classes. Not only objects but also subjects belong to trust zones. If a subject accesses an object, not only the information on the access but also the relation between the trust zones of the subject and object are checked.

In this paper, we evaluate the TZMAC model compared with the TZ model. Here, subjects randomly issue operations to objects. In the TZ model, access rules are given based on the many types of features of subjects and the trust zones of objects. In the evaluation, we assume that the departments where subjects belong are considered to decide the access rules. On the other hand, in the TZMAC model, access rules are given based on the relation between trust zones of subjects and objects. In both models, operations which do not follow each access rule are rejected to protect the networks from illegal accesses. The evaluation results show that there are operations rejected but illegal information flow occurs in the TZ model. Although the number of operations rejected in the TZMAC model is larger than the TZ model, illegal information flow does not occur.

In Sect. 2, we discuss the TZ model and MAC model. In Sect. 3, we present the TZMAC model. In Sect. 4, we evaluate the TZMAC model compared with the TZ model.

## 2   System Model

### 2.1   TZ Model

In a ZT model [8, 19], since it is assumed that all the accesses are not trustworthy, every access can be allowed is checked. Hence, the ZT model is regarded as the method to protect networks from any threats. However, to change the present system model to the ZT model is not easy. Since it is critical to make the shift from the present system model to the ZT model smooth, a TZ model [2] is proposed.

Here, every object is in a trust zone. First, an access from a subject arrives at a PEP (Policy Enforcement Point). Next, the access is forwarded to the PDP (Policy Decision Point) from the PEP. Then, the access is checked and an authorization decision is made in a trust algorithm by the PDP. Finally, the decision is sent to the PEP from the PDP. If the PDP decides to allow the access, the connection between the subject and object is enabled by the PEP. Otherwise, the connection is terminated by the PEP, i.e. the access is rejected. For each trust zone, a PEP is placed. Therefore, the PDP makes an authorization decision and the decision is used for a trust zone.

Authorization decisions are made by a trust algorithm in the PDP. The trust algorithm is the process to check each access is allowed. Here, many types of information on the access is used to protect networks. In this paper, we consider the trust algorithm based on criteria [19]. Here, for each access, information on the access is compared with the criteria. The access is allowed only if all the criteria are met.

## 2.2 MAC Model

In information systems, subjects need to manipulate objects according to access control models [4,21,22]. Here, a subject is issued access rights. Subjects can manipulate objects in only operations which are allowed by access rights issued to the subjects. We consider subjects $sb_i$ and $sb_j$ which manipulate objects $o_1$ and $o_2$. $sb_i$ is allowed to read information from $o_1$ and write information to $o_2$. On the other hand, $sb_j$ is allowed to read information from $o_2$. First, $sb_i$ reads information of $o_1$ and write the information to $o_2$. Then, $sb_j$ reads information from $o_2$. Here, $sb_j$ can read the information of $o_1$ written to $o_2$ by issuing a read operation to $o_2$. As above example, a subject may get information of an object which the subject is not allowed to get in the access control model via other subjects and objects, i.e. illegal information flow occurs [1].

The prevention of the illegal information flow is critical. In a MAC model [5,20], subjects and objects belong to security classes. Let an entity $e_k$ be of a security class $sc_k$. A precedent relation "→" between a pair of security classes means that information can flow between the security classes. For example, $sc_k \rightarrow sc_l$ means that information can flow from an entity $e_k$ to an entity $e_l$.

Security classes are partially ordered by the precedent relation "→" in a lattice. Access rules for a subject $sb$ to access an object $o$ are given based on the precedent relation "→" as follows:

**Definition 1.** $sb$ of $sc_k$ can write data to $o$ of $sc_l$ iff (if and only if) $sc_k \rightarrow sc_l$.

**Definition 2.** $sb$ of $sc_k$ can read data from $o$ of $sc_l$ iff $sc_l \rightarrow sc_k$.

**Definition 3.** $sb$ of $sc_k$ can read and write data from and to $o$, i.e. modify data of $o$, of $sc_l$ iff $sc_k \rightarrow sc_l$ and $sc_l \rightarrow sc_k$.

## 3    TZMAC Model

In the TZ model, every object is in a trust zone. The PDP makes an authorization decision and the decision is used for a trust zone. The decision is made in the trust algorithm and sent to the PEP. If the PDP decides to allow the access, the connection between the subject and object is enabled by the PEP. Otherwise, the connection is terminated by the PEP, i.e. the access is rejected.

Many types of information on an access is used in the trust algorithm. However, the prevention of the illegal information flow is not discussed. Therefore, we consider a MAC model. Here, subjects and objects belong to security classes. Access rules are given based on the precedent relation "→" between a pair of security classes. If subjects access objects following the access rules in the MAC model, illegal information flow does not occur.

In our previous studies [17], a TZMAC model is proposed. Here, the trust zones are used as the security classes. Not only objects but also subjects belong to trust zones. If a subject accesses an object, not only the information on the access but also the precedent relation between the trust zones of the subject and object are checked. In this paper, we assume that subjects issue *read* and *write* operations to objects.

Suppose a subject $sb_i$ accesses an object $o$. Let $e.tz$ be a trust zone to which an entity $e$ belongs. In the trust algorithm, the trust zones $sb_i.tz$ and $o.tz$ of $sb_i$ and $o$ are used. How $sb_i$ is allowed to access $o$ is decided based on the precedent relation "→" between the trust zones $sb_i.tz$ and $o.tz$. The access is rejected if the precedent relation "→" is not satisfied. Therefore, illegal information flow does not occur.

## 4   Evaluation

In this paper, we evaluate the TZMAC model compared with the TZ model in the simulation. Here, we consider fifty subjects and one hundred objects. In the TZ model, we assume that there are three departments, "Laboratory", "Committee", and "Club". Every subject belongs to at least one department. On the other hand, every object is in a trust zone. In the evaluation, we consider eight trust zones, $\{\phi\}$, $\{Laboratory\}$, $\{Committee\}$, $\{Club\}$, $\{Laboratory,Committee\}$, $\{Laboratory,Club\}$, $\{Committee,Club\}$, and $\{Laboratory,Committee,Club\}$. We assume access rules are given based on the relation between the departments where the subjects belong and trust zones of the objects. In the TZMAC model, not only objects but also subjects belong to trust zones. The trust zones are used as the security classes. Figure 1 shows the lattice of trust zones. An arrow means that the precedent relation "→" among a pair of trust zones is satisfied. For example, $\{Committee\} \rightarrow \{Committee,Club\}$ holds. The numbers of subjects and objects are assumed as shown in Table 1. The column "departments" means the departments where the subjects belong in the TZ model. The column also means the departments in each trust zone where the objects belong in the TZ model and the subjects and objects belong in the TZMAC model, respectively. For example, five subjects and ten objects belong to the pair of departments "Committee" and "Club" and the trust zone $\{Committee,Club\}$ in the TZ model. On the other hand, five subjects and ten objects belong to the trust zone $\{Committee,Club\}$ in the TZMAC model.

Let $st$ be the simulation steps. One simulation step is as follows:

1. The number $n$ of subjects are randomly selected. Here, the number $n$ is randomly decided ($1 \leq n \leq 50$).
2. The number $n$ of subjects issue operations. Each subject issues one operation to an object. The operation is randomly decided to be read or write. The order in which the number $n$ of subjects issue an operation is randomly decided.

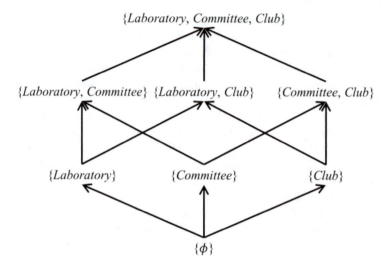

**Fig. 1.** Lattice of trust zones.

**Table 1.** Numbers of subjects and objects.

| departments | number of subjects | number of objects |
|---|---|---|
| Laboratory | 10 | 20 |
| Committee | 10 | 20 |
| Club | 10 | 20 |
| Laboratory, Committee | 5 | 10 |
| Laboratory, Club | 5 | 10 |
| Committee, Club | 5 | 10 |
| Laboratory, Committee, Club | 5 | 10 |

3. Each operation is allowed or rejected based on the relation between a subject and an object.

In the second procedure, an object $o$ is randomly selected. The subject $sb_i$ randomly selects the object $o$ where the departments in $o.tz$ are the same as the set of departments where the subject $sb_i$ belongs with the probability 0.7 in the TZ model. On the other hand, the subject $sb_i$ randomly selects the object $o$ whose trust zone $o.tz$ includes at least one department where the subject $sb_i$ belongs with the probability 0.3. In the TZMAC model, the subject $sb_i$ randomly selects the object $o$ from the same trust zone with the probability 0.7. On the other hand, the subject $sb_i$ randomly selects the object $o$ from other trust zones including at least one common department with the probability 0.3. For example, in the TZ model, a subject $sb_i$ belonging to the department "Committee" randomly selects an object $o$ from objects belonging to the trust zone {*Committee*} with the probability 0.7. On the other hand, the subject $sb_i$ randomly selects an object $o$ from objects belonging to {*Committee*}, {*Laboratory,Committee*},

{*Committee,Club*}, or {*Laboratory,Committee,Club*} with the probability 0.3. In the TZMAC model, a subject $sb_i$ belonging to the trust zone {*Committee*} randomly selects an object $o$ from objects belonging to the same trust zone {*Committee*} with the probability 0.7. On the other hand, the subject $sb_i$ randomly selects an object $o$ from objects belonging to {*Committee*}, {*Laboratory,Committee*}, {*Committee,Club*}, or {*Laboratory,Committee,Club*} with the probability 0.3. The above procedures for $st$ are iterated five hundred times and operations are counted. Finally, the average values are calculated.

The numbers of operations in the TZ model for the simulation steps $st$ are shown in Fig. 2. The line "all" shows the total number of operations issued by subjects. In the TZ model, an operation issued by a subject $sb_i$ to an object $o$ is allowed or rejected based on the relation between the departments where the subject $sb_i$ belongs and the trust zone $o.tz$. We assume that the following access rules are given in the TZ model:

- If $sb_i$ belongs to every department in $o.tz$, both read and write operations are allowed.
- If all the departments where $sb_i$ belongs are in $o.tz$ but $o.tz$ includes at least one department that is not in $sb_i.tz$, only write operation is allowed.
- Otherwise, every operation is rejected.

The line "read_rejected" shows the number of read operations rejected. On the other hand, the line "write_rejected" shows the number of write operations rejected. The total number of operations rejected increases as the simulations steps $st$ increase. For example, about 310 operations are rejected for $st = 100$ and about 1,560 operations are rejected for $st = 500$. The line "illegal information flow" shows the number of operations occurring illegal information flow. In the TZ model, operations which do not follow the above access rules are rejected to protect the network from illegal accesses. However, illegal information flow occurs.

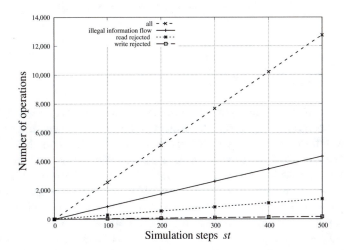

**Fig. 2.** Number of operations in the TZ model.

The numbers of operations in the TZMAC model for the simulation steps $st$ are shown in Fig. 3. The line "all" shows the total number of operations issued by subjects. In the TZMAC model, an operation issued by a subject $sb_i$ to an object $o$ is allowed or rejected based on the precedent relation "$\rightarrow$" between the trust zones $sb_i.tz$ and $o.tz$. The line "read_rejected" shows the number of read operations rejected. On the other hand, the line "write_rejected" shows the number of write operations rejected. The total number of operations rejected increases as the simulations steps $st$ increase. For example, about 420 operations are rejected for $st = 100$ and about 2,080 operations are rejected for $st = 500$. Although the number of operations rejected in the TZMAC model is larger than the TZ model, illegal information flow does not occur.

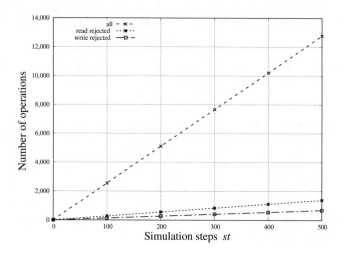

**Fig. 3.** Number of operations in the TZMAC model.

## 5    Concluding Remarks

A ZT model is proposed and discussed these days. Here, since it is assumed that all the accesses are not trustworthy, every access can be allowed is checked. Since it is critical to make the shift from the present system model to the ZT model smooth, a TZ model is proposed. Here, every object is in a trust zone and an authorization decision is made for each trust zone. In the TZ model, the prevention of the illegal information flow is not discussed. Hence, in our previous studies, a TZMAC model is proposed. Here, not only objects but also subjects belong to trust zones. The trust zones are used as the security classes. Not only the information on an access but also the precedent relation between trust zones of the subject and object are checked. In this paper, we evaluate the TZMAC model. In the TZ and TZMAC models, operations which do not follow each access rule are rejected to protect the networks from illegal accesses. The evaluation results show that there are operations rejected but illegal information flow occurs in the TZ model.

Although the number of operations rejected in the TZMAC model is larger than the TZ model, illegal information flow does not occur.

**Acknowledgements.** This work is supported by Japan Society for the Promotion of Science (JSPS) KAKENHI Grant Number JP23K16887.

# References

1. Denning, D.E.R.: Cryptography and Data Security. Addison Wesley, Boston (1982)
2. Department of Homeland Security, Cybersecurity & Infrastructure Security Agency: Trusted Internet Connections (TIC) 3.0 core guidance documents (2021). https://www.cisa.gov/resources-tools/resources/trusted-internet-connections-tic-30-core-guidance-documents
3. Enokido, T., Barolli, V., Takizawa, M.: A legal information flow (LIF) scheduler based on role-based access control model. Int. J. Comput. Stand. Interfaces **31**(5), 906–912 (2009)
4. Fernandez, E.B., Summers, R.C., Wood, C.: Database Security and Integrity. Adison Wesley, Boston (1980)
5. Ferraiolo, D.F., Kuhn, D.R., Chandramouli, R.: Role-Based Access Control, 2nd edn. Artech, Norwood (2007)
6. Gusmeroli, S., Piccione, S., Rotondi, D.: A capability-based security approach to manage access control in the internet of things. Math. Comput. Model. **58**(5–6), 1189–1205 (2013)
7. Hernández-Ramos, J.L., Jara, A.J., Marín, L., Skarmeta, A.F.: Distributed capability-based access control for the internet of things. J. Internet Serv. Inf. Secur. **3**(3/4), 1–16 (2013)
8. Kindervag, J., Balaouras, S., Mak, K., Blackborow, J.: No more chewy centers: the zero trust model of information security. Technical report, Forrester Research (2016)
9. Nakamura, S., Duolikun, D., Enokido, T., Takizawa, M.: A read-write abortion protocol to prevent illegal information flow in role-based access control systems. Int. J. Space-Based Situated Comput. **6**(1), 43–53 (2016)
10. Nakamura, S., Enokido, T., Takizawa, M.: Information flow control based on the CapBAC (capability-based access control) model in the IoT. Int. J. Mob. Comput. Multimedia Commun. **10**(4), 13–25 (2019)
11. Nakamura, S., Enokido, T., Takizawa, M.: Information flow control in object-based peer-to-peer publish/subscribe systems. Concurr. Comput. Pract. Exp. **32**(8), e5118 (2020)
12. Nakamura, S., Enokido, T., Takizawa, M.: Implementation and evaluation of the information flow control for the internet of things. Concurr. Comput. Pract. Exp. **33**(19), e6311 (2021)
13. Nakamura, S., Enokido, T., Takizawa, M.: Information flow control based on capability token validity for secure IoT: implementation and evaluation. IoT J. **15**, 100423 (2021)
14. Nakamura, S., Enokido, T., Takizawa, M.: Traffic reduction for information flow control in the IoT. In: Proceedings of the 16th International Conference on Broad-Band Wireless Computing, Communication and Applications, pp. 67–77 (2021)
15. Nakamura, S., Enokido, T., Takizawa, M.: Capability token selection algorithms to implement lightweight protocols. IoT J. **19**, 100542 (2022)
16. Nakamura, S., Enokido, T., Takizawa, M.: Assessment of energy consumption for information flow control protocols in IoT devices. IoT J. **24**, 100992 (2023)
17. Nakamura, S., Takizawa, M.: Trust zone model with the mandatory access control model. In: Proceedings of the 12th International Conference on Emerging Internet, Data, and Web Technologies (2024, accepted)
18. Oma, R., Nakamura, S., Duolikun, D., Enokido, T., Takizawa, M.: An energy-efficient model for fog computing in the internet of things (IoT). IoT J. **1–2**, 14–26 (2018)

19. Rose, S., Borchert, O., Mitchell, S., Connelly, S.: Zero trust architecture. Technical report, National Institute of Standards and Technology (2020). https://csrc.nist.gov/publications/detail/sp/800/207/final
20. Sandhu, R.S.: Lattice-based access control models. Computer **26**(11), 9–19 (1993)
21. Sandhu, R.S., Coyne, E.J., Feinstein, H.L., Youman, C.E.: Role-based access control models. Computer **29**(2), 38–47 (1996)
22. Yuan, E., Tong, J.: Attributed based access control (ABAC) for web services. In: Proceedings of the IEEE International Conference on Web Services, ICWS 2005, p. 569 (2005)

# PRVC: A Novel Vehicular Ad-Hoc Network Caching Based on Pre-trained Reinforcement Learning

Ziyang Zhang[✉], Lin Guan, Yuanchen Li, Seth Gbd Johnson, and Qinggang Meng

Loughborough University, Loughborough, UK
{Z.Zhang,L.Guan,Y.Li,Q.Meng}@lboro.ac.uk,
S.Johnson-18@student.lboro.ac.uk

**Abstract.** In recent years, network caching in Vehicular Ad-hoc Network (VANET) has gained significant interest, with particular interest around high mobility nodes. Traditional methods, based on Mobile Ad-hoc Network (MANET) caching, face limitations due to MANET's computational constraints and struggle to effectively address VANET-specific caching needs, leading to subpar performance. Addressing this, this paper introduces the novel Vehicular Ad-hoc Network Caching Method based on Pre-trained Reinforcement Learning (PRVC). This new approach uses pre-trained reinforcement learning to enhance VANET caching Quality of Service (QoS) and adapt to changing cache interests. Our empirical experiments show that PRVC outperforms benchmarks in cache hit rate, latency, and link load.

## 1 Introduction

Vehicular Ad-Hoc Network (VANET), a specialized form of a Mobile Ad-Hoc Network (MANET), is a networking methodology that focuses on vehicles equipped with wireless transceivers as the node base. The communication between nodes allows for the formation of decentralized, mobile networks, which do not rely on fixed infrastructure to operate. Although VANET shares many similarities with MANET, such as decentralized wireless connectivity and mobile nodes, it holds many unique characteristics that differentiate it. For example, compared to MANET, vehicle-based nodes exhibit a reduced requirement for energy efficiency and an increased demand for mobility and computational resources. As a critical element of the evolving Intelligent Transportation System (ITS), VANET enhances traffic efficiency, safety, and quality. Its unique architecture and characteristics have made VANET technology and solutions an emerging research area. In this context, network caching is essential for improving data transmission and service quality in VANETs. Effective caching reduces content retrieval time and network congestion, making the development of innovative caching methods crucial for optimal network performance and Quality of Service (QoS).

L. Barolli (Ed.): AINA 2024, LNDECT 199, pp. 43–54, 2024.
https://doi.org/10.1007/978-3-031-57840-3_5

Current VANET caching methodsare mainly based on static networks or MANET methodologies. When applied to VANET scenarios, these methods struggle due to the high mobility of vehicles, leading to lower cache hit rates and reduced user content access speeds. These methods are also usually reliant on extensive broadcasting. This places undue stress on network bandwidth, which is limited in a VANET contexts. Moreover, VANET has a vast amount of content to be cached and nature of user needs is far more dynamic. This makes it challenging to optimize caching effectively when using traditional methods, which tend to lack customization for individual nodes.

Recognizing these issues, there is a need for a VANET-specific caching method that can adapt to the high mobility and tailor caching to individual nodes of networks. This has led to the exploration of reinforcement learning (RL) for caching in VANET. However, RL in VANET faces challenges due to the long learning times and slow convergence, which may not align well with the rapidly changing network status and user interests in VANETs.

To address these, we propose a novel vehicular ad-hoc network caching based on pre-trained reinforcement learning method, named PRVC. This method enhances the efficiency of RL convergence, whilst retaining its effectiveness. Empirical simulations demonstrate that the proposed PRVC offers improved performance results. This includes, higher cache hit ratios, reduced content delivery times, and lower link loads, making it a more suitable solution for the unique challenges of VANET caching.

The original contributions of this paper are summarized below: 1) Enhancing VANET caching using a pre-trained reinforcement learning approach. 2) Dynamic optimization of cache content tailored to the individual interests of each node, addressing the unique and variable requirements within VANET. 3) The proposed PRVC method demonstrates superior network performance in simulations, outperforming conventional VANET caching techniques in several aspects, including higher cache hit rate, reduced latency, lower link load, and improved convergence efficiency. 4) PRVC exhibits a faster convergence speed in comparison to traditional mathematical algorithms and standard RL methods, indicating its effectiveness in rapidly changing VANET environments.

The paper is organised as follows: Sect. 2 reviews related work and highlights the advantages of the proposed PRVC method. Section 3 elaborates on the methodology of using pre-trained reinforcement learning-based approaches for VANET caching. Section 4 describes the experimental environment including setting up configuration and relevant parameters etc. Section 5 presents a comprehensive summary of the results, validating the performance improvements brought about by PRVC. Section 6 concludes the contributions and outlines potential future research in the proposed research domain.

# 2   Related Work

As the data content could be stored in multiple sources in VANET, the usage of the caching mechanisms became significant in enhancing the VANET QoS [1]. VANETs are highly dynamic in their topology, with rapid node changes. The hardware within nodes performs well, but does have limitations. Decision-making relies on monitoring global network conditions for applications like traffic management [2,3].

Traditional caching policies like Least Recently Used (LRU) [4], LRU-2 [5], and Most Recently Used (MRU) [6] manage cache by monitoring data access patterns. LRU eliminates the longest unused data, LRU-2 uses a request history for caching decisions, and MRU discards the most recent data when needed. These algorithms suit vehicle network caching due to low computational demands, yet they often fall short in adapting to VANET's dynamic environments, impacting network Quality of Service (QoS).

Enhancing VANET QoS with traditional methods has been an ongoing research focus. This includes predicting node trajectories using current locations and velocities to proactively cache content at future nodes along vehicle routes [7]. Various content caching criteria have been proposed, such as friendship [8], based on communication frequency and duration, and social similarity [9], calculated using the Jaccard similarity coefficient for cached content. While these content evaluation approaches improve VANET QoS, they cannot fully address the challenges posed by VANET's high dynamics and geographical constraints.

Because of the VANET high dynamic, the communications would face sever link loss issues, and the machine or RL approach could predict the consumers traffic trajectory to reduce the network traffic load and make prediction of the future content requirement based on local historical requests record to improve the QoS generally [10]. Alotaibi et al. [11] applied Deep Q-Learning in making caching decision process by minimizing the difference between the ideal profit and the real gain. The system would slice one day into several time periods and group them with peak and off-peak label to apply different calculating formulations. This scheme could be deployed in vehicle nodes and decreased the retrieve cost. However, these algorithms should update the learning parameters online.

Blasco and Gunduz [12] proposed the Learning-Based Optimization of Cache Content algorithm for the small base stations (sBS) which optimization the placed cache content by jointly considering the already known popular content and learning the historical data requests via machine learning. This algorithm showed short convergence time and increase the cache hit rate. Jiang et al. [13] used the Multi-Agent Reinforcement Learning (MARL) in only using the historical requests record without the needs of knowing the data popularity to make caching decisions. The proposed scheme could reduce the download latency and improve cache hit rate. However, as the complexity of the proposed algorithm, these algorithms could only be applied on the sBS, the vehicles cannot handle the decision process with the limited computation capacity.

Conventional caching on all nodes is limited by VANET's high dynamics. To improve QoS, machine learning methods are employed to adjust caching decisions in response to evolving network topology. However, due to high computation costs, many algorithms can only be deployed on sBS. In this paper, we proposed a novel algorithm capable of deployment on vehicular nodes, requiring only limited computational resources. The proposed method exhibits adaptability in a dynamic environment, effectively responding to the challenges of high mobility and the variability of cache states within network nodes. This adaptability is achieved through the pre-trained reinforcement learning techniques, demonstrating its efficacy in VANET caching scenarios through cache hit ratio, latency and link load.

## 3   Vehicular Ad-Hoc Network Caching Method Based on Pre-trained Reinforcement Learning

VANET, a vehicle-based ad-hoc network, comprises vehicles, Roadside Units (RSUs), and data servers. Vehicles, as mobile nodes, have limited cache space and high mobility. RSUs are stationary with larger cache capacity. Data servers, static nodes with unlimited cache, store all content and connect directly to RSUs. The proposed PRVC method, works on both vehicles and RSUs. It dynamically adjusts the cache through replacement control, boosting cache performance.

### 3.1   Cached Content Forward of PRVC

PRVC functions through two primary stages: processing cache requests and replies. In the vehicle stage, when a vehicle receives a cache request (self-generated or from others), it first verifies if the request is a duplicate. Duplicate requests are ignored to prevent broadcast loops. The vehicle then checks its cache for the requested content; if available, a reply is sent, and PRVC rewards this positively. If not, a negative reward is noted, and the request is broadcast to adjacent nodes. At the RSU stage, the process starts by identifying duplicate requests. Should the RSU's cache hold the requested content, a reply is issued. If the content isn't in the RSU cache, the RSU seeks it from the data server instead of further broadcasting. This procedure is depicted in Fig. 1 of the paper.

In the cache reply phase, when a vehicle receives a cache reply, it firstly checks if it's already received a similar reply for the same target, ignoring duplicates. Next, it determines if it is the intended recipient of the reply. If it is, the vehicle then assesses whether cache replacement is necessary. If the vehicle is not the target, it forwards the reply before checking if the content is already in its cache. If not present, it considers cache replacement; if already cached, the process ends. When replacement is deemed necessary, the decision on what to replace is guided by the RL algorithm. If no replacement is needed, the content is simply stored in the cache.

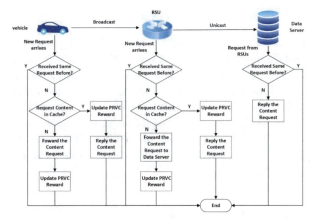

**Fig. 1.** PRVC request process

For RSUs, the process starts by checking for duplicate replies, which are ignored. Subsequently, the reply is forwarded to nearby vehicles. The RSU then either replaces or discards new cache content based on the RL algorithm's decision.

This cache reply process is detailed in Fig. 2 of the paper.

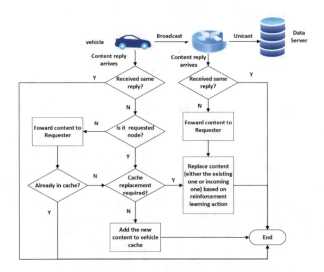

**Fig. 2.** PRVC reply process

## 3.2 Content Replacement Selection of PRVC

In the proposed cache replacement algorithm for VANET, the process begins by assessing the current cache contents and their locations. This is then fed as state information to the RL agent and a pre-trained algorithm. This pre-trained component informs the RL agent, hastening convergence by influencing the rewards. The RL agent determines the action for cache replacement, controlling which content is replaced. This process, including the cache status before and after replacement, is recorded to calculate rewards. Positive or negative rewards are given based on whether a new cache request can be satisfied with the current cache, influencing future cache replacement decisions through the reward function (Fig. 3).

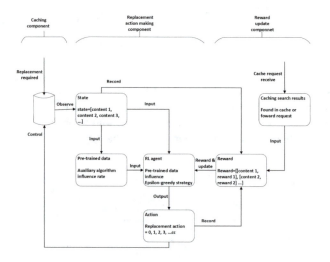

**Fig. 3.** PRVC RL update

In the VANET caching algorithm, Q-learning is chosen for its simplicity and quick convergence. This decision stems from the limitations of deep learning, which, despite potentially better performance, requires more computing power, data, and time to converge - resources that vehicles in VANET often lack. Additionally, the high mobility of VANET means longer convergence times could hinder reaching an optimal solution. Therefore, Q-learning is used and augmented with pre-trained data. This speeds up the learning process, making it a more practical choice for VANET's unique environment.

In this system, the 'state' is the cache content at the node and is always marked as full due to the algorithm being triggered only when cache space is at capacity. The 'action' is defined as the cache location being replaced, a practical choice considering the infinite potential cache contents and finite cache locations at a node.

The reward system in the algorithm assesses the efficacy of cache replacements. However, since the quality of these replacements is not immediately apparent in network performance, an asynchronous reward method is employed. Rewards are assigned based on whether new requests can be satisfied with existing cache content.

VANET characteristics often lead to limited time and requests, potentially hindering the algorithm from achieving an optimal solution. To counter this, pre-trained data guided by the Least Recently Used (LRU) algorithm is used to enhance convergence speed. Over time, the influence of LRU on rewards reduces, permitting the RL algorithm to converge to the optimal solution independently. Furthermore, reactivation of the LRU is possible for aiding convergence to new optimal solutions as network cache preferences shift.

## 4   Simulation Configuration of PRVC

In this paper, we utilized a unified framework integrating Simulation of Urban Mobility (SUMO), OMNET++, and the Vehicle in Network Simulation (VEINS) framework to assess the performance of the proposed caching system. The key parameters used in these simulations are outlined in Table 1.

**Table 1.** Parameters of the simulation

| Parameters | Value of the parameters |
| --- | --- |
| Node Acceleration | $2.6\,\mathrm{ms}^{-2}$ |
| Node Deceleration | $4.5\,\mathrm{ms}^{-2}$ |
| Maximum Node Speed | 120 mph |
| Minimum Node Gap | 2.5 m |
| Node beacon Interval | 0.1 s |
| Header Length | 80 bit |
| Transmitter Power | 20 mW |
| Bit Rate | 6 Mbps |
| Noise floor | $-98\,\mathrm{dBm}$ |
| Minimum Power Level | $-110\,\mathrm{dBm}$ |
| Antenna Height | 1.895 m |
| Decider Center Frequency | 5.89 GHz |
| Time between requests | 1 s |

Simulations were conducted in two settings: urban and highway, to evaluate the proposed method across different vehicle densities and speeds. Urban scenarios featured heavy traffic with slower speeds, while highway scenarios had fewer vehicles and higher speeds. In both environments, 3 and 15 Roadside Units (RSUs) were used to monitor performance changes under varying vehicle-to-vehicle (V2V) caching conditions.

For the urban scenario, the selected area was south of Cambridge Railway Station, encompassing the region from Hills Road and Cherry Hinton Road in the north to Long Road in the south. This network, as detailed in NETEDIT, consists of 3220 edges and 1227 junctions, depicted in Fig. 4.

**Fig. 4.** Urban scenario

The highway scenario focused on the stretch between junctions 35 and 36 on the A14, a principal route connecting urban and rural areas in the simulation. NETEDIT describes this network as having 289 edges and 160 junctions, which is illustrated in Fig. 5.

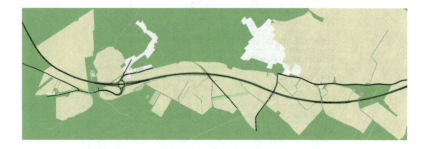

**Fig. 5.** Highway scenario

## 5    Simulation Results and Performance Evaluation

The effectiveness of the proposed method has been assessed using several key metrics: Cache Hit Ratio (CHR), average Content Delivery Time (CDT), and link load. CHR is the ratio of cache hits to the sum of cache hits and misses. CDT measures the time from a vehicle's content request to a nearby RSU to its receipt of content, with average CDT being the mean of all such durations. Link load is the caching-related bandwidth usage compared to total bandwidth. These simulation results are derived from the average across 10 simulations.

## 5.1   Urban Scenario

Figure 6 illustrates the effectiveness of the proposed PRVC method in an urban traffic setting. The specific outcomes for the three evaluation metrics - cache hit rate, latency, and link load - are depicted in Figs. 6(a), 6(b), and 6(c), respectively. Initially, the performance improvement of PRVC might not be immediately noticeable in urban traffic conditions. However, the method shows a significant enhancement after an initial learning period, particularly notable after 1200 s of simulation time.

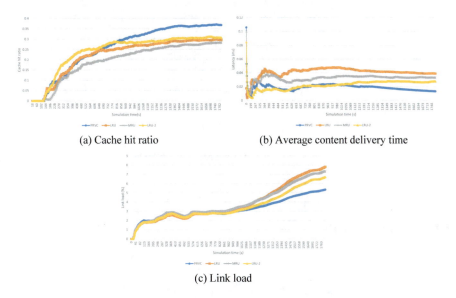

(a) Cache hit ratio                    (b) Average content delivery time

(c) Link load

**Fig. 6.** Results for urban scenario 3 RSUs

In scenarios involving 15 RSUs, the caching performance of the evaluated scheme is detailed in Figs. 7(a), 7(b), and 7(c). When compared to baseline schemes, the proposed PRVC method demonstrates superior performance. As the number of RSUs rises, the convergence rate of the PRVC algorithm also increases. While the baseline solution shows improvement too, the rapid convergence and enhanced performance in later stages maintain the position of PRVC as the most effective method among those tested.

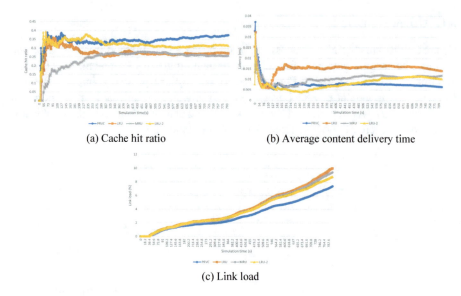

(a) Cache hit ratio

(b) Average content delivery time

(c) Link load

**Fig. 7.** Results for urban scenario 15 RSUs

## 5.2   Highway Scenario

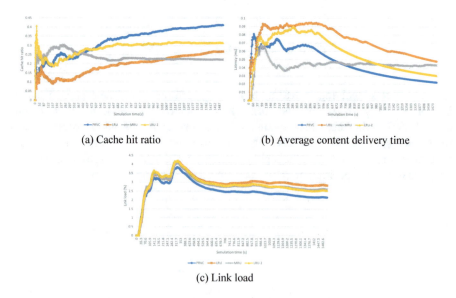

(a) Cache hit ratio

(b) Average content delivery time

(c) Link load

**Fig. 8.** Results for highway scenario 3 RSUs

Figure 8 presents the performance of PRVC in a highway environment with lower vehicle density. The effects on cache hit rate, latency, and link load are specifically shown in Figs. 8(a), 8(b), and 8(c). Unlike urban areas, highways see fewer requests between vehicles due to less dense traffic, leading to a slower initial convergence for PRVC. However, despite this initial delay, PRVC eventually catches up and surpasses the three benchmarks in all key metrics, illustrating its adaptability and efficiency in different traffic scenarios.

As depicted in Fig. 9, an increase in the number of RSUs notably accelerates the convergence speed of the proposed PRVC method. The results showcased in this figure indicate that PRVC not only converges rapidly but also significantly surpasses the three baseline schemes in performance.

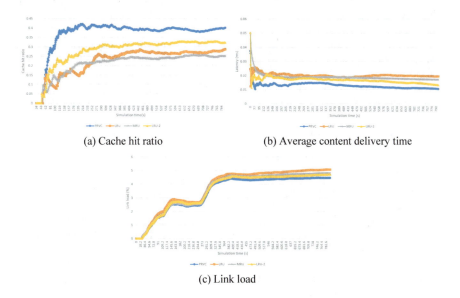

(a) Cache hit ratio                    (b) Average content delivery time

(c) Link load

**Fig. 9.** Results for highway scenario 15 RSUs

## 6    Conclusions and Future Work

This paper introduces PRVC, a novel VANET caching method based on pre-trained reinforcement learning. The proposed PRVC algorithm uses a hybrid approach to overcome the challenges faced by traditional caching methods in addressing the specific requirements of VANET nodes. It utilizes the LRU algorithm to generate pre-training data, which subsequently guides the RL process. Empirical simulations experiments clearly demonstrate that the proposed PRVC enhances network performance across various scenarios and adapts well to evolving network caching demands.

Future work will focus on enhancing the proposed algorithm for large-scale network environments. Investigating more complex mathematical and RL algorithms in such contexts will further confirm the advantages of the pre-training-based reinforcement learning approach.

# References

1. Glass, S., Mahgoub, I., Rathod, M.: Leveraging MANET-based cooperative cache discovery techniques in VANETs: a survey and analysis. IEEE Commun. Surv. Tutorials **19**(4), 2640–2661 (2017). https://doi.org/10.1109/COMST.2017.2707926
2. Lee, M., Atkison, T.: VANET applications: past, present, and future. Veh. Commun. **28**, 100310 (2021). https://doi.org/10.1016/j.vehcom.2020.100310
3. Al-Heety, O.S., Zakaria, Z., Ismail, M., Shakir, M.M., Alani, S., Alsariera, H.: A comprehensive survey: benefits, services, recent works, challenges, security, and use cases for SDN-VANET. IEEE Access **8**, 91028–91047 (2020). https://doi.org/10.1109/ACCESS.2020.2992580
4. Tian, H., Mohri, M., Otsuka, Y., Shiraishi, Y., Morii, M.: LCE in-network caching on vehicular networks for content distribution in urban environments. In: 2015 Seventh International Conference on Ubiquitous and Future Networks, pp. 551–556. IEEE, Sapporo (2015). https://doi.org/10.1109/ICUFN.2015.7182605
5. Boyar, J., Ehmsen, M.R., Kohrt, J.S., Larsen, K.S.: A theoretical comparison of LRU and LRU-K. Acta Informatica **47**(7–8), 359–374 (2010). https://doi.org/10.1007/s00236-010-0123-6
6. Guan, N., Lv, M., Yi, W., Yu, G.: WCET analysis with MRU cache: challenging LRU for predictability. ACM Trans. Embed. Comput. Syst. **13**(4s), 1–26 (2014). https://doi.org/10.1145/2584655
7. Ud Din, I., Ahmad, B., Almogren, A., Almajed, H., Mohiuddin, I., Rodrigues, J.J.P.C.: Left-right-front caching strategy for vehicular networks in ICN-based Internet of Things. IEEE Access **9**, 595–605 (2021). https://doi.org/10.1109/ACCESS.2020.3046887.
8. Yi, B., Wang, X., Huang, M.: Content delivery enhancement in Vehicular Social Network with better routing and caching mechanism. J. Netw. Comput. Appl. **177**, 102952 (2021). https://doi.org/10.1016/j.jnca.2020.102952
9. Yao, L., Wang, Y., Wang, X., Wu, G.: Cooperative caching in vehicular content centric network based on social attributes and mobility. IEEE Trans. Mobile Comput. **20**(2), 391–402 (2021). https://doi.org/10.1109/TMC.2019.2944829
10. Sheraz, M., et al.: Artificial intelligence for wireless caching: schemes, performance, and challenges. IEEE Commun. Surv. Tutorials **23**(1), 631–661 (2021). https://doi.org/10.1109/COMST.2020.3008362
11. Alotaibi, F., Eryilmaz, A., El Gamal, H.: Dynamic pricing and proactive caching with unknown demand profile. In: 2016 54th Annual Allerton Conference on Communication, Control, and Computing (Allerton), pp. 711–716. IEEE, Monticello (2016). https://doi.org/10.1109/ALLERTON.2016.7852301
12. Blasco, P., Gunduz, D.: Learning-based optimization of cache content in a small cell base station. In: 2014 IEEE International Conference on Communications (ICC), pp. 1897–1903. IEEE, Sydney (2014). https://doi.org/10.1109/ICC.2014.6883600
13. Jiang, W., Feng, G., Qin, S., Liu, Y.: Multi-agent reinforcement learning based cooperative content caching for mobile edge networks. IEEE Access **7**, 61856–61867 (2019). https://doi.org/10.1109/ACCESS.2019.2916314

# ElasticWISP-NG: Towards Dynamic Resource Provisioning for WISP Access Networks

Duncan E. Cameron[✉], Murugaraj Odiathevar, Alvin C. Valera, and Winston K. G. Seah

School of Engineering and Computer Science, Victoria University of Wellington, Wellington, New Zealand
{duncan.cameron,murugaraj.odiathevar,alvin.valera, winston.seah}@ecs.vuw.ac.nz

**Abstract.** *ElasticWISP-NG (Next Generation)* presents an innovative approach to dynamic resource provisioning in Wireless Internet Service Provider (WISP) access networks. This paper introduces the early stages of our novel scheme, which builds on the foundational concepts of our previous *ElasticWISP* model. We focus on the utilisation (and the implicit constraints) of renewable energy sources, and identify where opportunities to reduce network-wide energy consumption exist. Our findings outline the potential of *ElasticWISP-NG* to transform the accessibility of the Internet in underserved areas, promising significant advances in sustainable and scalable network management.

## 1 Introduction

The physical growth of the Internet since its inception has been remarkable. By 2025, the number of Internet of Things (IoT) devices alone is expected to exceed 37 billion, far surpassing the human population of our planet [20]. The Internet, through physical telecommunications networks, has brought billions of people across the globe closer together, enabling countless new industries to emerge, consequently changing the way we live forever. Looking toward the future, the proliferation of devices connected to the Internet is expected to increase, especially as we enter the IoT and Internet of Vehicles (IoV) era.

Today, we collectively take fast and accessible Internet connectivity for granted. It is easy to forget that in remote and rural areas throughout the world, Internet access is not equal and there is an all-too-real digital divide. Even in well-developed nations such as New Zealand, experiencing the displeasure of poor or non-existent connectivity often only requires a short drive out of any major urban centre. Furthermore, during the COVID-19 pandemic, we saw that for those living outside of well-connected areas, the ability to work remotely was not always possible, further degrading economic and health outcomes for those with a connectivity disadvantage. Although services such as Starlink have since

L. Barolli (Ed.): AINA 2024, LNDECT 199, pp. 55–66, 2024.
https://doi.org/10.1007/978-3-031-57840-3_6

become available in many parts of the world, inequality and cost of service still remain a major concern.

We must also consider that simply having Internet connectivity is no longer enough. Much of the modern Internet is media rich and not built considering remote and rural users with constrained connections. Having an Internet connection does not guarantee the ability to benefit from remote work tools such as Zoom, or the ability to stream media through YouTube or Netflix. The promise of telemedicine, which arguably benefits remote and rural communities the most, is far from a reality when many last mile access networks do not survive beyond their pilot phase [14]. Unfortunately, the path forward to improve remote and rural Internet infrastructure is not definitive throughout much of the world.

In 2019, a study put the number of people worldwide without basic Internet access at 400 million [4], many of them living outside urban centres. With so many lacking access to basic Internet infrastructure worldwide, it is reasonable to ask what is preventing further expansion of service. From the perspective of an Internet Service Provider (ISP), the density of potential subscribers in remote and rural areas generally means that there is little financial incentive to build infrastructure. The reality is often unfortunate. Without the incumbent ISPs taking on greater financial risk to build modern Internet infrastructure in these sparsely populated areas, there will be no *modern* infrastructure, if any at all.

The demand for better physical Internet infrastructure, combined with the inaction of slow-moving incumbent ISPs, has led to the development of community networks, often wireless, throughout the world [19]. Sometimes, these networks are run entirely by the community, such as Guifi.net and Freifunk. In many other cases, networks are relatively small commercial ventures and are referred to as Wireless Internet Service Providers (WISPs). These wireless networks are of particular interest, as they have the potential to offer high-speed services that can often exceed the commercial offerings of the incumbent ISPs and have been recognised as an important part of remote and rural Internet infrastructure [5]. However, WISPs face challenges, both technical and social, that have seen little improvement over the past decade.

By nature, many of the communities WISPs serve are in remote and rural areas, with limited other connectivity options available [13]. To reduce Capital Expenditure (CAPEX), WISPs deploy infrastructure in high, sometimes mountainous areas to provide the best coverage with the least number of physical sites required, much the same as any Mobile Network Operator (MNO). However, while high ground is generally advantageous for wireless network coverage, being located outside of urban centres often means that there is no available electricity grid [11]. Consequently, WISPs must often use renewable energy sources, typically solar and occasionally wind. Renewable energy sources present their own challenges, particularly in terms of system planning and ensuring site reliability. The possibility of power interruptions requires adequate detection and monitoring of faults, an area that still requires improvement to be suitable for the particular challenges encountered [12]. The use of renewable energy sources also creates a significant cost incentive to minimise the energy consumption of devices, where possible [5,11,14].

In New Zealand, more than 30 regional WISP networks play a crucial role, connecting more than 70,000 remote and rural homes and businesses to the Internet, underscoring their importance as infrastructure within the country [18]. Like in New Zealand, WISPs around the world do not always individually operate large networks, but collectively connect millions to the Internet. The next generation of WISP networks will need to be robust, energy efficient, and scalable enough to meet the demands of the future. For WISPs to remain an important part of last-mile Internet access, we **must** determine how we can create affordable, scalable, and maintainable networks. The following sections will introduce our contribution, *ElasticWISP-NG*, which expands on our earlier *ElasticWISP* work, and will enable WISP operators to build energy-efficient and scalable access networks.

## 2   Related Work and Motivation

Perhaps unsurprisingly, researchers have identified that many WISPs build their networks out of necessity to provide Internet access where connectivity is poor or non-existent, even if they lack background knowledge and experience in operating wireless networks [13]. As a result, many research efforts have focused on improving the outcomes of these wireless networks, often through improved ease of network management [7]. Several important advances have also been made in understanding and monitoring traditional off-grid energy systems used [12]. Despite advances in energy monitoring at WISP sites, and their subsystems, little progress has been made to reduce the cumulative energy consumption of network devices at these sites [1].

The implications of poor efficiency in energy-constrained WISP networks are significant. Microwave radios, often used for backhaul, can account for more than half the total energy consumption of the network [1]. In efforts to improve this energy efficiency, our previous research demonstrated the potential for up to 65% energy reduction within backhaul networks, through strategic deactivation of redundant links [1]. Furthermore, designing networks around ever-changing weather conditions can be a challenge. Known weather patterns at particular geographic locations can determine which frequency radios (e.g., mmWave or otherwise) are suitable for use, as well as the type and size of energy harvesting systems that are required. The design of network resource optimisation schemes that can manage high environmental variability requires the appropriate planning and sizing of energy harvesting systems. Underbuilding risks creating unreliable networks that cannot adapt to undesirable conditions. Overbuilding will improve reliability, with the potential implication of fewer sites being built due to resource constraints.

## 2.1   Location

Unfortunately, for MNOs and WISPs alike, finding suitable land will always be a challenge. Mountainous areas often provide excellent coverage potential, but come at a price. Take New Zealand WISP Venture Networks [15] as a case study. Venture Networks' Heights site is an example of an all too common situation, especially among WISPs. Lower areas of the Tararua Range, located around the Lower Central North Island of New Zealand, are ideal for building wireless infrastructure. Venture Networks, and other MNOs, have built wireless network infrastructure in the lower and more accessible areas of the Tararua Range. In any case, the grid electricity in such mountainous terrain is limited almost exclusively to network operators that have the financial ability to invest millions in their most important sites, as power infrastructure must be built to the location.

Consequently, Venture Networks, like many other WISPs around the world, is in a situation where renewable energy sources must be used due to financial constraints. As a result, planning and maintaining off-grid energy harvesting systems is critical to the long-term success of the network. Although solar and wind systems can be cost-effective compared to installing expensive power cables from the closest source, often kilometres away, building very remote WISP sites means that additional expertise is required of network operators. In addition to being able to plan and configure the wireless network infrastructure itself, operators also need to be comfortable managing low- to medium-voltage energy harvesting systems, which in some countries requires certification. In many cases, vehicle access to sites is not viable throughout the year. Transporting equipment and personnel to new or existing sites can mean using off-road vehicles, hiking, or, in some cases, using a helicopter.

Although MNOs and WISPs do not always face such a difficult terrain, there is strong motivation to design networks that are robust enough to survive challenging environments. As power systems are a fundamental part of any wireless infrastructure, it is necessary to emphasise energy efficiency. Furthermore, with more energy-efficient wireless infrastructure, greater financial savings can be achieved with reduced spending on energy harvesting and battery systems. In summary, the natural landscape is an invaluable resource, and wireless networks will continue to benefit as technology evolves and energy efficiency increases. However, considering the challenges created by terrain, existing grid-connected structures that can provide reasonable coverage to areas, such as guyed towers or grain silos, are comparatively accessible alternatives.

## 2.2   Energy Harvesting, Management, and Monitoring

The nature of operating remote and rural access networks often means that renewable energy sources outside the grid, such as solar or wind, must be harvested [1]. Grid electricity in developing nations, where it is available, is often poor quality, raising concerns about the reliability of the network [7]. Despite the fact that energy constraints within wireless access networks are well understood, few research efforts have produced tangible improvements [1]. In general, the energy consumption of wireless access networks has been a widely studied area, often with the objective of reducing the operational cost of energy per kWh consumed. Although important research, the fundamental issue of reliably providing power must also be considered.

Additionally, the operational cost of powering remote and rural access networks through renewable sources makes energy efficiency very important, meaning that related cost reduction initiatives are often applicable. Furthermore, our previous work found that traffic in WISP networks follows diurnal patterns [1]. Importantly, our findings are consistent with related work [6]. The distribution of daily traffic, as seen in Fig. 1, means that intelligent use of network resources, namely access radios, is possible. Our *ElasticWISP* scheme is one of such examples, but it is limited to point-to-point backhaul networks. Related research has identified that point-to-multipoint Long-Term Evolution (LTE) cellular base stations can almost halve their off-grid energy harvesting and storage requirements by implementing a sleep mode during low load periods [10]. For WISPs, we found no comparable point-to-multipoint energy conservation schemes, presenting an important opportunity for the development of our *ElasticWISP-NG* scheme.

**Fig. 1.** Access Network Traffic (4 x Access Points).

Planning off-grid energy systems and properly sizing them is another important task that can be exacerbated by weather variations. Solar Photovoltaic (PV) systems are widely used, but the limited sunlight hours during the winter months mean that energy harvesting and storage capacity planning must be carried out carefully. Unfortunately, there is no "one-size-fits-all" approach for capacity planning. The daytime hours of sunshine vary depending on the geographic location and the intensity of the solar radiation. Even with an appropriately sized off-grid energy system in place, monitoring and ensuring year-round availability are critical. To support this, our earlier work studied machine learning for automatic monitoring of off-grid energy systems and preemptively detecting faults in the power system before they occur [12]. Figure 2 shows the degradation of the battery bank at the Venture Networks Heights site during an extreme weather event, illustrating how quickly things can go wrong.

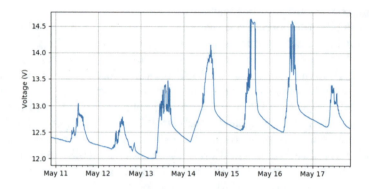

**Fig. 2.** Battery degradation and recovery during a storm.

## 3  Design

We have argued why energy efficiency in WISP networks is of utmost importance. Operators face challenges that we are not familiar with in daily life, even as engineers, and every incremental step we can make towards alleviating these pains is a major benefit. The following subsections describe the high-level formulation of our *ElasticWISP-NG* scheme. Although we are in the early stages of fully implementing *ElasticWISP-NG*, and seeing it through to becoming a production ready system, we believe it is important to highlight our contribution and what the next key steps are.

### 3.1  Model Formulation (Original)

Let us first review our *ElasticWISP* [1] model formulation, which is an adaptation of the *ElasticTree* [6] formulation. We define a flow network $G = (V, E)$ with edges $(u, v) \in E$ and with capacity $c(u, v)$. We have $k$ different commodities $K_1, K_2, ..., K_k$, where $K_i = (s_i, t_i, d_i)$. For the commodity $i$, $s_i$ is the source, $t_i$ is the sink, and $d_i$ is the demand. We establish that the flow of commodity $i$ along the edge $(u, v)$ is $f_i(u, v)$. We add standard multi-commodity flow constraints and find a flow assignment that can satisfy them [2,3]. Next, our original *ElasticWISP* formulation builds "elastic" wireless networks through energy minimisation constraints. First, we ensure that flows are restricted to radio pairs that are powered on (in a Point to Point (PTP) backhaul context). Bidirectional power constraints are also added so that both radios (at either end of the PTP backhaul pair) are powered on. Lastly, we prevent flow splitting, eliminating the possibility of problematic Transmission Control Protocol (TCP) packet reordering [8]. The objective function of our original model was to minimise the overall energy consumption of the network, subject to the mentioned constraints. For a complete refresh, please refer to our original *ElasticWISP* publication [1].

## 3.2   Extended Notation for PTMP Access Networks (New)

We now extend our original *ElasticWISP* formulation by adding constraints that consider the network beyond simple PTP backhauls, as well as including an updated objective function (definitions in Table 1).

**Table 1.** PTMP energy minimisation notation.

| Notation | Definition |
|----------|------------|
| $A$ | Set of access points |
| $S$ | Set of subscribers |
| $P(u)$ | Set of subscribers connected to access point $u$ |
| $Y_u$ | Binary decision variable for the power state of access point $u$ |
| $L_{u,s}$ | Binary decision variable indicating if subscriber $s$ is connected to access point $u$ |
| $g_s$ | Demand of subscriber $s$ |
| $b(u)$ | Power cost of access point $u$ |

**Access Point-Subscriber Assignment.** Ensure that each subscriber is assigned to one and only one powered-on access point. For each subscriber $s$,

$$\sum_{u \in A} L_{u,s} = 1, \tag{1}$$

and to ensure that the access point $u$ is on if subscriber $s$ is connected to it, we have

$$L_{u,s} \leq Y_u, \quad \forall s \in S, \forall u \in A. \tag{2}$$

**Access Point Capacity.** The total demand of all subscribers connected to an access point must not exceed its capacity.

$$\sum_{s \in P(u)} g_s \times L_{u,s} \leq c(u), \quad \forall u \in A. \tag{3}$$

**Updated Objective Function.** In addition to the original *ElasticWISP* objective function, during low load periods, ensure that only the minimum set of access points necessary to serve all subscribers is powered on. This can be viewed as a Steiner Tree problem, where Steiner points are the access points, and terminals are the subscribers:

$$\textbf{minimize} \sum_{(u,v) \in E} X_{u,v} \times a(u, v) + \sum_{u \in A} Y_u \times b(u), \tag{4}$$

where $X_{u,v}$ is the binary decision variable for the power state of the link $(u, v)$ and $a(u, v)$ is the power cost of the link $(u, v)$, as given by the original *ElasticWISP* formulation [1]. As detailed in our original publication, solvers such as Gurobi can be used to find solutions. For reference, we have uploaded an example using Gurobipy [17]. Note that a valid licence is required. Heuristics for our model are currently in development.

## 4   Proof of Concept

Steiner Tree problems have long been known to be NP-Complete [9], as are multi-commodity flow problems, such as our original *Elastic WISP* [3] scheme. While we continue to work on the evaluation of *Elastic WISP-NG* against larger network topologies, we should first consider the possible energy reduction possible with the scheme under worst-case conditions. Figure 3 and Fig. 4 demonstrate how *Elastic WISP-NG* could save energy over a 24-hour period, using real-world data from our experimental testbed. In this case, we consider 4 access points, each of which consume at most 15 W of energy. For our Proof of Concept (PoC), we add a fifth access point, which is used to offload traffic from the other four access points during periods of low load.

The process of identifying the energy saving potential in Figs. 3 and 4 is as follows. We denote the energy saving potential as $E_{savings}$. First, we set a safe capacity threshold, $C_{threshold} = C_{AP5} \cdot U_{threshold}$, where $C_{AP5}$ is the capacity of the fifth access point, and $U_{threshold}$ is the utilisation threshold (85% in this context). The utilisation threshold should be set appropriately to avoid saturating queues and inducing unnecessary queuing delay, or "bufferbloat". We then calculate the rolling average of total throughput, $T_{rolling}(t)$, over time $t$ and identify the points, $P_i$, where offloading is possible. Functionally, this means whenever $T_{rolling}(t_i) \leq C_{threshold}$, subject to constraints such as radio boot time. In our case, we considered rolling averages of 5 min. The offload period extends from each $P_i$ to the subsequent time when the rolling average exceeds $C_{threshold}$. Finally, we give the total offload time as $T_{offload}$, which is aggregated across all recorded periods. We then use $T_{offload}$ to quantify the energy saving potential, as a percentage:

$$E_{savings} = \left( \frac{E_{main} - E_{5th}}{E_{main}} \right) \cdot \left( \frac{T_{offload}}{T_{total}} \cdot 100 \right), \tag{5}$$

where $E_{main}$ and $E_{5th}$ are the energy consumption metrics of the four main access points and the fifth access point, respectively, and $T_{total}$ is the total observed time.

Figure 3 shows the potential energy savings of offloading the 4 access points when their cumulative throughput drops below 50 Mbps of downstream traffic. Network resources (access points) will begin to scale up when we reach 85% of the maximum offloading capacity, to avoid introducing excessive queueing delays when running too close to the capacity ceiling of the single access point. The percentage of time during which the offloading to the fifth access point is active is approximately 70% of the total observed time during this 24-hour period. The energy savings achieved under these conditions are approximately 53% of the total energy that would have been consumed if the four main access points had run simultaneously.

This is the worst-case scenario; as we could save more energy by offloading as few as two or more access points during low load. However, considering clusters of access points is not foolish, as high cumulative traffic across them is a better indicator of increased demand than from a single access point alone. Offloading is also not initiated until we fall below 85% of the target offload capacity, to prevent undesirable situations from occurring where excessive load is forced onto the single access point. The events highlighted in green indicate periods when only the fifth access point is operational, offering substantial energy savings, although less than with the higher capacity threshold shown next in Fig. 4.

Figure 4 shows the same approach, with the fifth access point supporting 75 Mbps of offload capacity. The percentage of time during which the offloading to the fifth access point is active is now approximately 93%. This indicates that most of the time, the network conditions are suitable for offloading to the fifth access point, and we only need to scale up resources under peak load. The energy savings achieved under these conditions are approximately 70% of the total energy that would otherwise have been consumed. Effectively, these results demonstrate that *ElasticWISP-NG* can enable network operators to run

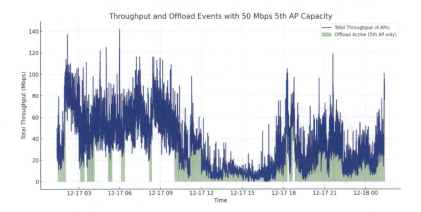

**Fig. 3.** Offloading at less than 50 Mbps of throughput.

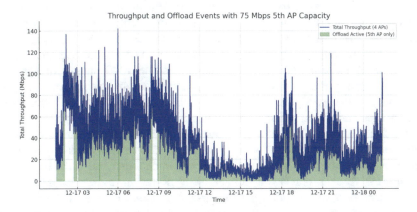

**Fig. 4.** Offloading at less than 75 Mbps of throughput.

**Fig. 5.** Radio Power Control Platform.

a reduced capacity access network, with a minimal power footprint, and scale resources to support peak demand, allowing better energy consumption today and scalability for the demands of tomorrow.

## 5   Future Work

Our contribution goes beyond an updated *ElasticWISP* model. We have designed hardware to enable the real-world deployment of *ElasticWISP* and *ElasticWISP-NG*, and we are pleased to announce that we will make it open source for all. Figure 5 shows an ultrawide voltage input power control platform that we developed as part of this work. The platform accepts 9–75 VDC in with a fixed 24 VDC output, supplying a load of up to 150W. The system can currently be adapted to operate at 9–75 VDC in with a fixed 48 VDC output by swapping the power supply module. We intend to extend this in the future to support both output voltages simultaneously. The platform has integrated management via Ethernet and has an affordable bill of materials.

Another important consideration for *ElasticWISP-NG* is how to react to, or preempt, network-wide subscriber demand. *ElasticTree* took a preemptive approach, using an autoregressive model to predict demand days in advance. *ElasticWISP* took a reactive approach, only powering up additional backhaul radios when there was a consistent increase in demand. Future work will look at how we can best anticipate and manage demand in the combined context of PTP backhaul and Point to Multi Point (PTMP) access networks. We know from our *ElasticWISP-NG* PoC that a reactive approach would still save a substantial amount of energy, although we also know that we can push this limit even further.

Furthermore, we know that flexible traffic engineering is mandatory to experience all the benefits of a scheme such as *ElasticWISP-NG*. Intelligent routing decisions within dynamic WISP access networks will help ensure that Quality of Service (QoS) remains acceptable and that network subscribers remain satisfied. However, we know that traditional shortest-path routing protocols are not, at least on their own, suitable for traffic engineering [1]. During our original *ElasticWISP* work, we encountered this exact problem: steering specific flows over specific links was not easily achievable using traditional shortest-path routing protocols, such as Open Shortest Path First (OSPF). With this in mind, targeted solutions are needed.

## 6 Conclusion

Our preliminary findings show great promise for dynamic resource provisioning in WISP access networks. Although still in the early stages of development, our PoC shows that there is a great opportunity to improve the energy efficiency of low-load or idle access networks. Our next steps will dive deeper into refining our formal model, as well as proposing less computationally complex heuristics. We must also assess the challenges of real-world implementation on a wider scale, as well as considering a broader dataset for evaluation. Practical deployment will help validate the effectiveness of the scheme and reveal any unforeseen challenges that we are yet to face. When completed, we can fully realise the path to more sustainable, energy efficient, and inclusive remote and rural connectivity solutions. Finally, to allow other researchers to build on our work, we have released real-world subscriber throughput data from our wireless testbed that can be used to assess how similar schemes could perform [16].

**Acknowledgement.** We thank Victoria University of Wellington for providing a Wellington Doctoral Scholarship that has funded this research, including the ongoing development of *ElasticWISP-NG*.

## References

1. Cameron, D., Valera, A., Seah, W.K.G.: ElasticWISP: energy-proportional WISP networks. In: Proceedings of the 2020 IEEE/IFIP Network Operations and Management Symposium (NOMS 2020), pp. 1–9 (2020). https://doi.org/10.1109/NOMS47738.2020.9110384
2. Cormen, T.H., Leiserson, C.E., Rivest, R.L., Stein, C.: Introduction to Algorithms, Third Edition. The MIT Press, 3rd edn. (2009)
3. Even, S., Itai, A., Shamir, A.: On the complexity of time table and multicommodity flow problems. In: Proceedings of the 16th Annual Symposium on Foundations of Computer Science (SFCS 1975), pp. 184–193 (1975). https://doi.org/10.1109/SFCS.1975.21

4. Hasan, S.: Towards Scalable Community Networks. Ph.D. thesis, EECS Department, University of California, Berkeley (2019)
5. Hasan, S., Ben-David, Y., Bittman, M., Raghavan, B.: The challenges of scaling WISPs. In: Proceedings of the 2015 Annual Symposium on Computing for Development (DEV 2015) (2015)
6. Heller, B., et al.: ElasticTree: saving energy in data center networks. In: Proceedings of the 7th USENIX Conference on Networked Systems Design and Implementation (NSDI 2010), p. 17 USENIX Association, USA (2010)
7. Jang, E., et al.: Crowdsourcing rural network maintenance and repair via network messaging. In: Proceedings of the 2018 Conference on Human Factors in Computing Systems, pp. 1–12. (CHI 2018), Association for Computing Machinery, New York, NY, USA (2018). https://doi.org/10.1145/3173574.3173641
8. Kandula, S., Katabi, D., Sinha, S., Berger, A.: Dynamic load balancing without packet reordering. SIGCOMM Comput. Commun. Rev. **37**(2), 51–62 (2007). https://doi.org/10.1145/1232919.1232925
9. Karp, R.M.: Reducibility among combinatorial problems. In: Miller, R.E., Thatcher, J.W., Bohlinger, J.D. (eds.) Complexity of Computer Computations. The IBM Research Symposia Series. Springer, Boston, MA (1972). https://doi.org/10.1007/978-1-4684-2001-2_9
10. Marsan, M.A., Bucalo, G., Di Caro, A., Meo, M., Zhang, Y.: Towards zero grid electricity networking: powering BSs with renewable energy sources. In: Proceedings of the 2013 International Conference on Communications Workshops (ICC 2013), pp. 596–601 (2013)
11. Nungu, A., Olsson, R., Pehrson, B.: On powering communication networks in developing regions. In: Proceedings of the 2011 IEEE Symposium on Computers and Communications (ISCC 2011), pp. 383–390 (2011). https://doi.org/10.1109/ISCC.2011.5983868
12. Odiathevar, M., Cameron, D., Seah, W.K.G., Frean, M., Valera, A.: Humans learning from machines: data science meets network management. In: Proceedings of the 2021 International Conference on Communication Systems Networks (COMSNETS 2021), pp. 421–428 (2021)
13. Pötsch, T., Schmitt, P., Chen, J., Raghavan, B.: Helping the lone operator in the vast frontier. In: Proceedings of the 15th ACM Workshop on Hot Topics in Networks (HotNets 2016) (2016)
14. Surana, S., et al.: Beyond Pilots: keeping rural wireless networks alive. In: Proceedings of the 5th USENIX Symposium on Networked Systems Design & Implementation (NSDI 2008) (2008)
15. Venture Networks: Super-fast Rural Broadband (no date). https://www.venture.net.nz/. Accessed 21 Aug 2019
16. WiNe Research Group: ElasticWISP Information (no date). https://elasticwisp.wine.ac.nz/. Accessed 19 Dec 2023
17. WiNe Research Group: ElasticWISP-NG PoC Code (no date). https://files.wine.ac.nz/elasticwisp-ng.py. Accessed 19 Dec 2023
18. Wireless Internet Service Providers Association of New Zealand Inc: Don't Waste Our Spectrum: A message to rural New Zealand (2019). Accessed 8 Feb 2020
19. Yaacoub, E., Alouini, M.-S.: A key 6G challenge and opportunity—connecting the base of the pyramid: a survey on rural connectivity. Proc. IEEE **108**(4), 533–582 (2020). https://doi.org/10.1109/JPROC.2020.2976703
20. Yao, H., Guizani, M.: Introduction. In: Yao, H., Guizani, M. (eds.) Intell. Internet Things Networks, pp. 1–21. Springer, Cham (2023). https://doi.org/10.1007/978-3-031-26987-5_1

# BBR-R: Improving BBR's RTT Fairness by Dynamically Adjusting Delay Detection Intervals

Zewei Han[1(✉)] and Go Hasegawa[2]

[1] Graduate School of Information Sciences, Tohoku University, Aoba Ward, Sendai 980-0845, Japan
han.zewei.p1@dc.tohoku.ac.jp
[2] Research Institute of Electrical Communication, Tohoku University, 2-1-1 Katahira, Aoba-ku, Sendai 980-8577, Japan
hasegawa@riec.tohoku.ac.jp

**Abstract.** Proposed by Google in 2016, the Bottleneck Bandwidth and Round-trip propagation time (BBR) congestion control mechanism has been widely used on the Internet. BBR reduces packet loss and significantly minimizes end-to-end packet latency, garnering substantial attention in recent years. However, recent studies have shown that BBR suffers from severe fairness problems when flows with different Round Trip Times (RTTs) share a bottleneck link. Existing methods for fairness improvement mostly limit the packet sending rate of longer RTT flows, which is not a good way since the root of the problem is not considered and their results still have room for improvement. In this paper, we propose a simple but effective modification to BBR, named BBR Refined (BBR-R), to improve throughput fairness among flows with different RTTs. Based on the root causes of unfairness, BBR-R adaptively changes the delay detection intervals for draining the queues in the bottleneck link buffer while maintaining link utilization. We evaluated the proposed method with some existing works including BBRv2 and confirmed that while ensuring high link utilization and lower latency, BBR-R improved the fairness index by up to 40%.

## 1 Introduction

Transmission Control Protocol (TCP) [1], which was proposed in the 1980s, is still one of the key protocols in Internet communications and is widely used in various network applications. A congestion control algorithm (CCA) is an important function of TCP. The concept of congestion control was first proposed by Jacobson [2] in the 1980s and remains an important topic in network communications.

The *operating point* of a Congestion Control Algorithm (CCA) denotes its dynamic adjustment of transmission rate and parameters to sustain network performance especially in the face of congestion. CCAs seek an optimal operating point, balancing stability and throughput. Kleinrock proposed the optimum operating point by adjusting transmission rates based on the Bandwidth-Delay Product (BDP) [3]. Unfortunately, Jaffe [4] later proved the impracticality of achieving the optimum operating point through

© The Author(s), under exclusive license to Springer Nature Switzerland AG 2024
L. Barolli (Ed.): AINA 2024, LNDECT 199, pp. 67–77, 2024.
https://doi.org/10.1007/978-3-031-57840-3_7

distributed algorithms, redirecting research toward congestion avoidance strategies. During network development, the loss-based CCAs like CUBIC [5] have played crucial roles but contribute to a problem called bufferbloat [6]. Bufferbloat highlighted the existence of large latency in bottleneck link buffers. Since modern routers and switches equipped with substantial buffers, and loss-based CCAs tend to fill the whole buffer, thus increasing the latency [7].

To improve the bufferbloat and achieve the optimum operating point of Kleinrock, in 2016, Google proposed a new CCA called Bottleneck Bandwidth and Round-trip propagation time (BBR) [8]. Unlike traditional loss-based CCAs, BBR does not react to packet losses. Instead, it decides its sending rate based on the observed throughput. It also adjusts the inflight packets to prevent filling the bottleneck link buffer and maintain optimal network performance. Therefore, BBR is expected as a CCA to avoid bufferbloat and has been used in servers of Google, YouTube, Netflix, and Amazon Web Services [8–10]. The use of BBR in the current Internet is around 20% [11].

However, several recent researches have reported fairness issues in BBR. In 2017, Hock [12] presented the first independent study on BBR and found fairness problems. [13–15] analyzed BBR in experimental and mathematical ways. These studies show that when the BBR flows with different Round Trip Times (RTTs) competing on the same bottleneck link, the RTT fairness problem will happen. Especially with large buffers at the bottleneck link, longer RTT flows will occupy more bandwidth than flows with shorter RTT.

RTT fairness is a serious problem. As affirmed by [16]. A direct approach to address this is suppressing the sending rate of longer RTT flows [17], but this doesn't resolve the root problem and can be improved [18]. Google's release of BBR version 2 (BBRv2) in 2019 aimed to tackle certain concerns [19], yet the RTT fairness challenge persists [20,21].

Hence, in this paper, we propose a modification to BBR, BBR-R, to improve the fairness between flows with different RTTs, while keeping the link utilization and lower latency. Based on the root causes for the unfairness, that are first described in this paper, BBR-R incorporates delay information to consider the link utilization and the queue formed in the bottleneck link buffer. Then it dynamically adjusts the delay detection interval, which is constant in the original BBR, to drain the queue at the bottleneck link buffer in a timely manner. This modification can reduce excessive data transmission while ensuring link utilization and avoiding high tail latency. Note that the modification to the original BBR is limited to one additional control mechanism and two parameters' settings to maintain its fundamental characteristics.

We implemented BBR-R based on BBR in Linux kernel v5.4. We conduct extensive experiments to validate the performance of BBR-R under the emulated network environment. The comparative experimental results show that BBR-R outperforms BBR and other methods with improved fairness while keeping low latency.

In summary, this paper makes three key contributions: 1) Root cause analysis of BBR RTT-unfairness, 2) Introduction of a new idea, BBR-R, enhancing fairness and reducing latency through dynamic detection delay interval adjustments, and 3) Extensive experiments to confirm that BBR-R outperforms other existing CCAs.

The remainder of this paper is organized as follows. Section 2 summarizes the previous works related to this study. In Sect. 3, we briefly explain the BBR CCA and present our analysis of the cause of the RTT fairness problem. In Sect. 4, we present our proposed BBR-R. We discuss the performance of BBR-R through experiments in Sect. 5. Finally, we conclude this paper and mention future works in Sect. 6.

## 2    Related Work

Ma et al. proposed BBQ [17]. The BBQ constantly detects the presence of redundant queues in the buffer. When queues form, the BBQ will prevent longer RTT flow from sending excess packets. However, as pointed out by Yang et al. in paper [18], BBQ does not fundamentally solve the RTT fairness problem, the share of different flows on the bottleneck is not considered and flows will experience high latency.

G. H. Kim in paper [22] proposed a bottleneck queue buildup suppression method to enhance the fairness between different RTT flows. They proved this method is not stable while reducing the link utilization and proposed an improved version BBR-E [23]. Later, they proposed Delay-Aware BBR [16] by reducing the excessive data transmission of the original BBR and adjusting the sending rate based on each flow's RTT to improve fairness. Proposed by Yang et al., the Adaptive-BBR [18] adjusts the probe bandwidth phase and a parameter relative to sending rate of BBR to control the sending rate and maintain network fairness. Mahmud et al. [24] proposed BBR-Advanced (BBR-A). BBR-A will reduce the cwnd by half when congestion is detected and packet loss occurs in flows with varying RTTs to improve RTT fairness. Cardwell et al. [19] proposed BBRv2 to mitigate the fairness problem of BBR. However, the modification leads to a decline in the competitiveness of BBR, and in some cases, the fairness of BBRv2 is worse than the original BBR [20,21].

Recently, Yuan et al. developed FEBBR [25]. FEBBR deals with the aggressiveness of BBR by introducing delay deviations to solve the fairness problem. Sun et al. suggest that the BBR can detect the delay more frequently and offer a method to modify the delay detection interval [26].

Most of the BBR variants above try to improve RTT fairness based on suppressing the packet sending rate of longer RTT flows by dynamically tuning the parameters relative to the cwnd or sending rate, or by changing them to some static values. However, we believe that such methods are not particularly good ways to address the RTT-fairness issue at its roots, as described in Sect. 3. That is, there is a lack of detailed justification and explanation for the existing mechanisms. Their methods may also degrade some fundamental characteristics and advantages of the original BBR and their results still have room for improvement.

## 3    BBR Overview and Fairness Analysis

We briefly introduce the BBR algorithm [8] and analyze the cause of its RTT-unfairness.

## 3.1 BBR Overview

BBR executes Startup phase and Drain phase at the start of a flow. Then it enters ProbeBW phases and ProbeRTT phases repeatedly. In each phase, the sender determines the packet pacing rate $x(t)$ and the congestion window size $cwnd(t)$ at the time $t$ when the sender receives an ACK packet with Eqs. (1) and (2), respectively.

$$x(t) = g_p \cdot BtlBw(t) \tag{1}$$

$$cwnd(t) = g_w \cdot BtlBw(t) \cdot RTprop(t) \tag{2}$$

$x(t)$ and $cwnd(t)$ are determined based on $BtlBw(t)$ and $RTprop(t)$, calculated by Eqs. (3) and (4), respectively.

$$BtlBw(t) = \max(d(t)) \quad \forall t \in [t - W_B, t] \tag{3}$$

$$RTprop(t) = \min(RTT(t)) \quad \forall t \in [t - W_R, t] \tag{4}$$

$BtlBw(t)$ is the base of packet pacing rate at the time $t$, which is the maximum value of data delivery rate, denoted by $d(t)$. $d(t)$ means the data receiving rate at time $t$ calculated from received ACK packets, in the recent past. It is updated every time when an ACK packet is received. $RTprop(t)$ is the minimum RTT recorded by the sender in the recent past. $RTT(t)$ is the RTT observed by the sender at the time $t$. $W_B$ and $W_R$ are set to $10 \cdot RTT(t)$ and 10 [sec], respectively. $g_p$ in Eq. (1) is the parameter to determine the packet pacing rate $x(t)$ relative to $BtlBw(t)$. $g_w$ in Equation (2) is the parameter to determine the target congestion window size $cwnd(t)$. $g_p$ and $g_w$ are called as pacing gain and cwnd gain, respectively.

In a stable state, a BBR flow primarily stays in the ProbeBW and ProbeRTT phases. In ProbeBW phase, $g_p$ is dynamically changed in every RTT to follow the pattern [1.25, 0.75, 1, 1, 1, 1, 1, 1]. $g_w$ is constantly set to two. When entering ProbeBW state, BBR starts a 10-second timer for checking whether $RTprop(t)$ is updated by a new smaller $RTT(t)$ with Eq. (4). When it is updated before the timer expires, the timer is reset to 10 s. Otherwise, after the timer expires, BBR enters ProbeRTT phase to actively measure minimum RTT without queuing delay. In ProbeRTT phase, BBR sets cwnd to four packets for 200 ms to drain the queue and then back to ProbeBW phase.

## 3.2 Operating Point of the Original BBR

Figure 1 depicts the typical changes in the delivery rate and the RTT as a function of the inflight packets when a single flow is on the bottleneck link. We utilize this figure for explaining *operating point* of BBR. In the graph, *BDP* is the bandwidth-delay product (BDP) for the flow, which is calculated by the product of the bottleneck link bandwidth and the round-trip delay without queuing at the bottleneck link buffer. When the inflight packets are smaller than *BDP*, the delivery rate is proportional to the inflight packets while the RTT remains minimum (as $RTT_1$ in the graph) without the queuing delay. When the inflight packets are larger than *BDP*, the delivery rate equals to the bottleneck link bandwidth, while the RTT increases because of the queuing delay at the bottleneck link buffer. Finally, when the inflight packets reach the sum of the BDP and the buffer size, $BDP + B$ in the graph, packet losses occur due to buffer overflow. The optimum operating point of Kleinrock is where the inflight packets are equal to $BDP$, where the delivery rate is equal to the bottleneck link bandwidth, while the RTT remains the minimum. A BBR flow continuously observes $BtlBw$ and $RTprop$ by Eqs. (1) and (2) to regulate the packet transmission. Therefore, when $BtlBw$ is equal to the bottleneck link bandwidth and $RTprop$ is equal to the round-trip delay without the queuing delay, the BBR flow keeps the operating point optimal. In what follows, we use $BDP_{bbr}$ to represent the product of $RTprop$ and $BtlBw$.

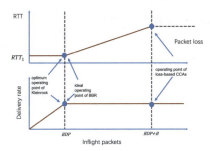

**Fig. 1.** Operating point based on the inflight packets

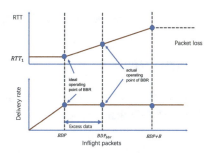

**Fig. 2.** Ideal and actual operating point of a BBR flow

### 3.3   Root Causes of BBR's RTT-Unfairness

Figure 2 shows the operating point of a BBR flow when it is competing with other BBR flows in a bottleneck link with a large buffer ($B > BDP$). [12] reported that when multiple BBR flows share a bottleneck link, the queuing delay becomes approximately 1–1.5 times the RTT due to overestimating the bottleneck link bandwidth. Then BBR flows transmit excessive data to the bottleneck link, resulting in that the actual operating point of the BBR flow being shifted right from the optimum operating point of Kleinrock. This generates a persistent queue at the bottleneck link buffer, causing a large queuing delay. Figure 3 depicts operating points for two BBR flows (flow 1 and flow 2) with different RTTs without queuing delay ($RTT_1$ and $RTT_2$ ($> RTT_1$), respectively) when they share a bottleneck link. In the figure, $BDP_1$ and $BDP_2$ are the BDPs of flow 1 and flow 2, respectively. $BDP_{bbr1}$ and $BDP_{bbr2}$ are the product of $BtlBw$ and $RTprop$ of flow 1 and flow 2, respectively. As explained above in Fig. 2, a persistent queue is formed at the bottleneck link when two BBR flows share the bottleneck. This causes $RTprop$ of the flow to be larger than the round-trip delay without queuing. On the other hand, because the $BDP_{bbr}$ is determined by the $RTprop$ of each flow, $BDP_{bbr2}$ becomes larger than $BDP_{bbr1}$, that makes inflight packets gap between flow 1 and flow 2, as depicted in Fig. 2. Then flow 2 has a larger cwnd than flow 1. So, flow 2 has a larger share of the queue at the bottleneck link buffer than flow 1. According to the queuing theory, the increased queue share of flow 2 allows it to operate at a higher delivery rate than flow 1, so flow 2 can occupy more bandwidth than flow1. These are the root causes of the RTT-unfairness in BBR.

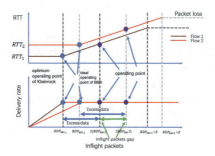

**Fig. 3.** Operating points of BBR flows with different RTTs

## 4    BBR Refined

In this section, we propose a modification to the original BBR to improve RTT fairness, named BBR Refined (BBR-R).

### 4.1   Dynamically Adjusting Delay Detection Interval

As described above, the root causes of the RTT-unfairness are (1) a persistent queue is formed at the bottleneck link buffer, and (2) a longer RTT flow occupies the larger share of the queue. So, a straightforward idea to alleviate the RTT unfairness is to decrease the queue at the bottleneck link buffer. It also means that the operating point becomes near to the optimum one. The original BBR has a mechanism to drain the queue at the bottleneck link, but it does not work well especially when multiple BBR flows compete with the bottleneck link bandwidth.

In the original BBR, the minimum duration of ProbeBW phase is fixedly set to 10 s. We believe that it has two disadvantages. One is that 10 s is too long. When BBR fails to behave ideally and the sending rate becomes larger than the ideal rate, too many packets are injected into the bottleneck link for 10 s. Probing for available bandwidth for 10 s may not perform well especially when low latency is needed in heterogeneous networks [26]. The other is that it is fixed to 10 s, regardless of the degree and frequency of changes in the network condition. In the 10 s, various situations could happen on the bottleneck link such as packet losses and queuing delay fluctuation. On the other hand, when the network condition is quite stable, the 10 s may be too short. Each after ProbeBW phase, the BBR flow enters ProbeRTT phase, in which, the flow decreases the inflight packets to four during 200 ms, which significantly decreases the flow throughput in many cases. So, we need to avoid entering ProbeRTT phase if it is not necessary.

In order to solve this problem, we propose a new mechanism to dynamically adjust delay detection intervals. In ProbeBW phase, when the observed RTT in ProbeBW phase is larger than $k \cdot RTprop$, the BBR flow immediately enters ProbeRTT phase without waiting for expiring the 10-sec timer. Otherwise, our method remains the same as the original BBR. We set $k$ to two in the evaluations in Sect. 5. We expect two effects of this mechanism. One is that when a BBR flow sends excess packets in ProbeBW phase due to some reasons, the flow immediately detects it by observing RTTs and enters ProbeRTT phase to drain the queue. The other is that when the network condition remains stable and RTTs do not so much increase, we avoid entering ProbeRTT phase more frequently compared with the original BBR.

## 4.2  Adaptively Set the Packet Inflight in ProbeRTT Phase

Due to the same reason in Subsect. 4.1, we consider that it is not a good idea that the cwnd in ProbeRTT phase is fixed to four packets. However, the above-mentioned proposed mechanism would make the duration of ProbeBW phase only dependent on the network condition without considering the link utilization of the ProbeBW and ProbeRTT phase. So, a new mechanism is introduced to determine the inflight packets of ProbeRTT phase and the minimum duration of ProbeBW phase based on the target utilization of the bottleneck link bandwidth.

Here, we propose that the inflight packets in ProbeRTT phase are set to $\alpha \cdot BDP_{bbr}$, instead of four in the original BBR, and the minimum duration of ProbeBW phase is set to $T_{PB}$. Note that the values of $\alpha$ and $T_{PB}$ would determine the bottleneck link utilization during steady states. That is, we set these values under the following conditions based on the target bottleneck link utilization, $\sigma$:

$$g_p \cdot \alpha \le 1.0 \tag{5}$$

$$\frac{T_{PB} + 0.2 \cdot \alpha}{T_{PB} + 0.2} \ge \sigma \tag{6}$$

Equation (5) ensures that the inflight packets in ProbeRTT phase are smaller than $BDP_{bbr}$. The left hand of Eq. (6) means the estimated utilization of the bottleneck link bandwidth, where 0.2 means the duration of ProbeRTT phase.

# 5  Evaluation

In this section, we validate the performance of our BBR-R through the experimental evaluations, using Mininet [27] for emulating the network environment.

## 5.1  Evaluation Settings

Figure 4 shows the network topology used in the evaluations. The network consists of $n$ hosts (h1...h$n$), one receiver (R1) with $n$ ports, one switch (SW1), $n$ access links, and one bottleneck link. Each sender is equipped with the original BBR [8], BBRv2 [19], CUBIC [5] or BBR-R. The four algorithms are implemented based on the source codes for Linux kernel, the literature [5,8,19] and Sect. 4. The data and ACK packet sizes are 1500 and 40 [bytes], respectively. All flows between the senders and the receiver start the data transmission simultaneously at 0 [sec]. Mininet version 2.3.1 was run on Ubuntu with Linux kernel version 5.4. Each sender uses iPerf2, version 2.0.13, to send packets to the different ports of the receiver.

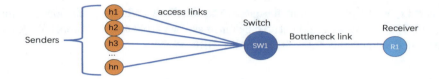

**Fig. 4.** Network topology

We assess the throughput fairness, the RTT, and the bottleneck link utilization of flows. We use Jain's fairness index [28] to measure the fairness when equal to or more than four flows co-exist in the network. For two flows, we directly give the throughput values.

## 5.2  Fairness Between Two Flows

We first evaluate throughput fairness among different RTT flows for the original BBR, BBRv2, CUBIC and BBR-R. Figure 5 shows the average throughput during 50-second experiments when two flows with identical algorithms co-exist in the network. The bottleneck link has 10 Mbps bandwidth, a 5 ms one-way propagation delay, and a buffer of 500 packets. We use five combinations of the round-trip delays without queuing at the bottleneck link of the two flows: (50 ms, 50 ms), (60 ms, 40 ms), (70 ms, 30 ms), (80 ms, 20 ms), and (90 ms, 10 ms) to confirm the effect of RTT difference. From Fig. 5, we can see that the throughput gap between the two original BBR flows becomes larger when the RTT difference between the two flows becomes larger. When BBR-R is used, the fairness is improved. CUBIC keeps a high fairness even when the 10 ms RTT flow competes with the 90 ms RTT flow. When BBRv2 is used, the gap between the two flows' average throughput is large even when the two flows' RTTs are close or equal.

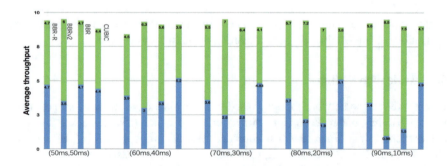

**Fig. 5.** Average throughput of two flows with different RTTs

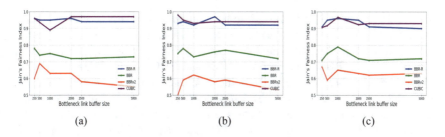

**Fig. 6.** Effect of bottleneck link buffer size

## 5.3 The Effect of Buffer Size

Figures 6 (a), (b), and (c) show the comparison of the fairness with many flows, as a function of the buffer size at the bottleneck link. We set the composition of the RTTs for the flows as: four 10 ms-flows and four 50 ms-flows in Fig. 6 (a), eight flows with 10, 15, 20, 25, 30, 35, 40, and 45 ms in Fig. 6 (b), and ten flows with 10, 15, 20, 25, 30, 35, 40, 45, 50, and 55 ms in Fig. 6 (c). We use a bottleneck bandwidth of 100 Mbps, and for 10 flows we use a bottleneck bandwidth of 200 Mbps, for eight and ten flows experiments, respectively. The one-way propagation delay is set to 3 ms for all these experiments. The fairness indexes are calculated from the average throughput in the 50-sec experiment of each flow. From the results, the fairness of BBR-R is close to CUBIC and it remains high regardless of the buffer size. BBR's fairness is worse compared with BBR-R and CUBIC. BBRv2's fairness is very low no matter how the buffer size changed. This may be because BBRv2 is more sensitive to network congestion than BBR, and the adjustment of sending rate is too conservative and the bandwidth estimation is not accurate enough due to the increase of queuing delay, which has a greater impact on shorter RTT flows than longer RTT flows. We can see from these figures that when BBR-R is used, the fairness index among those flows is improved. Specifically, BBR-R improves the fairness index to 40% or even more.

## 5.4 The Effect of the Number of Competing Flows

We use Fig. 7 to present the effect of the number of competing flows on the four methods' performance. We use 4, 8, 12, 16, and 20 flows in total. For four flows, we have one flow with 10 ms, one flow for 20 ms, one flow for 40 ms, and one flow for 80 ms. For eight flows, we have two flows with 10 ms, two flows for 20 ms, two flows for 40 ms, and two flows for 80 ms, and the same settings is used for 12, 16, and 20 flows. We use the bottleneck link with 200 Mbps bandwidth, 3 ms one-way propagation delay, and 2000 packet buffer. The experiment time is 30 s. Figures 7 (a), (b), and (c) plot the results of the fairness, the utilization of the bottleneck link, and the average RTT of competing flows. We can see from Fig. 7 (a) that BBR-R achieves better fairness than BBR and BBRv2, especially when the number of flows increases. On the other hand, CUBIC has good fairness regardless of the number of competing flows. With the number of flows increasing, BBRv2's fairness becomes worse. This is because BBRv2 is more sensitive to packet losses and when packet losses occur, it adjusts the sending rate more conservatively, which can lead to unfairness especially when the number of flows is large. From Fig. 7 (b), we observe that BBR-R maintains enough high bottleneck link utilization when compared with other methods. Especially when the number of flows increases, BBR-R outperforms BBR and BBRv2. This is because, with the increase in the number of flows in the bottleneck link, the network becomes more congested, and queuing latency increases, but BBR-R is more aware of network conditions and responds to congestion timely. Finally, from the results of average RTTs in Fig. 7 (c), we can see that BBR-R has smaller RTTs than BBR and BBRv2, regardless of the number of competing flows. This is because BBR-R drains the queues adaptively and is more sensitive to the queue in the bottleneck link buffer compared with BBR and BBRv2. On the other hand, we can see that CUBIC has quite large RTTs, that is the sacrifice for good fairness in Fig. 7 (a).

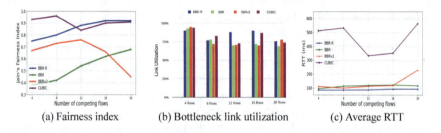

(a) Fairness index          (b) Bottleneck link utilization          (c) Average RTT

**Fig. 7.** Effect of the number of competing flows

From all results in Fig. 7, we conclude that our BBR-B has the best performance balance among the RTT fairness, bottleneck link utilization, and average RTT.

## 6   Conclusion

In this paper, we proposed a modification to the congestion control algorithm for BBR, named BBR-R, to improve throughput fairness among flows with different RTTs. We introduced a new mechanism to dynamically set the duration of delay detection based on the network condition. We evaluated the performance of BBR-R through extensive experiments. Our results indicated that BBR-R flows can improve throughput fairness and end-to-end latency when different RTT flows share the bottleneck link, regardless of the bottleneck link buffer size and the number of competing flows. One of our future works is evaluations in large-scale networks with a larger number of competing heterogeneous flows and in cases where there are multiple bottlenecks. Another future work is to solve the unfairness problem when BBR flows share a bottleneck link with loss-based CCAs.

# References

1. Postel, J.: Transmission control protocol. Technical report (1981)
2. Jacobson, V.: ACM SIGCOMM Comput. Commun. Rev. **18**(4), 314 (1988)
3. Kleinrock, L.: ICC 1979: International Conference on Communications, vol. 3, pp. 43–1 (1979)
4. Jaffe, J.M.: Networks **14**(1), 95 (1984)
5. Ha, S., Rhee, I., Xu, L.: ACM SIGOPS Oper. Syst. Rev. **42**(5), 64 (2008)
6. Gettys, J.: IEEE Internet Comput. **15**(3), 96 (2011)
7. Utsumi, S., Hasegawa, G.: 2021 IEEE International Workshop Technical Committee on Communications Quality and Reliability (CQR 2021), pp. 1–6 (2021). https://doi.org/10.1109/CQR39960.2021.9446211
8. Cardwell, N., Cheng, Y., Gunn, C.S., Yeganeh, S.H., Jacobson, V.: Commun. ACM **60**(2), 58 (2017)
9. Cohen, A.: Netflix is killing my other streaming services, Compira Labs Blog (2020)
10. A. Admin: TCP BBR congestion control with Amazon CloudFront (2019)
11. Mishra, A., Sun, X., Jain, A., Pande, S., Joshi, R., Leong, B.: Proc. ACM Meas. Anal. Comput. Syst. **3**(3), 1 (2019)
12. Hock, M., Bless, R., Zitterbart, M.: 2017 IEEE 25th International Conference on Network Protocols (ICNP), pp. 1–10 (2017). https://doi.org/10.1109/ICNP.2017.8117540
13. Dong, M., et al.: 15th USENIX Symposium on Networked Systems Design and Implementation (NSDI 2018), pp. 343–356 (2018)
14. Scholz, D., Jaeger, B., Schwaighofer, L., Raumer, D., Geyer, F., Carle, G.: 2018 IFIP Networking Conference (IFIP Networking) and Workshops, pp. 1–9. IEEE (2018)
15. Turkovic, B., Kuipers, F.A., Uhlig, S.: arXiv preprint arXiv:1903.03852 (2019)
16. Kim, G.H., Cho, Y.Z.: IEEE Access **8**, 4099 (2019)
17. Ma, S., Jiang, J., Wang, W., Li, B.: arXiv preprint arXiv:1706.09115 (2017)
18. Yang, M., Yang, P., Wen, C., Liu, Q., Luo, J., Yu, L.: 2019 IEEE Wireless Communications and Networking Conference (WCNC), pp. 1–6. IEEE (2019)
19. Cardwell, N., et al.: Proceedings of IETF 106th Meeting, pp. 1–32 (2019)
20. Song, Y.J., Kim, G.H., Mahmud, I., Seo, W.K., Cho, Y.Z.: IEEE Access **9**, 37131 (2021). https://doi.org/10.1109/ACCESS.2021.3061696
21. Kfoury, E.F., Gomez, J., Crichigno, J., Bou-Harb, E.: Comput. Commun. **161**, 212 (2020)
22. Kim, G.H., Mahmud, I., Cho, Y.Z.: 2019 International Conference on Electronics, Information, and Communication (ICEIC), pp. 1–2 (2019). https://doi.org/10.23919/ELINFOCOM.2019.8706412
23. Kim, G.H., Song, Y.J., Mahmud, I., Cho, Y.Z.: 2019 Eleventh International Conference on Ubiquitous and Future Networks (ICUFN), pp. 358–360 (2019). https://doi.org/10.1109/ICUFN.2019.8806064
24. Mahmud, I., Cho, Y.Z.: ICT Express **6**(4), 343 (2020)
25. Yuan, H., Han, Y., Zhong, Y., Liao, M., Yuan, Z., Muntean, G.M.: 2023 IEEE International Symposium on Broadband Multimedia Systems and Broadcasting (BMSB), pp. 1–6 (2023). https://doi.org/10.1109/BMSB58369.2023.10211562
26. Sun, W., Meng, K., Wang, A.: International Conference on Wireless Algorithms, Systems, and Applications, pp. 142–154. Springer, Cham (2022)
27. De Oliveira, R.L.S., Schweitzer, C.M., Shinoda, A.A., Prete, L.R.: 2014 IEEE Colombian Conference on Communications and Computing (COLCOM), pp. 1–6. IEEE (2014)
28. Jain, R., Durresi, A., Babic, G.: ATM Forum Contribution, vol. 99 (1999)

# Reducing Signaling Overhead in 5G Mobile Network for IoT Device Mobility

Takahiko Kato$^{(\boxtimes)}$, Chikara Sasaki, and Atsushi Tagami

KDDI Research, Inc., Saitama, Japan
{kh-katou,ch-sasaki,at-tagami}@kddi.com

**Abstract.** Internet of Things (IoT) services over 5G mobile networks will have unique communication characteristics of massive connectivity and sporadic small-scale data transfer. These characteristics increase the signaling overhead of C-plane message exchange associated with connection management in RAN and mobile core. 3GPP has standardized Small Data Transmission with RRC-INACTIVE that reduces this overhead. However, a new type of IoT service accompanied by device mobility is emerging. The 3GPP standard approach does not adequately account for the frequent device mobility and does not sufficiently reduce the overhead for these new IoT services. This paper proposes a new communication method for the mobility of IoT devices, which suppresses the signaling overhead of RAN and mobile core. It releases the device status management of RAN by introducing simplified status information. Our simulation results showed that the proposed method suppresses the signaling overhead compared with the 3GPP approaches. Specifically, the proposed method could reduce the signaling message exchange up to 28%.

## 1 Introduction

Internet of Things (IoT) services, such as smart cities, factories, and agriculture, are expected to be realized on top of the 5G mobile networks [1]. The various sensors in these services intermittently report measurement data to the server, and the individual data sizes are relatively small. The IoT communication characteristics of massive connectivity and sporadic small data transfer are different from traditional human communications, where the data transmission continues for a while. These characteristics of IoT may bring changes in the architecture and functionalities to the current 5G network.

IoT services are commonly known to increase the control plane (C-plane) signaling overhead associated with connection management (CM) mechanisms in 5G mobile networks [2,3]. This overhead includes the number of signaling in the network, the state retention resources, and the computing resources, etc. The CM mechanism allows the RAN to release the status of a user equipment (UE), called the UE context, and to release the radio resources allocated for the UE when the UE has not been communicating for a predefined period, e.g., 10 s. Then, the UE changes its state from the connected state to the idle state.

When the UE attempts to resume data transmission, the RAN needs to allocate a new radio resource for the UE and request the corresponding UE context to the core network. This mechanism was introduced to ensure the efficient use of radio resources and RAN computational resources. In IoT communications, however, the sporadic data transfer makes the state of a UE transit to the idle state before it sends the next data. The state changes in a huge number of IoT devices invoke the request and release storms for UE contexts. As a result, the 5G network functions (NFs) supporting C-plane signaling may cause congestion that induces severe problems in the network.

As for the small IoT data transfer over the 5G network, there are several proposals. 3GPP has introduced three mechanisms: the C-Plane CIoT Optimization (CP-CIoT) [4], the U-Plane CIoT Optimization (UP-CIoT) [4], and the Small Data Transmission in RRC-INACTIVE (Inactive-SDT) [5]. CP-CIoT transfers small IoT data in C-Plane messages and reduces the processing load of UEs and RANs. UP-CIoT and Inactive-SDT allow a RAN to maintain the UE context even after it releases the radio resources. Inactive-SDT also enables a UE to send data over the C-plane in order to reduce the signaling overhead related to the allocation of U-plane radio resources between the UE and the RAN. R. Bhatia et al. proposed introducing a new NF dedicated to IoT communications [6]. IoT data is transferred to this NF anyway. Although it reduces the power consumption of UEs, the new NF becomes the performance bottleneck. Among the conventional mechanisms, Inactive-SDT is the most suitable for reducing the signaling overhead of IoT communications.

On the other hand, some new IoT services have emerged in recent years: eHealth and Smart Logistics. The eHealth allows people to monitor their health through wearable devices in daily lives, and the Smart Logistics tracks the entire supply chain to help inventory management and optimize logistics. These new IoT services require device mobility in addition to massive connectivity and sporadic small data transfer.

When a UE moves in the 5G network adopting Inactive-SDT, the following signaling overheads will be newly increased.

- Retrieval of UE context by RAN: When a UE moves and resumes communication under a new RAN that does not have the UE context, the new RAN retrieves the UE context from the RAN where the UE was located. If the retrieved UE context is transferred to the new RAN, the RAN needs to perform a handover procedure with 5GC.
- Location management of UEs in TA/RNA granularity: The 5G mobile network generally performs location management of UEs with the idle state in the tracking area (TA) granularity, where a TA comprises multiple RANs. When a UE moves to another TA, it performs the location management signaling procedure [7]. Inactive-SDT introduces its own location management area called the RAN-based Notification Area (RNA), which is a subset of TA. A UE is required to invoke the location management signaling procedure when it moves to another RNA [5].

This paper proposes a new mechanism to reduce C-plane signaling overhead for IoT communications with the device mobility as well as the massive connectivity and the sporadic small data transfer. The proposed method, we called Idle-SDT, releases both the UE context and radio resources in the RAN to transit to the idle state when a UE has been silent for a predetermined time. When the UE attempts to resume data transmission, it provides the RAN with the information about the corresponding U-plane function (UPF) to which the data is transferred. The RAN relays the data to the UPF in a newly introduced idle state with limited context information. Three key points of Idle-SDT are as follows. First, in the idle state with limited context information, the RAN communicates with the UE through the C-plane and establishes a temporal session with the UPF. This allows the UE and the UPF to exchange small data. In this state, the 5GC regards the RAN as being idle, and so the RAN does not require the UE context retrieval and the RNA based location management accompanied by the UE moving. Secondly, in the UE context at the RAN, only the UPF information varies in the actual communication environment because small data are transferred over the C-plane between the UE and the RAN. The other parameters are unnecessary or set to the predetermined values. The UPF information is given to the UE when it is turned on. Thirdly, Idle-SDT allows the UPF to select the RAN side endpoint identifier used in the downlink data transfer when the UE is turned on, and to provide the value for the RAN through the UE. In the original 5G system, this parameter is determined dynamically by the RAN and given to the UPF through the C-plan signaling. The approach in Idle-SDT saves this signaling overhead.

The remainder of the paper is organized as follows. Section 2 explains the 5G connection management and location management related to moving IoT device communications. Section 3 describes the proposed method. Section 4 evaluates the performance of the proposed method and discusses these results. Section 5 concludes the paper.

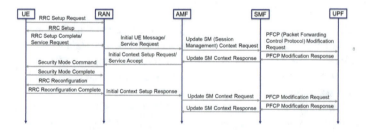

**Fig. 1.** Service Request Procedure

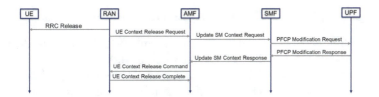

**Fig. 2.** AN Release Procedure

# 2   Connection Management and Location Management in 5G Network

## 2.1   C-plane Signaling Related to Connection Management

The access management function (AMF) of the original 5G system has two connection management states with a UE: CM-IDLE and CM-CONNECTED. When a UE is in CM-CONNECTED, the corresponding RAN allocates the radio resources and the UE context for the UE. In this state, the UE can send data over the PDU session established among the UE, the RAN, and the UPF. To transit to CM-CONNECTED, a UE invokes the Service Request Procedure [7] as shown in Fig. 1. This procedure starts from the radio resource setup, followed by C-plane message exchanges among the UE, the RAN, the AMF, the session management function (SMF) and the UPF. When a UE is silent for some period, that is, the inactivity timer expires, the RAN invokes the AN Release Procedure [7] as shown in Fig. 2. Then, the UE is shifted to CM-IDLE, and the RAN releases the radio resources for the UE and its UE context. When a UE in CM-IDLE state tries to send data, it needs to invoke the Service Request Procedure again.

In order to save the signaling overhead of those repeated procedures in a sporadic small data transfer, another state, CM-CONNECTED with RRC-INACTIVE, is introduced. Only the UE and the RAN recognize this state, not the AMF. In Inactive-SDT, when the inactivity timer expires, the RAN sends RRC release message to the UE, and releases the radio resource while maintaining the UE context. The UE and RAN transit to CM-CONNECTED with RRC-INACTIVE. When the UE in CM-CONNECTED with RRC-INACTIVE tries to send data, it sends a C-plane message requesting the radio resource setup with conveying the data in it. These procedures are shown in Fig. 3.

Inactive-SDT proposes two methods for the case that a UE accesses a new RAN, not maintaining the UE context. They are shown in Fig. 4 and Fig. 5. One is a method to relocate the UE context to the new RAN. In this method, the RAN performs the handover process with the 5GC in addition to searching for the UE context. The other method does not relocate the UE context to the new RAN, but the previous RAN keeps working as an anchor. This method can reduce the C-plane signaling overhead of the handover with the 5GC. However, it increases the delay because the new RAN transfers user data to UPF via the RAN that maintains the UE Context. The red messages shown in Fig. 4 and Fig. 5 are additionally generated by the UE moving.

## 2.2    C-plane Signaling Related to Location Management

The 5G mobile network manages the location of a UE in order to deliver user data to it. Since a UE in CM-IDLE has released the radio resources, the 5GC does not know which RAN the UE is located under. Therefore, the AMF associates a UE with a TA, an area where the UE can move without notifying the AMF. Only when a UE moves to another TA, it invokes the Mobility Registration Procedure [7] to obtain the new TA from the AMF, as shown in Fig. 6. When the 5GC sends C-plane messages or user data to a UE in CM-IDLE, it sends paging to all the RANs in the TA where the UE is located, which makes the transition of the UE state to CM-CONNECTED.

In addition, Inactive-SDT introduces the location management using the RNA so that the RAN can track the location of UEs in CM-CONNECTED with RRC-INACTIVE. When a UE moves to another RNA, it invokes the RNA Update Procedure shown in Fig. 6.

**Fig. 3.** Procedure of Inactive-SDT

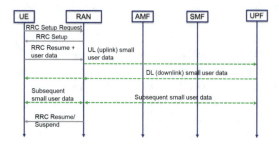

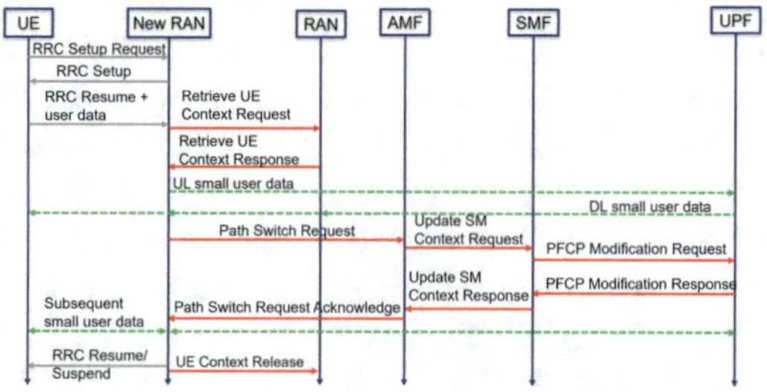

**Fig. 4.** Procedure of Inactive-SDT with UE Context Relocation in UE Movement

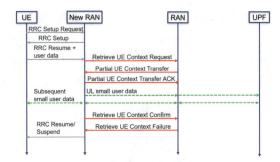

**Fig. 5.** Procedure of Inactive-SDT without UE Context Relocation in UE Movement

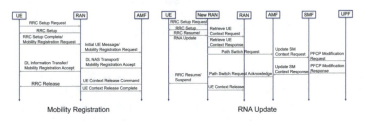

**Fig. 6.** Procedure of Location Management

# 3    Proposed Method

## 3.1    State Transition of Idle-SDT

In order to reduce C-plane signaling overhead for IoT communications with the device mobility, our proposed method, Idle-SDT, introduces a new state, CM-IDLE with Limited Context, and modifies the C-plane processing in the existing states. Figure 7 shows the state transition of Idle-SDT related to the connection management. When a UE is turned on, it first invokes the Initial Registration Procedure for authentication and location registration and the PDU Session Establishment Procedure for U-plane communication. The proposed method allows a UE to negotiate with the AMF to use Idle-SDT in the Initial Registration Procedure. In the PDU Session Establishment Procedure, the RAN and UPF establish a GTP tunnel to exchange user data. The UPF allocates the GTP tunnel ID, the tunnel endpoint identifier (TEID) for uplink (UL) communication, and the RAN allocates the TEID for downlink (DL) communication. Then, the RAN and the UPF tell each other their tunnel IDs via the AMF and the SMF. As described below, the UL TEID value selected by the UPF can be used as the DL TEID for sporadic small data transfer. Idle-SDT makes the SMF inform the UE of the corresponding UPF IP address and the UL TEID used by the RAN. This is the CM-CONNECTED state shown in the figure.

When the inactivity timer expires, the state shifts to CM-IDLE through the AN Release Procedure. In this procedure, the SMF informs the UPF that the UE is transitioning to CM-IDLE with the proposed method applied.

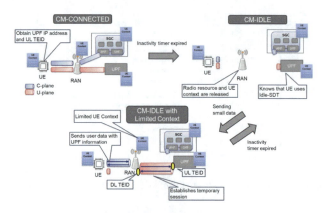

**Fig. 7.** State Transition of Proposed Method

When a UE in CM-IDLE tries to transfer UL small data, it also sends a part of its UE context to the RAN, specifically the UPF IP address and the UL TEID, together with the user data over the C-plane session. Then, the RAN establishes a GTP Tunnel for UL in a temporary session with the UPF using the provided UPF information and transfers user data to the UPF. At the same time, the RAN and the UPF establish a GTP Tunnel for DL in the temporary session. The state is shifted to CM-IDLE with Limited Context. After the data transfer, the inactivity timer expiration shifts the state to CM-IDLE again.

### 3.2 Building Limited UE Context at RAN

This subsection describes how the limited UE context is built in the CM-IDLE with Limited Context state. Figure 8 lists up the parameters included in a UE context maintained by a RAN. These are used both to coordinate the radio section between the UE and the RAN and to establish sessions with AMF and UPF. Since small data are exchanged over the C-plane in the radio section, the values of the parameters except the UPF information are determined in the following way, as specified in Fig. 8 in detail.

- AMF identifier and C-plane session information with AMF: unnecessary.
- Radio security information including the security capability and the key over the radio section: unnecessary.
- PDU session identifier: set to predetermined value.
- Slice identifier: set to predetermined value.
- Maximum data rate (the data rate limit determined by the subscription contract): unnecessary.
- PDU session type (IPv4, IPv6, unstructured, or Ethernet): set to predetermined value.
- QoS flow information specifying the QoS classes: set to default QoS.

Only the values of the UPF information, including the UPF IP address and the TEID, vary in the actual communication environment. They are provided by the UE together with user data as described in the previous subsection.

## 3.3   Allocating Downlink and Uplink GTP Tunnel IDs for Small Data Transfer

In the original 5G system, the UL TEID is allocated by the UPF and the DL TEID by the RAN. The allocating side, either the UPF or the RAN, needs to inform the other side of the TEID through the C-plane signaling. In Idle-SDT, the UL TEID is allocated by the UPF as a unique value within it and is given to the RAN by the UE when it sends user data in the CM-IDLE with Limited Context state. The RAN will no longer need to send C-plane messages to the 5GC.

Idle-SDT uses the following approach to determine a DL TEID value that is unique within the RAN and to share the value between the UPF and the RAN without any C-plane signaling in the CM-IDLE with Limited Context state.

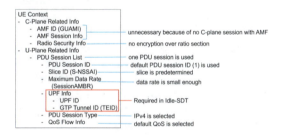

**Fig. 8.** Structure of UE Context at RAN and How Parameters Are Specified

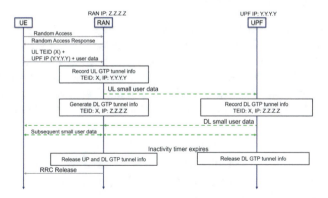

**Fig. 9.** Procedure of Idle-SDT

- When the UPF allocates the UL TEID, it determines it as a unique value within itself as well as within any RANs. That is, the UL TEID value is determined so that it can be used as the DL TEID value.
- The unique value can be determined by combining the IP address of the UE assigned by SMF/UPF, which is unique within the UPF, and the identifier of the UPF.
- When a UE sends user data together with the UPF information including the UL TEID, the RAN picks up the UL TEID value and uses it as the DL TEID between the RAN and the UPF.
- When the UPF detects user data transferred from the RAN belonging to a UE in CM-IDLE with Limited Context, it records the value of DL TEID as the UL TEID in order to transfer downlink user data to the UE.

### 3.4   Procedure in CM-IDLE with Limited Context

The procedure for user data transfer by a UE in CM-IDLE with Limited Context is shown in Fig. 9. As described above, the UE sends the UL TEID $X$, and UPF IP $Y.Y.Y.Y$, provided during the PDU Session Establishment Procedure, together with user data to the RAN over the C-plane. The RAN records $X$ and $Y.Y.Y.Y$ as the UL GTP tunnel information and sends the user data to the UPF. At the same time, the RAN uses $X$ as the DL TEID. When the UPF receives user data, it checks which session the data belongs to. If the UPF determines that the data belongs to the session of the UE in CM-IDLE with Limited Context, it sets the UL TEID ($X$) and the IP address of the originating RAN ($Z.Z.Z.Z$) as the DL GTP tunnel information. As a result, the UL and DL communications between the UE and the UPF become possible.

## 4   Performance Evaluation

### 4.1   Signaling Overhead Model

To validate the effectiveness of the proposed method for the sporadic data transfers of moving IoT devices, we focus on the number of signaling messages exchanged among UEs, RANs, and 5GC NFs, as an indicator of overhead, which is a similar method adopted in [8]. This is because we emphasize the C-plane signaling overhead involved in connection management processing and location management. We define the total overhead $C_{total}$ in the C-plane for moving IoT devices by

$$C_{total} = C_{CM} + C_{LM}. \tag{1}$$

Here, $C_{CM}$ indicates the signaling overhead associated with connection management that occurs as the UEs transfer user data. $C_{LM}$ indicates the signaling overhead of the location management for moving UEs.

The connection management overhead $C_{CM}$ takes different values in the following two cases. One is that the RAN receiving data from a UE has its UE context (Case 1) and the other is that the RAN does not have one (Case 2). So, $C_{CM}$ is given by

$$C_{CM} = c_1 \cdot \sum_{i=1}^{n} x_i + c_2 \cdot \sum_{i=1}^{n} y_i. \qquad (2)$$

Here, the values of $n$, $c_1$ and $c_2$ indicate the number of RANs, the signaling overhead of a single user data transfer by a UE in Case 1, and that in Case 2, respectively. The symbols $x_i$ and $y_i$ indicate the number of user data transmitted from UEs to $RAN_i$ in Case 1 and that in Case 2, respectively.

The signaling overhead for the location management $C_{LM}$ arises when a UE moves to another TA or RNA. So, $C_{LM}$ is given by

$$C_{LM} = c_3 \cdot \sum_{i=1}^{n} \sum_{j=1, j \neq i}^{n} z_{i,j} \cdot a(i, j). \qquad (3)$$

Here, the value $c_3$ and the symbol $z_{i,j}$ indicate the signaling overhead related to the location management for one moving and the number of UEs that move from $RAN_i$ to $RAN_j$, respectively. The function $a(i, j)$ returns 0 if the TA (or RNA) of $RAN_i$ is the same as that of $RAN_j$ and returns 1 otherwise.

From Eqs. (1), (2) and (3), the total overhead, $C_{total}$, is obtained by

$$C_{total} = c_1 \cdot \sum_{i=1}^{n} x_i + c_2 \cdot \sum_{i=1}^{n} y_i + c_3 \cdot \sum_{i=1}^{n} \sum_{j=1, j \neq i}^{n} z_{i,j} \cdot a(i, j). \qquad (4)$$

Table 1 shows the signaling overhead with one data transfer or one UE moving, i.e., the values of $c_1$, $c_2$ and $c_3$, for each method. They are obtained by counting the number of message exchanges in the procedures presented in Sects. 2 and 3. The signaling overhead of the connection management in the Original 5GC is the sum of messages in the Service Request and AN Release Procedures. Since Original 5GC and Idle-SDT do not maintain a UE context during CM-IDLE, the value of $c_1$ in these methods is zero. In our simulation, we assume each method performs either the Mobility Registration or RNA Update Procedure for the location management. The original 5GC also assumes that the UE moves to another TA only when it is in CM-IDLE.

## 4.2   Evaluation Environment

We evaluate the performance of the proposed method using NS-3 [9]. We suppose a topology as shown in Fig. 10. RANs are deployed homogeneously in a square area of 12 km by 12 km. Each RAN has a square communication area of 1 km by 1 km. RNA consists of nine RANs and TA consists of four RNAs. There are four TAs in the topology. The number of UEs is 10,000 in this area. The UEs are randomly located initially and move according to the Random Work Model [10]. We compare the proposed method with Original 5GC and Inactive-SDT with/ without Context Relocation regarding signaling overhead. With a simulation time of 4 h, we evaluate each performance with changing the mobility speed of UEs and the time interval for transferring user data.

### 4.3   Evaluation Results

Figure 11 shows the signaling overhead for each method when the mobility speed is 0 km/h (UE does not move), 4 km/h, 10 km/h and 40 km/h. The UEs transfer user data every 10 min. When the mobility speed is 0 km/h, the signaling overhead for Idle-SDT is the same as that for Inactive-SDT with Context Relocation and Inactive-SDT without Context Relocation. This is because there was no processing of the UE context retrieval by the RAN, and no processing of the location management by the RAN, since UEs do not move. Idle-SDT achieves the lowest signaling overhead at the other mobility speeds. The signaling overhead for Idle-SDT decreases to approximately 32-17% compared to the original 5GC, 48-30% compared to the Inactive-SDT without Context Relocation. When mobility speed is 40 km/h, Inactive-SDT with/without Context Relocation have a higher signaling overhead than Original 5GC. This is because these methods perform RNA Update more frequently than Mobility Registration.

Figure 12 shows the results when the interval of user data transfer is 1 min, 10 min and 60 min. Idle-SDT achieves the lowest signaling overhead at any interval. When the interval of user data transfer is 1 min, the signaling overhead of

**Table 1.** Signaling Overhead with One Behavior for Each Method

|  | Connection management | | Location management | |
|---|---|---|---|---|
|  | Overheads with Context in RAN ($c_1$) | Overheads without Context in RAN ($c_2$) | Trigger | Overheads ($c_3$) |
| Original 5GC | 0 | 26 (Fig. 1 + Fig. 2) | move to another TA | 9 (Fig. 6) |
| Inactive-SDT with Context Relocation | 4 (Fig. 3) | 13 (Fig. 4) | move to another RNA | 13 (Fig. 6) |
| Inactive-SDT without Context Relocation | 4 (Fig. 3) | 9 (Fig. 5) | move to another RNA | 13 (Fig. 6) |
| Idle-SDT | 0 | 4 (Fig. 9) | move to another TA | 9 (Fig. 6) |

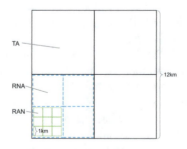

**Fig. 10.** Evaluation topology

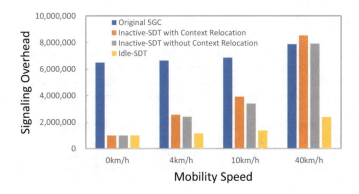

**Fig. 11.** Signaling Overhead with Changing Mobility Speed

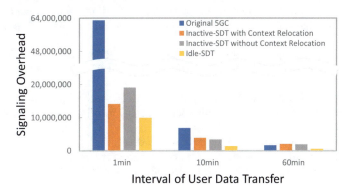

**Fig. 12.** Signaling Overhead with Changing Interval of User Data Transfer

Inactive-SDT without Context Relocation is higher than that of Inactive-SDT with Context Relocation. This is because there are many times when UEs send user data to RANs that do not maintain a UE Context. These evaluations confirmed that Idle-SDT reduces the number of C-Plane signaling messages even when the UE moves.

## 5   Conclusion

In this paper, we propose a new IoT communication method with low signaling overhead, even for moving IoT devices, which we named Idle-SDT. Idle-SDT introduces the new state, called CM-IDLE with Limited Context, in which UE can transfer user data to UPF without any C-plane signaling between RAN and 5GC. Idle-SDT allows UE to transfer UPF IP address and UL TEID to RAN together with user data, and this enables RAN and UPF to establish a temporary session. The simulation results showed that Idle-SDT reduced the signaling message exchange up to 28% compared with Inactive-SDT proposed by 3GPP.

# References

1. Dahlman, E., Parkvall, S., Skold, J.: 5G NR: The Next Generation Wireless Access Technology, 2nd edn. Elsevier Academic Press, London (2021)
2. Vittal, S., Franklin, A.A.: HARNESS: high availability supportive self reliant network slicing in 5G networks. IEEE Trans. Netw. Service Manag. **19**(3), 1951–1964 (2022)
3. Kahn, C., Viswanathan, H.: Connectionless access for mobile cellular networks. IEEE Commun. Mag. **53**(9), 26–31 (2015)
4. 3GPP Technical Specification Group Services and System Aspects, "System architecture for the 5G System (5GS) (Release 18)," 3GPP TS 23.501, V 18.3.0, September 2023
5. 3GPP Technical Specification Group Radio Access Network, "NR and NG-RAN Overall Description (Release 17)," 3GPP TS 38.300, V 17.6.0, September 2023
6. Bhatia, R., et al.: Massive machine type communications over 5G using lean protocols and edge proxies. In: 2018 IEEE 5G World Forum (5GWF), pp. 462–467, July 2018
7. 3GPP Technical Specification Group Services and System Aspects, "Procedures for the 5G System (5GS) (Release 18)," 3GPP TS 23.502, V 18.3.0, September 2023
8. Oliveira, L.A.N., Alencar, M.S., Lopes, W.T.A., Madeiro, F.: On the performance of location management in 5G network using RRC inactive state. IEEE Access **10**, 65520–65532 (2022)
9. ns-3 network simulator. https://www.nsnam.org/
10. Camp, T., Boleng, J., Davies, V.: A survey of mobility models for ad hoc network research. Wireless Commun. Mobile Comput. **2**(5), 483–502 (2002)

# On Network Design and Planning 2.0 for Optical-Computing-Enabled Networks

Dao Thanh Hai[1(✉)] and Isaac Woungang[2]

[1] School of Science, Engineering and Technology, RMIT University Vietnam, Ho Chi Minh City, Vietnam
hai.dao5@rmit.edu.vn

[2] Department of Computer Science, Toronto Metropolitan University, Toronto, ON, Canada
iwoungan@torontomu.ca

**Abstract.** In accommodating the continued explosive growth in Internet traffic, optical core networks have been evolving accordingly thanks to numerous technological and architectural innovations. From an architectural perspective, the adoption of optical-bypass networking in the last two decades has resulted in substantial cost savings, owning to the elimination of massive optical-electrical optical interfaces. In optical-bypass framework, the basic functions of optical nodes include adding (dropping) and cross-connecting transitional lightpaths. Moreover, in the process of cross-connecting transiting lightpaths through an intermediate node, these lightpaths must be separated from each other in either time, frequency or spatial domain, to avoid unwanted interference which deems to deteriorate the signal qualities. In light of recently enormous advances in photonic signal processing/computing technologies enabling the precisely controlled interference of optical channels for various computing functions, we propose a new architectural paradigm for future optical networks, namely, *optical-computing-enabled networks*. Our proposal is defined by the added capability of optical nodes permitting the superposition of transitional lightpaths for computing purposes to achieve greater capacity efficiency. Specifically, we present two illustrative examples highlighting the potential benefits of bringing about in-network optical computing functions which are relied on optical aggregation and optical XOR gate. The new optical computing capabilities armed at optical nodes therefore call for a radical change in formulating networking problems and designing accompanying algorithms, which are collectively referred to as *optical network design and planning 2.0* so that the capital and operational efficiency could be fully unlocked. To this end, we perform a case study for network coding-enabled optical networks, demonstrating the efficacy of *optical-computing-enabled networks* and the challenges associated with greater complexities in network design problems compared to optical-bypass counterpart.

## 1 Introduction

From an architectural perspective, optical networking has been shifted from optical-electrical-optical mode to optical-bypass operations so that transiting lightpaths could

L. Barolli (Ed.): AINA 2024, LNDECT 199, pp. 91–102, 2024.
https://doi.org/10.1007/978-3-031-57840-3_9

be optically cross-connected from one end to the other end rather than being undergone unnecessary optical-electrical-optical conversions [1]. Optical-bypass networking has then gone a long way from a conceptual proposal to a widely adopted technology by network operators in the last two decades [2]. However, one of the major challenges in scaling up the optical networks to support explosive traffic growth in a greater efficiency manner is the fact that signal processing functions are mostly implemented in electrical domain based on digital signal processing. This involves a number of the well-established procedures including mainly optical-to-electrical conversion, electronic sampling, digital signal processing and finally, back-conversion to optical domain. Here, the critical bottleneck lies in the electronic sampling rate and in circumventing this major concern, solutions are directed towards migrating certain signal processing functions to the optical domain. Indeed, photonic signal processing/computing technologies offer a new powerful way for handling high speed signals thanks to its inherent merits of wide bandwidth, transparency and energy-efficiency [3].

In optical-bypass networking, the basic functions of optical nodes are to add (drop) and cross-connect the transiting lightpaths. Moreover, in cross-connecting transiting lightpaths through an intermediate node, these lightpaths must be maximally separated from each other in either time, frequency or spatial domain [2] to avoid unwanted interference which deems to deteriorate the signal qualities. This turns out to be a fundamental limitation as various optical computing operations could be performed between such transitional lightpaths to generate the output signals which are spectrally more-efficient than its inputs. In light of recent enormous advances in photonic computing technologies enabling the controlled interference of optical channels for various computing capabilities [4–7], we propose a new architectural paradigm for future optical networks, namely, optical-computing-enabled networks. Our proposal is defined by the added capability of optical nodes permitting the superposition of transitional lightpaths for computing purposes to realize a greater capacity efficiency. Specifically, we present two illustrative examples highlighting the potential benefits of bringing about in-network optical computing which are relied on optical aggregation [4,5] and optical XOR gate [6,7]. The new optical computing capabilities armed at optical nodes calls for a radical change in optical network design and planning in order to fully reap spectral and cost benefits as well as operational efficiency. To this end, we perform a case study for network coding-enabled optical networks, demonstrating the efficacy of optical-computing-enabled networks and challenges associated with greater complexities in network design problems compared to optical-bypass counterpart.

The paper is structured as followed. In Sect. 2, we introduce a new concept of optical-computing-enabled paradigm. We also highlight the applications of two optical computing operations, namely, optical aggregation and optical XOR whose enabling technologies have been progressing fast and their integration to future optical networks could be foreseen. We also address the computational impact and intricacies for network design and planning in the paradigm of optical-computing networking. Next, as a case study to reveal more complicated network design problems arisen in the optical-computing-enabled network, we focus on the network coding-enabled scenarios and formulate the routing, wavelength and network coding assignment problem in Sect. 3. Section 4 is dedicated to showcase the numerical evaluations comparing our proposal that leverages the use of optical XOR encoding within the framework of optical-computing-enabled networks to the traditional optical-bypass networking. The comparison is drawn on a realistic COST239 and NSFNET network topologies. Finally, Sect. 5 concludes the paper.

## 2 Optical-Computing-Enabled Paradigm

The optical-computing-enabled paradigm is characterized by the key property that optical nodes are empowered with the optical computing capability. Specifically, two or more optical channels could be optically mixed together to compute a new optical channel and thus, to attempt achieving a greater capacity efficiency [8–11]. In light of tremendous progresses in optical computing technologies permitting precisely controlled interference between optical channels, in-transit lightpaths traversing the same optical node are offered unique opportunities to optically superimpose to each other to generate the output signals which are spectrally more-efficient than their inputs. Such optical operations involving the interaction of transitional optical channels pave the way for redefining the optical network architecture, disrupting the conventional assumption of keeping transitional lightpaths untouched. In this context, optical-computing-enabled framework is foreseen to be the next evolution of optical-bypass networking. In this section, we highlight the efficient use and impact of introducing two optical computing operations, namely, optical aggregation (de-aggregation) and optical XOR gate to optical networks. It is noticed that the enabling technologies for realizing such two optical computing capabilities have been accelerating and therefore technical readiness of integration to optical nodes could be foreseen. Of course, there are many other ways for mixing two optical channels and in the future, as photonic computing technologies move forward, a wide range of computing functions could be technologically realized. These advances will be expected to have massive impacts to optical networks from both design, planning, operation and management.

### 2.1 Optical Aggregation and De-aggregation

Aggregation of lower-speed channels into a single higher-speed one has been a key function in the operation of optical networks. The goal of doing so is to achieve a greater capacity efficiency by freeing up the lightpaths of the lower wavelength utilization. Traditionally, this function has been performed in the electronic domain by terminating optical channels, re-assembling, re-modulating and finally back-converting to the optical domain. Clearly, there have been many limitations of doing so and thus, it is not scalable for higher bit-rate operations. In mitigating this major issue, the concept of optical aggregation has recently been proposed, implemented and pushed forward [12, 13]. The main industrial player for this revolutionary effort is INFINERA, whose the goal is to develop a new ecosystem of devices and components, with the capability of transforming the traditional operation of the optical nodes. In term of functionality, an optical aggregator can add two or more optical channels of lower bit-rate and/or lower-order modulation format into a single higher bit-rate and higher-order modulation format one. In this example, we consider the use of an optical aggregator and de-aggregator whose function is to combine two QPSK signals into a single 16-QAM channel and vice-versa. Figure 1 illustrates the schematic diagram for adding two QPSK channels of lower bit-rate into a single 16-QAM channel of higher bit-rate. In doing so, the spectral efficiency could thus by improved twice.

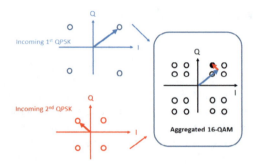

**Fig. 1.** Schematic illustration of the aggregation of two QPSK signals into a single 16-QAM one and vice-versa

In leveraging the use of the aforementioned aggregator for optical networks, we first consider the conventional way of accommodating traffic demands in optical-bypass networking. Figure 2 shows the routing and wavelength assignment for two demands $a$ and $b$ of the same line-rate 100G and format QPSK. Due to the wavelength uniqueness constraint on a link, two wavelengths are needed on link $XI$ and $IC$. Now, let us consider the case that at node $X$, the optical aggregation is enabled. By permitting the optical aggregation, two 100G QPSK transitional lightpaths crossing node $X$ could be optically added to generate the output signal of 200G, which is modulated on the 16-QAM format. In Fig. 3, it is clearly observed that by having a single wavelength channel of 200G capacity, a greater capacity efficiency has been realized. At the common destination node $C$, the aggregated lightpath could be decomposed into constituent ones and such decomposition operation could be performed either in an optical or electrical domain. It is important to note that in order to maximize the aggregation opportunities, new network design and planning algorithms should be developed to determine the pairing of demands for aggregation, the respective aggregation node and more importantly, the transmission parameters for aggregated lightpaths.

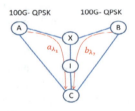

**Fig. 2.** Traffic Provisioning in Optical-bypass Networking

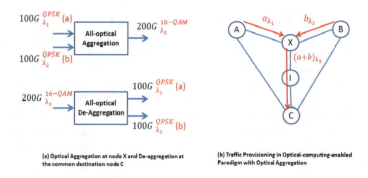

(a) Optical Aggregation at node X and De-aggregation at the common destination node C

(b) Traffic Provisioning in Optical-computing-enabled Paradigm with Optical Aggregation

**Fig. 3.** Optical-computing-enabled Paradigm with Optical Aggregation and De-aggregation

## 2.2   Optical XOR Encoding and Decoding

Technologies for realizing all-optical logic gates have been accelerating in recent years that permit performing the bit-wise exclusive-or (XOR) between optical signals of very high bit-rates and/or different modulation formats [6, 7]. Different from the aggregation operation, the optical XOR encoding output is kept at the same bit-rate and/or format as the inputs. A functional description of such device is shown in Fig. 5(a), where two optical signals of 100G QPSK on different wavelength are coded together to generate the output X of the same bit-rate and format.

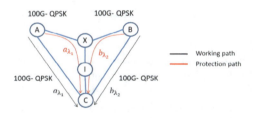

**Fig. 4.** Traffic Provisioning in Optical-bypass Networking

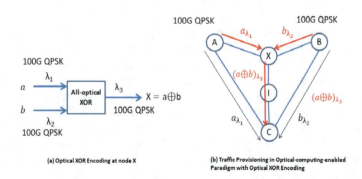

(a) Optical XOR Encoding at node X

(b) Traffic Provisioning in Optical-computing-enabled Paradigm with Optical XOR Encoding

**Fig. 5.** Optical-computing-enabled Paradigm with Optical XOR Encoding and Decoding

In exploiting the optical XOR gate, we focus on the protection scenario and we assume that there are two demands with dedicated protection. The provisioning of such two demands are shown in Fig. 4 for the optical-bypass framework. Because the protection lightpath of demand $a$ and demand $b$ cross the same links $XI$ and $IC$, it therefore requires at least two wavelengths on those links. Looking at Fig. 5(b) for the optical-computing-enabled paradigm when node $X$ is armed with the optical XOR encoding capability, the protection signal of demand $a$ and demand $b$ could thus be optically encoded to produce the signal $X = a \oplus b$. Such encoded output is routed all the way from node $X$ to the shared destination node $C$. By doing so, only one wavelength channel on link $XI$ and $IC$ is sufficient, resulting in a spectral saving of 50%. Although the protection signals of demand $a$ and demand $b$ is encoded, it is always possible to recover the original signal for both demand $a$ and demand $b$ in any case of a single link failure. The recovery is as simple as the encoding by making use of the XOR operation on the two remaining signals. Specifically, if the working signal of demand $b$ is lost, it can be retrieved in an another way by $b = (a \oplus b) \oplus a$.

The combination of optical encoding and dedicated protection appears to be matched to attain a greater capacity efficiency while keeping a near-immediate recovery speed. Nevertheless, in order to realize the encoding benefits, more complicated network design problems emerge. The more intricacies are related to the determination of the pair of demands for encoding and the selection of the routes and/or transmission parameters for the encoded lightpaths.

## 2.3   Impact for Network Design and Planning

It should be noted that the *optical-computing-enabled* paradigm introduces more networking flexibility by permitting the precisely controlled interference among two or more favorable lightpaths, and this poses important ramifications for the network design algorithms to maximize the potential benefits [14–19]. In optical-bypass networking, the central problem for the network design and planning is the routing and resource allocation. For solving such problem, the selection of the route and assigning transmission parameters, including the wavelength/spectrum and/or format, is determined for each individual demand. In optical-computing-enabled networking, more complicated network design problems arise due to the interaction of transitional lightpaths. Specifically, in addition to determining of the route and transmission parameters for each demand, the pairing of demands and subsequently, the selection of route and assigning transmission parameters for special lightpaths involving the interaction of two or many demands, must also be identified. This represents a radical change in the network design and planning, disrupting the conventional set of algorithms that have been developed for optical-bypass networking in many years. In recognizing this disruption, we call for a new framework, that is, optical network design and planning 2.0, which encompasses new problems emerging from various ways that transitional lightpaths could be optically mixed and accompanying algorithms including exact and heuristic solutions for solving them. In the subsequent section, we formulate the problem entitled, the routing, wavelength and network coding assignment problem arising in the application of optical XOR for optical-computing-enabled network and highlight how such problem is different from its counterpart, that is, routing and wavelength assignment in optical-bypass network.

## 3 Routing, Wavelength and Network Coding Assignment Problem Formulation in Optical-Computing-Enabled Network

In this section, we consider the design of network coding-enabled networks to support a set of traffic demands with minimum wavelength link cost. The optical encoding scheme to be used is the simple XOR, where the input signals of the same wavelength, line-rate and format, are XOR-coded to produce the output signal of the same wavelength and format. The main advantage of such scheme, XOR coding between signals of the same wavelength, is the elimination of a probe signal and therefore, could be highly cost-efficient [20]. Moreover, for ease of operations, the encoding is restricted only on the protection signals of demands having the same destination node and the decoding is only taken place at the destination. Furthermore, each demand is permitted to have maximum one encoding operation. In this framework, there are a set of sufficient constraints on the network coding assignment for any two code-able demands, namely: i) two demands must have common destination, ii) two demands must use the same wavelength, iii) the link-disjoint constraint between their working paths and between one's working path to the another's protection path, iv) their protection paths must have a common sub-path whose one end is the shared destination.

Inputs:

- $G(V,E)$: A graph representing the physical network topology with $|V|$ nodes and $|E|$ fiber links.
- $D$: A set representing the traffic demands, indexed by $d$. Each demand $d \in D$ requests *one wavelength capacity* (e.g., 100 Gbps)
- $W$: A set representing the wavelengths on each fiber link, indexed by $w$. The link capacity measured in number of wavelength is $|W|$

Outputs:

- Routing and wavelength assignment for each lightpath
- Determination of pair of demands for optical XOR encoding and determination of respective coding nodes.
- Routing and wavelength assignment for encoded lightpaths
- The usage of wavelength on each link

Objective: Minimize the wavelength link usage

The mathematical model is formulated in the form of integer linear programming (ILP). In addition to typical variables and constraints accounting for the selection of route and assigning wavelength for each demand, new variables and constraints emerge as the interaction of demands have been introduced and thus, causing the mathematical model one order of magnitude computationally harder than its counterpart, that is, the traditional routing and wavelength assignment in optical-bypass networking [21]. In acknowledging the NP-nature of the model, we therefore propose the following scalable heuristic (Algorithm 1) that could be used in large networks.

---

**Algorithm 1:** Heuristic Solution

---

**Input:** $G(V,E), D, W$

**Output:** $\alpha_{e,w}^d, \beta_{e,w}^d, \theta_w^d, z_{e,w}^{d,v}, \delta_v^d, f_{d_1}^{d_2}, \gamma_{e,w}$

1  **for** *node* $v \in V$ **do**

2  $\quad\lfloor$  Find demand $d \in D : r(d) = v$

3  $\quad\lfloor$  Insert $d$ into set $X_v$

4  **Sort** $X_v$ according to its size $|X_v|$ in descending order

5  **for** *demand* $d \in X_v$ **do**

6  $\quad\lfloor$  Find $k$ shortest cycles (modified Suurballe algorithm) including working and routing route for demand $d$

7  **Sort** demand $d \in X_v$ according to its length of $k$ cycles in descending order

    `/* Perform routing, encoding and wavelength assignment for all`
      `sorted demands                                             */`

8  **for** $d \in X_v$ **do**

9  $\quad\lfloor$  Select one cycle for demand $d$ out of $k$ cycles

10 $\quad\lfloor$  Perform encoding with a suitable demand

11 $\quad\lfloor$  First-fit Wavelength Assignment

---

## 4   Numerical Results

This section presents the numerical evaluations comparing our proposal that leverages the use of optical XOR encoding within the framework of optical-computing-enabled networks to the traditional optical-bypass networking. The comparison is drawn on a realistic COST239 and NSFNET network topologies as shown in Fig. 6. The metric for comparison is a traditional one, that is, the wavelength link cost. Two designs, namely, WNC and NC are performed, where WNC refers to the routing and wavelength assignment in optical-bypass networking and NC refers to the more advanced problem, i.e., routing, wavelength and network coding assignment in optical-computing-enabled networks.

We first test the solutions from solving the ILP models and heuristic on small-scale topology, 6 nodes, all of degree 3 as shown in Fig. 6(a). The traffic is randomly generated between node pairs with the unit capacity (i.e., one wavelength) and the fiber capacity is 40 wavelengths. We consider three increasing load corresponding to 30%, 70% and 100% (full mesh) node-pair traffic exchange. The result in Table 1 (except the full mesh) is averaged over 20 samples. For the network coding-based design in full mesh traffic, the computation is overly long and thus, the results is obtained after 10 h of running. The well-studied heuristic for WNC achieves optimal results which are on a par with its ILP model while the heuristic for NC produces reasonably good solutions with tight gaps compared to its ILP model, avoiding the overly long computational time. Due to the sub-optimal nature of the heuristic algorithms, the gain obtained by these algorithms is slightly reduced compared to the one obtained from the ILP.

We apply the heuristic algorithm for larger networks, NSFNET and COST239 topologies with the same setting as in the 6-node topology about traffic generation, fiber capacity and number of traffic samples. The results are presented in Table 2. It can

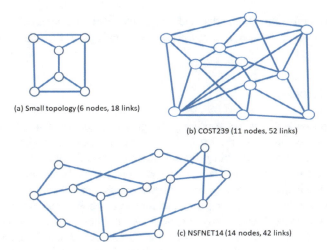

(a) Small topology (6 nodes, 18 links)

(b) COST239 (11 nodes, 52 links)

(c) NSFNET14 (14 nodes, 42 links)

**Fig. 6.** Network Topologies under Test

**Table 1.** Performance Comparison between Exact Solution and Heuristic One

| Load | ILP | | | Heuristic | | |
|------|------|------|------|------|------|------|
| | WNC | NC | Gain | WNC | NC | Gain |
| 30% | 32.6 | 30.6 | Max: 9%, Mean: 6% | 32.6 | 31.2 | Max: 9%, Mean: 4% |
| 70% | 75.8 | 67.9 | Max: 12%, Mean: 10% | 75.8 | 70.6 | Max: 9%, Mean: 7% |
| 100% | 108 | 99 | Max = Mean = 8% | 108 | 100 | Max = Mean = 7% |

be observed that up to about 8% gain could be achieved with the NSFNET network. For more densely connected COST239 network, the lower gain is obtained, up to 5%. It is evident that the solution from NC cases is always better than that from the WNC, resulting in improved capacity efficiency. Compared to the findings on O-E-O case [22], where the gain was reported up to 20%, there is reduced gain in all-optical case. This may be due to wavelength-related constraints for network coding assignments, curbing

**Table 2.** Numerical Results for Realistic topologies

| Topo | Load | WNC | NC | Gain | No Coding Operation |
|------|------|------|------|------|------|
| NSFNET | 30% | 318.3 | 299.5 | Max = 8%, Mean = 6% | 9.7 |
| | 70% | 730.4 | 682.4 | Max = 8%, Mean = 7% | 24.5 |
| | 100% | 1048 | 981 | Max = Mean = 6% | 35 |
| COST239 | 30% | 126.2 | 123.3 | Max = 3%, Mean = 2% | 1.4 |
| | 70% | 295.6 | 285.4 | Max = 5%, Mean = 3% | 5.1 |
| | 100% | 420 | 404 | Max = Mean = 4% | 8 |

the coding capability among the demands. Moreover, it should be noted that the gain is highly dependent on the structure of the network topology, traffic and network design algorithms.

## 5   Conclusion

This paper presents a new networking paradigm for future optical networks named, *optical-computing-enabled framework*. As a potential candidate for the next evolution of optical-bypass architecture, our proposal aims to exploit the optical computing capability at optical nodes so that greater capacity efficiency could be achieved. In highlighting the potential benefits of such *optical-computing-enabled framework*, we brought out two revealing examples leveraging the efficient use of optical aggregation and optical XOR gate. A numerical case study for the network coding-enabled design was provided to demonstrate the efficacy and the more complicated network design algorithm of *optical-computing-enabled networking* compared to the optical-bypass counterpart.

Albeit still primitive, the perspective of permitting optical mixing at intermediate nodes heralds a reinvention of optical networking, altering the way we think about optical network architecture. Although the experiments and practical demonstration of optical-computing-enabled networks are still ahead, the disconnection between the theoretical studies and the implementation realities are only just beginning to be rectified as enabling technologies have been more maturing and the huge profits have been foreseen. It should be pointed out that network design algorithms play a key role in achieving the operational efficiency of a network and thus, more robust and carefully designed algorithms should be developed to optimize the advantages of *optical-computing-enabled networking*. As for future works, we plan to develop an ecosystem of new research problems and the accompanying algorithms, collectively referred to as *optical network design and planning 2.0*, to capture a wide range of optical computing operations between in-transit lightpaths at their optimal usage scenarios.

## References

1. Saleh, A., Simmons, J.M.: Technology and architecture to enable the explosive growth of the internet. IEEE Commun. Mag. **49**(1), 126–132 (2011). https://doi.org/10.1109/MCOM. 2011.5681026
2. Saleh, A., Simmons, J.M.: All-optical networking: evolution, benefits, challenges, and future vision. Proc. IEEE **100**(5), 1105–1117 (2012). https://doi.org/10.1109/JPROC.2011. 2182589
3. Willner, A.E., et al.: All-optical signal processing techniques for flexible networks. J. Lightwave Technol. **37**(1), 21–35 (2019). https://doi.org/10.1109/JLT.2018.2873245
4. Wang, H., Pan, L., Ji, Y.: All-optical aggregation and de-aggregation of 4×BPSK-16QAM using nonlinear wave mixing for flexible optical network. IEEE J. Sel. Top. Quantum Electron. **27**(2), 1–8 (2021). https://doi.org/10.1109/JSTQE.2019.2943375
5. Li, Q., Yang, X., Yang, J.: All-optical aggregation and de-aggregation between 8QAM and BPSK signal based on nonlinear effects in HNLF. J. Lightwave Technol. **39**(17), 5432–5438 (2021). https://doi.org/10.1109/JLT.2021.3084353

6. Chen, L.K., Li, M., Liew, S.C.: Breakthroughs in photonics 2014: optical physical-layer network coding, recent developments, and challenges. IEEE Photonics J. **7**(3), 1–6 (2015). https://doi.org/10.1109/JPHOT.2015.2418264

7. Kotb, A., Zoiros, K.E., Guo, C.: 1 Tb/s all-optical XOR and AND gates using quantum-dot semiconductor optical amplifier-based turbo-switched Mach-Zehnder interferometer. J. Comput. Electron. (2019). https://doi.org/10.1007/s10825-019-01329-z

8. Hai, D.T.: On routing, wavelength, network coding assignment, and protection configuration problem in optical-processing-enabled networks. IEEE Trans. Netw. Serv. Manage. **20**(3), 2504–2514 (2023). https://doi.org/10.1109/TNSM.2023.3283880

9. Hai, D.T.: Optical-computing-enabled network: an avant-garde architecture to sustain traffic growth. Results Opt. **13**, 100504 (2023). https://doi.org/10.1016/j.rio.2023.100504. https://www.sciencedirect.com/science/article/pii/S2666950123001566

10. Hai, D.T.: Optical networking in future-land: from optical-bypass-enabled to optical-processing-enabled paradigm. Opt. Quantum Electron. **55**(864) (2023). https://doi.org/10.1007/s11082-023-05123-x

11. Hai, D.T.: Quo Vadis, optical network architecture? Towards an optical-processing-enabled paradigm. In: 2022 Workshop on Microwave Theory and Techniques in Wireless Communications (MTTW), pp. 193–198 (2022). https://doi.org/10.1109/MTTW56973.2022.9942542

12. Welch, D., et al.: Point-to-multipoint optical networks using coherent digital subcarriers. J. Lightwave Technol. **39**(16), 5232–5247 (2021). https://doi.org/10.1109/JLT.2021.3097163

13. Bäck, J., et al.: Capex savings enabled by point-to-multipoint coherent pluggable optics using digital subcarrier multiplexing in metro aggregation networks. In: 2020 European Conference on Optical Communications (ECOC), pp. 1–4 (2020). https://doi.org/10.1109/ECOC48923.2020.9333233

14. Hai, D.T.: Leveraging the survivable all-optical WDM network design with network coding assignment. IEEE Commun. Lett. **21**(10), 2190–2193 (2017). https://doi.org/10.1109/LCOMM.2017.2720661

15. Hai, D.T., Chau, L.H., Hung, N.T.: A priority-based multiobjective design for routing, spectrum, and network coding assignment problem in network-coding-enabled elastic optical networks. IEEE Syst. J. **14**(2), 2358–2369 (2020). https://doi.org/10.1109/JSYST.2019.2938590

16. Hai, D.T.: A bi-objective integer linear programming model for the routing and network coding assignment problem in WDM optical networks with dedicated protection. Comput. Commun. **133**, 51–58 (2019). https://doi.org/10.1016/j.comcom.2018.08.006

17. Hai, D.T.: On routing, spectrum and network coding assignment problem for transparent flex-grid optical networks with dedicated protection. Comput. Commun. (2019). https://doi.org/10.1016/j.comcom.2019.08.005

18. Hai, D.T.: Re-designing dedicated protection in transparent WDM optical networks with XOR network coding. In: 2018 Advances in Wireless and Optical Communications (RTUWO), pp. 118–123 (2018). https://doi.org/10.1109/RTUWO.2018.8587873

19. Hai, D.T.: Network coding for improving throughput in WDM optical networks with dedicated protection. Opt. Quantum Electron. **51**(387) (2019). https://doi.org/10.1007/s11082-019-2104-5

20. Porzi, C., Scaffardi, M., Potì, L., Bogoni, A.: All-optical XOR gate by means of a single semiconductor optical amplifier without assist probe light. In: 2009 IEEE LEOS Annual Meeting Conference Proceedings, pp. 617–618 (2009). https://doi.org/10.1109/LEOS.2009.5343425

21. Varvarigos, E., Christodoulopoulos, K.: Algorithmic aspects of optical network design. In: 2011 15th International Conference on Optical Network Design and Modeling (ONDM), pp. 1–6 (2011)
22. Øverby, H., Biczók, G., Babarczi, P., Tapolcai, J.: Cost comparison of 1+1 path protection schemes: a case for coding. In: 2012 IEEE International Conference on Communications (ICC), pp. 3067–3072 (2012). https://doi.org/10.1109/ICC.2012.6363928

# DEAR: DRL Empowered Actor-Critic ScheduleR for Multipath QUIC Under 5G/B5G Hybrid Networks

Pattiwar Shravan Kumar[1(✉)], Paresh Saxena[1], and Özgü Alay[2,3]

[1] BITS Pilani, Hyderabad, India
{p20190019,psaxena}@hyderabad.bits-pilani.ac.in
[2] University of Oslo, Oslo, Norway
ozgua@ifi.uio.no
[3] Karlstad University, Karlstad, Sweden

**Abstract.** Recently, the Internet has experienced a substantial increase in the use of bandwidth-intensive applications due to the introduction of fifth generation (5G) and beyond 5G (B5G) systems. Empirical evidence has shown that multipath transport layer protocols such as multipath TCP (MPTCP) and multipath QUIC (MPQUIC) are successful in addressing the increasing need for higher bandwidth in the existing Internet infrastructure. Nevertheless, multipath schedulers still face difficulties in efficiently handling significant amounts of variability in diverse network scenarios. This paper introduces, **D**eep reinforcement learning (DRL) **E**mpowered **A**ctor-critic schedule**R**, **DEAR**, a method designed for multipath QUIC in 5G/B5G hybrid networks. DEAR is developed utilising a DRL based actor critic methodology. This approach significantly improves the decision-making abilities of the scheduler in various rapidly changing network scenarios. We conducted experiments with the DEAR scheduler in several network settings, encompassing networks with rapidly fluctuating bandwidth, networks with rapid and short-term fluctuations, and networks experiencing progressive outages. We also conducted tests on the DEAR algorithm using the Lumos5G dataset, which consists of real network traces from two distinct service providers. We have performed a comparative analysis of DEAR with other state-of-the-art multipath schedulers, including another scheduler based on RL, Peekaboo, and other non-RL rule-based schedulers such as round robin (RR), earliest completion first (ECF), blocking estimation (BLEST), and minimum round trip time (min-RTT). Our evaluation demonstrates DEAR's superior performance compared to existing algorithms. In scenarios with fast-changing bandwidth, DEAR outperforms rule-based schedulers by 42.30% and Peekaboo by 17.77%. Similarly, in networks with fast short-scale variations, DEAR achieves gains of 22.22% over rule-based schedulers and 6.06% over Peekaboo. Moreover, in networks facing progressive outage and recovery, DEAR showcases gains of 13.90% over rule-based schedulers and 8.79% over Peekaboo.

**Keywords:** Multipath Networking · MPQUIC · Schedulers · 5G/B5G Networks

© The Author(s), under exclusive license to Springer Nature Switzerland AG 2024
L. Barolli (Ed.): AINA 2024, LNDECT 199, pp. 103–113, 2024.
https://doi.org/10.1007/978-3-031-57840-3_10

# 1 Introduction

The evolution of the Internet and network technologies has led to a growing need for quicker, more dependable, and effective data transfer [1]. Multipath transport protocols have emerged as a promising solution to meet these requirements. MPQUIC, a multipath extension of the widely-used QUIC (Quick UDP Internet Connections) protocol, is leading the way in this technological progress [2]. MPQUIC, functioning in the user space [3], is not restricted to stationary desktops, laptops, and mobile phones and hence it is more flexible to deploy as compared to muliptath transmission control protocol (MPTCP) [3]. It can also be incorporated into forthcoming fifth generation (5G) and beyond 5G (B5G) systems and applications, including connected automated vehicles (CAVs), unmanned aerial networks (UAVs) [4], and high-speed trains [5].

The MPQUIC uses a scheduler which is crucial for optimising the transmission of data over multiple paths, resulting in enhanced network speed and increased quality of service (QoS) for various applications. However, the multipath scheduler presents several significant challenges due to its intricate design. An important challenge is to develop a scheduler that can efficiently allocate packets across several networks, maximising the use of available resources. The complex interplay of various parameters and the ever-changing nature of networks pose a significant challenge in designing an optimal scheduler in a dynamic environment [6]. Researchers have recently focused on resolving problems related to multipath schedulers, including the head of line (HoL) blocking [7] and bufferbloat [8], which occurs when packets transmitted through a slower network experience delays, whereas packets transmitted through a faster network arrive at their destination early. Consequently, the quicker network packets may have to be held in the buffer until the slower network packets arrive. This can result in packets being received out of sequence, reduced data transfer rate, and a greater need for buffer capacity in the end-user device. Schedulers such as bufferbloat mitigation (BBM) [9] and blocking estimation (BLEST) [7] specifically address this concern and provide solution to mitigate HoL. Such schedulers focus mainly on HoL issue and they are not designed for scheduling packets, especially when the networks are dynamic and heterogeneous. In [10], researchers have made efforts to tackle this challenge using Peekaboo, a reinforcement learning (RL) based multipath scheduler. However, Peekaboo is based on mutli armed bandit, a subset of RL, and does not exploit extensively the capabilities of deep reinforcement learning (DRL). The DRL is used in ReLeS [11] to solve the scheduling problems. It relies on a single agent using DRL to decide how to manage all the network paths. However, according to [12], as the number of paths increases, this approach struggles to learn the best way to handle them effectively. MARS [13], an another DRL-based scheduler, aims to decrease buffer size usage at end devices. However, the computational burden increases in MARS because it relies on multiple agents for each path.

To address the aforementioned issues, we propose DRL Empowered Actor-critic scheduler (DEAR). Our proposed scheduler leverages the power of the actor-critic technique to train a neural network, and hence capable of learning and adapting to the frequent changes in the networks. We implemented the DEAR scheduler in MPQUIC using QUIC Go [3] and compared it with several state-of-the-art schedulers, chosen based on the availability of their open-source implementation. All software components of DEAR are provided as open-source to the community[1]. Furthermore, we extensively tested the proposed scheduler for diverse network scenarios, encompassing networks with rapidly fluctuating bandwidth, networks with rapid and short-term fluctuations, networks experiencing progressive outages and we also conducted experiments over 5G-based real network traces [14] to demonstrate the efficacy of the proposed approach.

## 2   System Model and Proposed Approach

In this section, we begin with a comprehensive explanation of the system model, offering valuable insights into the core components of our system and providing the necessary foundation for our proposed approach. Finally, we introduce DEAR, our innovative framework, and provide a detailed account of its implementation processes. These elements collectively contribute to a thorough understanding of our system model and the proposed approach, forming the essence of our research.

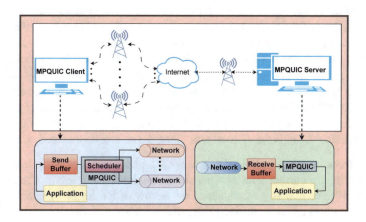

**Fig. 1.** System Model

Figure 1 presents the system model where both the client and the server use MPQUIC to send and receive data, respectively. The client may be connected to the multiple networks while the server is connected to only one network. At the client side, the application data is sent to the send buffer. The scheduler, positioned between the send buffer and networks, plays a crucial role in determining which packet from the send buffer is scheduled to be transmitted on which specific path, based on the scheduling policy. Finally, the data received from the transmission links at the server side is stored in the receive buffer. The receive buffer serves as a temporary storage space for received data before it is delivered to the relevant application. The MPQUIC module is positioned between the receive buffer and the application. Its responsibility includes reordering data packets to ensure the sequential delivery of data to the respective application.

---

[1] https://mutdroco.github.io/publications/dear/dear.html.

## 2.1   Proposed Approach: DRL Empowered Actor-Critic Scheduler (DEAR)

In this section we describe in detail the overall architecture of the proposed DEAR scheduler. From a methodological point of view, we address multipath scheduling as a decision-making problem and leverage RL to address it. Specifically, RL consists of an agent that interacts with its environment at time step $t$, observes state $s_t$, selects an action $a_t$, and transits to a new state $s_{t+1}$ with reward $r_t$. The agent selects the action based on the policy $\pi(a_t|s_t)$, i.e., the probability of taking an action $a_t$ at state $s_t$. The overall objective of RL agent is to find an optimal policy such that the discounted reward, $R_t = \sum_{t=0}^{\infty} \gamma^t r_t$, is maximized where $\gamma$ is a discount factor and $0 < \gamma < 1$. Additionally, the state value, which is the expected return starting from the state $s_t$, is defined as $V(s) = \mathbb{E}_\pi \left[ \sum_{k=0}^{\infty} \gamma^k r_{t+k+1} \mid s_t = s \right]$. Figure 2 shows the overall architecture of the DEAR scheduler. Since RL is utilized to solve the scheduling problem, we map the state, action and reward as follows.

- State $s_t$: It contains the relevant information regarding all $n$ available paths. For each available path $i$, the inputs are round trip time ($RTT_i$), congestion window ($CWND_i$) and in-flight packet ($Inf_i$) where in-flight packets refer to the packets that has been sent by the client but not yet acknowledged by the server.
- Action $a_t$: It contains the path $i$ to be selected to transmit the current packet.
- Reward $r_t$: The reward is the overall goodput and it is given by, $r_t = \frac{((l_t^s - l_t^{re}) \times MSS)}{\eta_t}$, where, $l_t^s$ is the total number of packets that the client has sent through all the available paths and received the acknowledgement, $l_t^{re}$ is the total number of packets that the sender has re-transmitted through all the available paths, $MSS$ is the maximum segment size, and $\eta_t$ is the elapsed time since the start of data transfer at time step $t$.

In this work, we propose DEAR that uses DRL based actor-critic methods to generate action given the current state with the aim to maximize the rewards. The actor-critic methods [15] have two key components: the **actor**, which learns to select actions to maximize expected rewards, and the **critic**, which evaluates the quality of these actions.

### 2.1.1   Actor

At each time step $t$, the actor neural network parameterized with parameters $\theta$, receives input $s_t$ and using this information it predicts a probability $p_i$ to send the data to each available path. The weighted random average method is used to select the action $a_t$ and the corresponding path $k$ to send the packet. The path $k$ is selected as follows: $k = \arg\min_i \left| \rho - \frac{p_i}{\sum_{j=1}^n p_j} \right|$ where $\rho$ is a randomly generated number within 0 to 1. After performing action $a_t$, the reward $r_{t+1}$ is calculated. The actor network updates its parameters $\theta$ using gradient descent by minimizing the following actor loss function $L_t(\pi_\theta(a_t|s_t)) = \frac{1}{D} \sum_{t=1}^{D} \left( -\log(\pi) \times \delta_t \right)$ where $D$ is the number of samples and $\delta_t = (r_{t+1} + \gamma \times V_\phi(s_{t+1})) - V_\phi(s_t)$ is the temporal difference (TD) error [16]. Specifically, TD error is estimated by the critic and sent to the actor. Hence, critic neural network, parameterized with parameters $\phi$, also estimates both state-value functions $V_\phi(s_t)$ and $V_\phi(s_{t+1})$. Finally, the actor updates its policy $\pi(a_t|s_t)$ and corresponding parameters as:

$$\theta_{t+1} = \theta_t + \alpha_\theta \cdot \delta_t \cdot \nabla_\theta \log \pi_\theta(a_t|s_t; \theta_t) \tag{1}$$

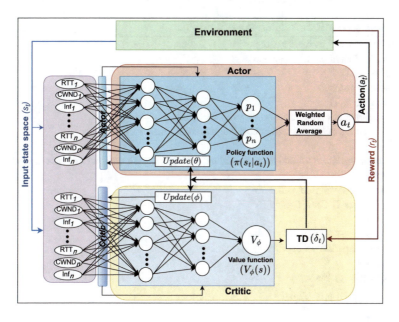

**Fig. 2.** Detailed architecture of DEAR

where $\alpha_\theta$ is the learning rate for actor neural network and $\nabla_\theta \log \pi_\theta(a_t|s_t;\theta_t)$ represents the gradient of the logarithm of the policy function with respect to the actor's parameters.

### 2.1.2 Critic

The critic neural network, parameterized with parameters $\phi$, estimates TD error $\delta_t$, and the state value functions $V_\phi(s_t)$ and $V_\phi(s_{t+1})$ and send them to the actor. The critic network is trained using the loss function $L(V_\phi) = \frac{1}{D} \sum_{t=1}^{D} \left( V_\phi(s_{t+1}) - V_\phi(s_t) \right)^2$ and using the loss function, the critic parameters are updated as follows:

$$\phi_{t+1} = \phi_t + \alpha_\phi . \delta_t . \nabla_\phi V(s_t) \tag{2}$$

where $\alpha_\phi$ is the learning rate for critic neural network. Finally, Algorithm 1 outlines the functioning of proposed DEAR scheduler. It takes $n$ available paths and output the selected path $i$ based on the methodology described above.

## 3   Experimental Setup

In this section, we provide details of our experimental setup, network setup, followed by the state-of-the-art schedulers used for the comparison with the proposed DEAR scheduler, performance metrics utilized in our study and hyperaparameters used for the proposed DEAR scheduler.

**Algorithm 1.** DEAR Scheduler: Algorithm

---

**Input**: Number of available paths $n$
**Output**: Selected path $k$ as an action $a_t$
**if**($paths == 1$)
    **return** $paths$
**else if**($paths > 1$)
    **if(actor critic neural network exist == true)**
        Load the actor and critic neural network with their weights
    **else**
        Initialize the actor and critic neural networks with their weights by random seed
    **end if**
    **for each** Available path $n$ **do**
        $(s_t)$: $RTT_i, CWND_i, Inf_i$
    **end for**
    **for each** Available path $n$ **do**
        Compute the probability $p_i$
    **end for**
    **for each** $p_i$ **do**
        Compute $p_i/(\sum_{j=1}^{n} p_j)$
    **end for**
    Generate $\rho \leftarrow \text{random}(0-1)$
    Find the closer match and select path $k$ as the ouput
    Get the **Reward**$(r_t) \leftarrow (l_t^s - l_t^{re}) \times \text{MSS}/\eta_t$
    Compute $V_\phi(s) \leftarrow \mathbb{E}_\pi \left[ \sum_{k=0}^{\infty} \gamma^k r_{t+k+1} \mid s_t = s \right]$
    Calculate TD Error $(\delta_t) \leftarrow (r_{t+1} + \gamma.V_{\phi(s_{t+1})}) - V_\phi(s_t)$
    Apply the loss function of critic $L(V_\phi) \leftarrow \frac{1}{D} \sum_{t=1}^{D} \left( V_\phi(s_{t+1}) - V_\phi(s_t) \right)^2$
    Update the critic: $\phi_{t+1} \leftarrow \phi_t + \alpha_\phi.\delta_t.\nabla_\phi V(s_t)$
    Apply the loss function of actor $L_t(\pi_\theta(a_t|s_t)) \leftarrow \frac{1}{D} \sum_{t=1}^{D} \left( -\log(\pi) \times \delta_t \right)$
    Update the actor: $\theta_{t+1} \leftarrow \theta_t + \alpha_\theta \cdot \delta_t \cdot \nabla_\theta \log \pi_\theta(a_t|s_t; \theta_t)$
    **return** selected path $k$ as action $a_t$
**end if**

---

## 3.1 Testbed Setup

The experimental setup is shown in Fig. 3. Note that the same setup we have used in our previous work [6] where we have integrated several state-of-the-art schedulers and provided the comprehensive comparison between them. The experimental setup includes the following components: (i) Intel NUC client and server. Both the client and the server run on Intel NUCs with i3 Processor, 64 GB RAM and Ubuntu 22.04.3 LTS operating system. The MPQUIC client and MPQUIC server programs are implemented in Go Language version 1.21.3. (ii) Two Raspberry Pi (Raspberry Pi 4 with 4 GB RAM) devices serves the purpose of emulating network traffic using Netem [17] on the separate paths within the multipath environment and (iii) a switch for forwarding of packets to their designated sender and receiver IP addresses.

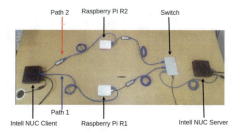

**Fig. 3.** Physical Testbed

## 3.2   Network Setup

Based on real-time observations of terrestrial network behaviour we derived 5G-based test cases using datasets from two operators. Specifically, we considered three predominant patterns identified in [18] through extensive network analysis. For all these patterns, the maximum, average and the minimum bandwidth values are derived from the 5G dataset [14] and a random delay between 1 and 2 s is introduced between each change in bandwidth.

- **Network with fast bandwidth changing**: Figure 4a presents the network pattern when the user walks without any environmental obstacles. Due to frequent user equipment (UE) movement, the bandwidth experiences frequent changes.
- **Networks with fast short-scale variations with high and low profiles**: Figure 4b presents the network pattern when the user is walking in environments where obstacles such as trees and high-rise buildings occuring frequently in the environment. Hence, UE faces the bandwidth fluctuates between high and low profiles when encountering obstacles.
- **Networks with progressive outage and subsequent recovery**: Figure 4c presents the network pattern when the user is on high-speed trains. As the train enters a tunnel, the bandwidth gradually drops, leading to an outage, and upon exiting the tunnel, the bandwidth recovers to its maximum level.
- **Real 5G traces:** Finally, we have also included real traces from the 5G dataset [14] and the corresponding pattern is shown in the Fig. 4d. We consider the bandwidth values from the traces in [14] and introduces a random delay (between 1 and 2 s) between each trace to change the network to make it realtime fast moving scenario 4d.

## 3.3   Existing MPQUIC Schedulers, Performance Metrics and Hyperparameters

We integrated several state-of-the-art multipath schedulers in addition to the proposed DEAR scheduler. The choice of considering these specific schedulers is influenced by the availability of their open-source implementations. Specifically, we have integrated RR, minRTT, ECF [19], BLEST [7] and Peekaboo [10] in our setup. We assume that client sends a file of size $\Omega$ (measured in MegaBytes) and the total time taken by the

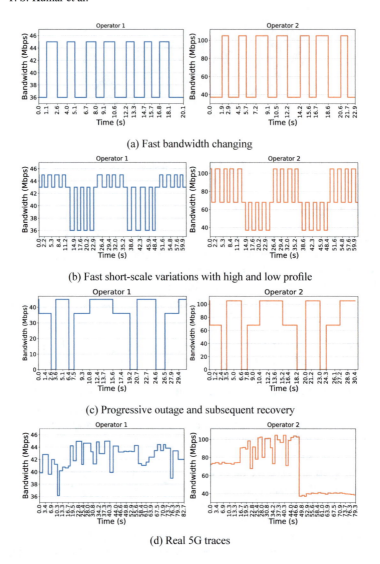

(a) Fast bandwidth changing

(b) Fast short-scale variations with high and low profile

(c) Progressive outage and subsequent recovery

(d) Real 5G traces

**Fig. 4.** Network patterns

client to transfer the file to the server, i.e., file transfer time, is represented by $T$ (seconds). We conduct 50 iterations to obtain each data point and collected data that allows us to calculate statistics such as the minimum, mean, median, and maximum to have box plots for the demonstration of results. To implement DEAR in the test setup, we have used discount factor $\gamma = 0.99$ and learning rate $\lambda = 0.001$. To speed up the convergence, we have used the momentum terms $\mu_1$ and $\mu_2$ as 0.9 and 0.999, respectively. To ensure the numerical stability, a small constant value of $\varepsilon = 1e - 8$ is used for optimization. The actor and critic networks contains one input layer and two hidden layers where the first hidden layer contains eight neurons and the second hidden layer contains four neurons. These selections are made after extensive experiments to strike a balance between model complexity and generalization.

## 4   Performance Evaluation

In this section, we undertake a comprehensive performance evaluation of the proposed DEAR scheduler along with other state of art schedulers. Based on the observations of terrestrial network behavior across various scenarios, we evaluate the performance of DEAR with different network patterns as shown in Sect. 3.2.

Figure 5a presents the performance of various schedulers for networks with fast bandwidth changing pattern shown in Fig. 4a. We observe that the round robin performs poorly as compared to others as it cyclically utilizes the two paths which fails to account for the heterogeneous characteristics of paths [20]. ECF, BLEST and MinRTT provide similar results whereas we observe substantial gains from RL based schedulers: Peekaboo and DEAR. Based on the actor-critic method, DEAR provides even better performance than Peekaboo. DEAR leverages the guidance of its critic, surpassing Peekaboo with superior performance in reducing data transfer time and facilitating adaptable policy learning for enhanced decision-making in dynamic network environments.

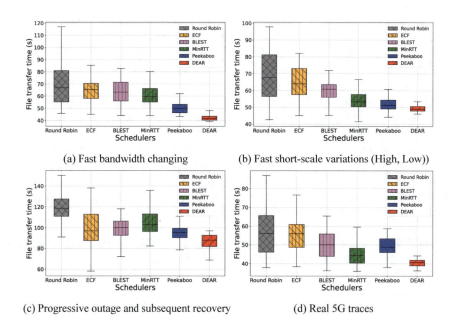

(a) Fast bandwidth changing

(b) Fast short-scale variations (High, Low))

(c) Progressive outage and subsequent recovery

(d) Real 5G traces

**Fig. 5.** Performance of DEAR and other multipath schedulers in the different 5G network patterns

Figure 5b presents the comparison of different schedulers for fast short-scale variations with high and low profiles network pattern shown in Fig. 4b. As expected, RR performs poorly and as the complexity of the pattern is increased, ECF is not able to estimate the path that provide early completion. BLEST also fails to utilize the slower path focusing only on HoL blocking. MinRTT scheduler performs better than BLEST and ECF because unlike BLEST it does not wait for the faster path. Peekaboo and DEAR performs better than other schedulers but DEAR outperforms even Peekaboo due its fast adaptability to the new patterns and quick decision making with the advantage of DRL.

Figure 5c presents the comparison of different schedulers in progressive outage and subsequent recovery network pattern shown in Fig. 4c. RR performed poorly as expected however ECF outperformed BLEST and MinRTT by estimating path availability and scheduling on it. BLEST performs better than MinRTT due to avoiding HoL blocking issues arising from outages and faster recovery. Peekaboo adapted well to new patterns, while DEAR demonstrates faster adaptability and enhanced performance.

Finally, Fig. 5d presents the comparison of different schedulers when real 5G traces are utilized. Interestingly, minRTT performs better than Peekaboo since the later suffers when the network changes with high frequency as pointed in [10]. However, DEAR still performs better than minRTT due to its fast adaptability with effective balance between exploration and exploitation.

## 5   Conclusion

This work aims to investigate the issue encountered by multipath schedulers while dealing with networks that are dynamically changing and heterogeneous. The identification of this problem is based on thorough investigation and analysis of current state-of-the-art schedulers. In order to address this issue, a novel scheduler named DEAR has been developed. DEAR utilises deep reinforcement learning techniques, namely actor-critic methods, to effectively adapt to the highly dynamic and diverse network conditions, thereby enhancing the overall throughput with limited CPU consumption and memory overhead. The DEAR algorithm's limitation lies in its inability to effectively operate in scenarios with very less data transmission sizes, primarily because of its reliance on online learning. Future work includes the integration of DEAR scheduler with 5G and B5G application scenarios including CAVs, UAVs, assisted driving, delivery robots, automated guided vehicles, and public safety applications.

**Acknowledgements.** This work has been supported by SERB, DST, Government of India's start-up research grant agreement SRG/2019/002027 (MUT-DROCO).

## References

1. Yang, W., et al.: Semantic communications for future internet: fundamentals, applications, and challenges (2022)
2. Zheng, Z., et al.: XLINK: QoE-driven multi-path QUIC transport in large-scale video services. In: Proceedings of the 2021 ACM SIGCOMM 2021 Conference, pp. 418–432 (2021)
3. De Coninck, Q., Bonaventure, O.: Multipath QUIC: design and evaluation. In: Proceedings of the 13th International Conference on Emerging Networking Experiments and Technologies, pp. 160–166 (2017)

4. Fatima, N., Saxena, P., Gupta, M.: Integration of multi access edge computing with unmanned aerial vehicles: current techniques, open issues and research directions. Phys. Commun. **52**, 101641 (2022)
5. Yang, W., Shu, S., Cai, L., Pan, J.: MM-QUIC: mobility-aware multipath QUIC for satellite networks. In: 2021 17th International Conference on Mobility, Sensing and Networking (MSN), pp. 608–615 (2021)
6. Kumar, P.S., Fatima, N., Saxena, P.: Performance analysis of multipath transport layer schedulers under 5G/B5G hybrid networks. In: 2022 14th International Conference on COMmunication Systems & NETworkS (COMSNETS), pp. 658–666 (2022)
7. Ferlin, S., Alay, Ö., Mehani, O., Boreli, V.: BLEST: blocking estimation-based MPTCP scheduler for heterogeneous networks. In: 2016 IFIP Networking Conference (IFIP Networking) and Workshops, pp. 431–439 (2016)
8. Jiang, H., Wang, Y., Lee, K., Rhee, I.: Tackling bufferbloat in 3G/4G networks. In: Proceedings of the 2012 Internet Measurement Conference, pp. 329–342 (2012)
9. Ferlin-Oliveira, S., Dreibholz, T., Alay, Ö.: Tackling the challenge of bufferbloat in multipath transport over heterogeneous wireless networks. In: 2014 IEEE 22nd International Symposium of Quality of Service (IWQoS), pp. 123–128 (2014)
10. Wu, H., Alay, Ö., Brunstrom, A., Ferlin, S., Caso, G.: Peekaboo: learning-based multipath scheduling for dynamic heterogeneous environments. IEEE J. Sel. Areas Commun. **38**(10), 2295–2310 (2020)
11. Zhang, H., Li, W., Gao, S., Wang, X., Ye, B.: ReLeS: a neural adaptive multipath scheduler based on deep reinforcement learning. In: IEEE Conference on Computer Communications, IEEE INFOCOM 2019, pp. 1648–1656 (2019)
12. He, B., et al.: DeepCC: multi-agent deep reinforcement learning congestion control for multipath TCP based on self-attention. IEEE Trans. Netw. Serv. Manage. **18**(4), 4770–4788 (2021)
13. Han, X., Han, B., Li, R., Ji, X.: MARS: an adaptive multi-agent DRL-based scheduler for multipath QUIC in dynamic networks. In: 2023 IEEE/ACM 31st International Symposium on Quality of Service (IWQoS), pp. 1–10 (2023)
14. Narayanan, A., et al.: A variegated look at 5G in the wild: performance, power, and QoE implications. In: Proceedings of the 2021 ACM SIGCOMM 2021 Conference, pp. 610–625 (2021)
15. Xu, Z., et al.: An actor-critic-based transfer learning framework for experience-driven networking. IEEE/ACM Trans. Netw. **29**(1), 360–371 (2020)
16. Sutton, R.S., Barto, A.G.: Reinforcement Learning: An Introduction. MIT Press (2018)
17. Jurgelionis, A., Laulajainen, J.-P., Hirvonen, M., Wang, A.I.: An empirical study of NetEm network emulation functionalities. In: 2011 Proceedings of 20th International Conference on Computer Communications and Networks (ICCCN), pp. 1–6 (2011)
18. Sackl, A., Casas, P., Schatz, R., Janowski, L., Irmer, R.: Quantifying the impact of network bandwidth fluctuations and outages on web QOE. In: 2015 Seventh International Workshop on Quality of Multimedia Experience (QoMEX), pp. 1–6 (2015)
19. Lim, Y., Nahum, E.M., Towsley, D., Gibbens, R.J.: ECF: an MPTCP path scheduler to manage heterogeneous paths. In: Proceedings of the 13th International Conference on Emerging Networking Experiments and Technologies, pp. 147–159 (2017)
20. Paasch, C., Ferlin, S., Alay, O., Bonaventure, O.: Experimental evaluation of multipath TCP schedulers. In: Proceedings of the 2014 ACM SIGCOMM workshop on Capacity sharing workshop, pp. 27–32 (2014)

# Integrated Vehicle Access Protocol with Priority-Based Messaging for VANETs

Mayssa Dardour[(✉)], Mohamed Mosbah, and Toufik Ahmed

Univ. Bordeaux, Bordeaux INP, CNRS, LaBRI, UMR5800, 33400 Talence, France
{mayssa.dardour,mosbah,tad}@labri.fr

**Abstract.** Enhancing the functionality of connected and automated vehicles (CAVs) hinges on the robust exchange of real-time safety messages. These communications are critical for conveying traffic information, aiding drivers in accident prevention. Ensuring the prompt and dependable transmission of such messages is vital, necessitating effective channel access strategies. This study introduces a novel hybrid method that integrates Vehicular Deterministic Access (VDA) and Carrier Sense Multiple Access with Collision Avoidance (CSMA/CA) protocols. This approach also incorporates a specialized algorithm for prioritizing messages, thereby optimizing the reliability and efficiency of safety message communication in vehicular networks. The hybrid model synergizes the advantages of both VDA, which is effective under high-traffic conditions, and CSMA/CA, which is preferable for lower traffic scenarios. The addition of a message prioritization approach guarantees priority for urgent communications like Decentralized Environmental Notification Messages (DENMs). Through simulations conducted using the Artery framework, our hybrid approach demonstrates a significant enhancement in performance, achieving over 93% delivery rates for DENMs, while also minimizing collision risks across various channel utilizations. This research contributes valuable insights into developing effective communication frameworks for CAVs, offering a detailed evaluation of the balance between different access protocols and message prioritization strategies in the context of vehicular networks.

**Keywords:** Access Protocols · CSMA/CA · DENM · Message Prioritization · VANET · VDA

## 1 Introduction

Connected and automated vehicles (CAVs) are revolutionizing transportation, bringing advancements in safety, efficiency, and convenience [1]. Key to this revolution is the exchange of safety-critical messages such as Cooperative Awareness Messages (CAMs) [2] and Decentralized Environmental Notification Messages (DENMs) [3], facilitating advanced driver assistance and traffic management.

© The Author(s), under exclusive license to Springer Nature Switzerland AG 2024
L. Barolli (Ed.): AINA 2024, LNDECT 199, pp. 114–128, 2024.
https://doi.org/10.1007/978-3-031-57840-3_11

However, the dynamic and heterogeneous vehicular networks pose significant communication challenges due to high mobility and variable topology, leading to potential safety risks [4]. Therefore, it is imperative to develop communication mechanisms that are not only efficient but also reliable, ensuring the safe functioning of CAVs. This paper introduces a novel approach that combines Vehicular Deterministic Access (VDA) - a protocol for vehicular communications that schedules wireless channel access to enhance reliability and reduce delays in high traffic situations [5] - with Carrier Sense Multiple Access with Collision Avoidance (CSMA/CA) and a message prioritization algorithm. This synergy is designed to facilitate reliable and efficient communication of safety messages within vehicular networks. Our approach capitalizes on the individual strengths of the VDA and CSMA/CA protocols, each tailored to specific traffic scenarios, to enhance channel utilization and minimize the likelihood of collisions. Concurrently, it ensures stringent adherence to Quality of Service (QoS) standards for messages of high priority, guaranteeing their reliable and timely delivery.

Our research introduces several contributions. Primarily, we propose a hybrid access scheme that efficiently toggles between VDA and CSMA/CA protocols based on the network load, utilizing VDA for high-load scenarios to ensure timely delivery of high-priority messages and CSMA/CA for low-load situations to adapt to traffic fluctuations and minimize collisions. We've also developed an analytical model to evaluate the performance of our hybrid scheme under diverse setups. Alongside this, we introduce a refined message prioritization algorithm for DENMs, which effectively organizes transmissions by urgency, importance, age, and deadline, incorporating both priority level and time to collision in determining transmission order. Our methodology's effectiveness is further validated through extensive simulations using the Artery framework, which align with our analytical model's findings. Additionally, we explore the impact of various factors on system performance across different network conditions. Finally, we benchmark our approach against current state-of-the-art schemes in terms of Emergency Notification Time (ENT), Packet Delivery Ratio (PDR), Latency (Lat) and False Alarm Rate (FAR).

The structure of this paper is outlined as follows: Sect. 2 offers a review of existing literature on access protocols and message prioritization algorithms. Our proposed hybrid access scheme and the accompanying analytical model are introduced in Sect. 3. Section 4 elaborates on the message prioritization algorithm. The simulation setup is detailed in Sect. 5. Moreover, Sect. 5 presents a validation of our analytical model through simulation results and compares our approach with previous research to demonstrate its effectiveness. The paper concludes in Sect. 6, where we discuss future directions and underscore the importance of our work in advancing connected and automated transportation.

## 2    Related Work

Reliable and efficient communication in vehicular networks is fundamentally anchored in the effectiveness of access protocols. Predominantly, CSMA/CA and

TDMA, are the protocols most frequently employed [6]. However, each of these protocols presents its own set of limitations.

In vehicular network protocols, the limitations of CSMA/CA and TDMA have spurred the development of hybrid protocols like BH-MAC [7]. It merges CSMA's slot reservation with TDMA's slot access. This protocol employs a bitmap system to minimize overhead and collisions. In fact, it features a fixed-size bitmap aiding in slot allocation and an error bitmap for rapid collision detection. Despite that, BH-MAC falls short in prioritizing safety-critical messages. Similarly, QCH-MAC [8], an extension of EDCA, combined with TDMA, offers improvements in transmission delay, packet loss rate and throughput. The protocol operates on a dual-period system: a Reservation Period for slot reservation and a Transmission Period for data transfer. It prioritizes safety messages via EDCA and optimizes channel sharing through TDMA. Yet, the study lacks a thorough analytical model, leaving its robustness less explored in the academic discourse. CS-TDMA [9], integrating CSMA and TDMA, shows promise in reliability and resource utilization. Moreover, it dynamically adjusts Transmission (TS) and Reservation (RS) periods according to vehicle traffic density. The RS period uses CSMA for new slot reservations, while the TS period facilitates reserved transmissions via TDMA. Nonetheless, its adaptability to varying network conditions needs further exploration. Furthermore, the research lacks a clear prioritization algorithm for different message types based on importance and urgency. Additionally, HER-MAC [10], combining TDMA and CSMA, targets improved reliability of safety messages and service channel efficiency. It segments the control channel transmissions into reserved and contention periods. Thus, it allows concurrent transmission of safety messages on the control channel and non-safety messages on the service channel. However, HER-MAC faces an elevated collision risk with increasing vehicle nodes, attributed to the rise in overhead packet transmission. On account of this, V. Nguyen et al. [11] proposed a hybrid TDMA/CSMA multichannel MAC protocol. Notably, the proposed protocol enhances control channel throughput by eliminating unnecessary overhead. In spite of this, it still falls short of adequately prioritizing safety messages. This could impact the QoS in Vehicle Ad-hoc Networks (VANETs).

In message prioritization within vehicular networks, various algorithms have been explored. The Priority-Based Scheduling (PBS) [12] algorithm assigns priority levels to messages based on their urgency and importance, arranging their transmission sequence accordingly. Nonetheless, it does not account for the fluctuating traffic load. Its simplistic approach leads to suboptimal channel utilization in high-traffic situations. Besides, the PBS algorithm does not account for message age or deadlines, risking delays in delivering time-critical messages. Likewise, the DCC-gatekeeper [13] sorts packets by traffic classes based on communication needs. Using a traffic class parameter from the facilities layer, it orders packets within DCC queues considering their urgency. Subsequently, it ensures consistent rate assignment. Despite these features, it increases end-to-end delays and message loss rates due to its impact on networking and facilities layer protocols. M. Kunibe et al. [14] developed an adaptive prioritization method based on optimal transmission intervals. This strategy dynamically adjusts the transmission priorities of CAMs and Cooperative Perception Messages (CPMs). The algorithm updates intervals to match the target Age of Information (AoI). This increases CPM priority when the AoI exceeds the target. Despite that, there is a risk of degrading the quality of lower-priority messages by focusing on critical-safety messages. This could affect the overall transmission quality.

To our knowledge, our approach is the first to combine VDA and CSMA/CA protocols along with a message prioritization algorithm. Our intention was to maximize channel utilization, reduce collisions, and ensure QoS for high-priority messages. In fact, the dynamic switch between VDA for high-traffic loads and CSMA/CA for low-traffic loads ensures stringent deadline assurances for high-priority messages and an effective adaptation to shifting traffic conditions. Our prioritization algorithm considers urgency, importance, age, and deadlines of safety-critical messages. Additionally, we present an analytical model for an in-depth performance analysis of the hybrid protocol, focusing on the key metric PDR. Referencing this model and using practical implementations in the Artery Framework, we examined other metrics such as ENT, Lat, and FAR, highlighting the impact of our approach on the future of CAVs.

## 3  Integrated Scheme for Enhanced Delivery of Prioritized Messages in Vehicular Communication Networks

Our proposed methodology introduces a vehicular infrastructure that emphasizes real-time data exchange, prioritizing two key processes: (1) message generation of CAMs and DENMs, and (2) message generation of different DENM types. Tailored for various scenarios, we primarily assess this infrastructure in a level-crossing zone, a common site for accidents. This specific evaluation highlights the broader applicability of our approach in other adequately equipped vehicular networks. The architecture, detailed in Fig. 1, supports multiple communication channels and is crucial for evaluating our method's efficacy.

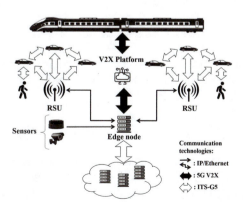

**Fig. 1.** Schematic of the Proposed Architecture for Real-Time Data Exchange.

Intelligent Transportation Systems-Stations (ITS-S) use VDA for exchanging CAMs and DENMs, a choice based on VDA's advantages over 802.11p schemes like EDCA or DCF [5]. VDA efficiently processes safety messages, balancing packet collision reduction and faster transmission. It also allows predetermined wireless medium access with less contention, effectively adjusting the Contention Free Period (CFP) duration, and reserving the Contention Period (CP) for private services. Regular CAM exchanges occur at 100 ms intervals, with priority shifting to DENMs in emergencies [15], as shown in our VDA scheduling in Fig. 2.

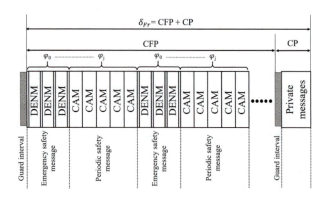

**Fig. 2.** Scheduling Scheme for Deterministic Medium Access in Vehicular Networks.

The wireless channel allocation in VDA follows a structured approach. ITS-S reserve the channel as multiple time slots ($T$), with all ITS-S within the same range being aware of the reservation due to VDA advertising messages from VDA Opportunities (VDAOP) requester and granter nodes [5]. Effective utilization of these $T$ for communication hinges on precisely calculating packet transmission durations in number of $T$. For determining the transmission duration ($\eta_p^T$) of a DENM/CAM packet $p$, we employ (1).

$$\eta_p^T = \frac{A + \text{DTrns}_p \times 10^3}{\Delta^T} \tag{1}$$

Where A is the Arbitration Inter-Frame Spacing (AIFS) ($\mu s$), $\text{DTrns}_p$ is the Duration of the Transmission of the packet (ms) (calculated in 2) and $\Delta^T$ represents the Duration of a single $T$ ($\mu s$).

$$DTrns_p = \frac{S_p \times 8}{\beta} \times 10^3 + Lat_p + Pt_p \tag{2}$$

With $\beta$ representing the Bandwidth (bps), $S_p$ the Size of the packet (byte), $Lat_p$ the Latency (ms) and $Pt_p$ the Processing time (ms).

The duration $\dot{d}$ of VDAOP, measured in time-slots, is defined as follows:

$$\dot{d} = \eta_p^T \times \lambda \tag{3}$$

In this model, $\lambda$ symbolizes the ITS-S message generation rate. We account for a maximum delay $\delta$ for each message type $x$ (CAM or DENM), considering a path's maximum number of hops $(Max_h)$ and the number of messages of type $x$ $(k_x)$. The generation rate is modeled as a continuous-time Markov chain with two states: high (H) and low (L) rates. The estimate of $\lambda$ involves calculating the expected number of messages produced per unit time in both H and L states, represented by $E[k_{x_H}]$ and $E[k_{x_L}]$:

$$\lambda = \frac{E[k_{x_H}]}{\delta} + \frac{E[k_{x_L}]}{\delta}, \delta > 0 \tag{4}$$

Here, $k_{x_H}$ and $k_{x_L}$ denote the number of messages generated by an ITS-S in states H and L, respectively. To calculate the expected values, we utilize the properties of the exponential distribution as follows:

$$E[k_{x_H}] = \lambda 1 \times \delta_H \tag{5}$$

$$E[k_{x_L}] = \lambda 2 \times \delta_L \tag{6}$$

In this context, $\lambda 1$ represents the high state message generation rate, while $\lambda 2$ corresponds to the low state generation rate. The terms $\delta_H$ and $\delta_L$ indicate the maximum delay $\delta$ for message type $x$ for $Max_h$ in a path in the high (H) and low (L) states, respectively.

We can infer that the frequency at which a vehicle generates messages follows a Poisson distribution, with the mean interval being influenced by the channel's current load status. $\rho$ represents the number of ITS-S in state H and $\gamma$ the number of ITS-S in state L. To depict the distribution of vehicles across various generation rates, we model a Markov Chain State Transition Diagram (Fig. 3).

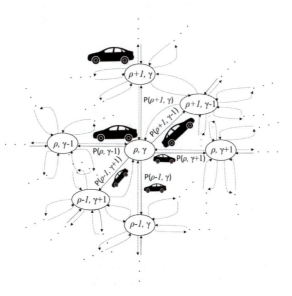

**Fig. 3.** Markov Chain State Transition Diagram for ITS-S Load States.

Based on this model, we can further infer that the overall message generation rate for all vehicles aligns with a Poisson distribution. This distribution is
characterized by an average generation rate calculated as follows:

$$\Phi_{\lambda|(H,\rho)|(L,\gamma)} = \rho \times \lambda_{(E[k_{x_H}] \geq E[k_{x_L}])} + \gamma \times \lambda_{(E[k_{x_H}] < E[k_{x_L}])} \tag{7}$$

Accordingly, the likelihood of generating $n$ messages within a specific $T$ during a successful transmission can be represented as follows:

$$P(n|(H,\rho)|(L,\gamma)) = \frac{\Phi_{\lambda|(H,\rho)|(L,\gamma)}^n \times T \times e^{-\Phi_{\lambda|(H,\rho)|(L,\gamma)} \times T}}{n!} \tag{8}$$

Based on that, the probability of a successful transmission, which permits
only a single packet to be transmitted per slot, is defined as follows:

$$P_s(1|(H,\rho)|(L,\gamma)) = \Phi_{\lambda|(H,\rho)|(L,\gamma)} \times T \times e^{-\Phi_{\lambda|(H,\rho)|(L,\gamma)} \times T} \tag{9}$$

Thus, the mean Packet Reception Rate (PRR) for successful transmissions
is calculated using (10):

$$\Theta_{RR|(H,\rho)|(L,\gamma)} = \frac{P_s(1|(H,\rho)|(L,\gamma))}{T} \tag{10}$$

To calculate PDR for a scenario, it is necessary to first ascertain the total
message generation rate $(\Phi_\lambda)$ and reception rate $(\Theta_{RR})$, dependent on the Probability Distribution Function (PDF). This calculation entails assessing the number
of vehicles in each state and their transition probabilities. For the low-offered
load state, PDF$(\gamma)$ (calculated in (11)) is determined using the binomial distribution, based on the probability of successful transmissions $(\eta^{STR})$ per vehicle
per slot $T$ and the total available transmission slots $(\eta^{AvT})$. Conversely, the PDF
for the high-offered load state (PDF$(\rho)$, given in (12)) is calculated using the
Poisson distribution, centered on the average vehicle count in that state.

$$PDF(\gamma) = \frac{\eta^{AvT}!}{\eta^{STR}!(\eta^{AvT} - \eta^{STR})!} \times P_s(1|(H,\rho)|(L,\gamma))^{\eta^{STR}} \times \tag{11}$$
$$(1 - P_s(1|(H,\rho)|(L,\gamma)))^{\eta^{AvT} - \eta^{STR}}$$

$$PDF(\rho) = \frac{\eta_{v|H}^{T^\rho} \times e^{-\eta_{v|H}^T}}{\rho!} \tag{12}$$

$\eta_{v|H}^T$ is the mean number of slots available to each vehicle under high-offered
load conditions.

The overall PDF is derived by merging the PDFs for each state. Specifically,
the joint PDF$(\rho,\gamma)$ is the product of PDF$(\rho)$ and PDF$(\gamma)$:

$$PDF(\rho,\gamma) = PDF(\rho) \times PDF(\gamma) \tag{13}$$

The calculation of the total message generation rate $\Phi_\lambda$ and the total message
reception rate $\Theta_{RR}$ can now be performed as follows:

$$\Phi_\lambda = \sum_{\rho=0}^{\rho_{\max}} \sum_{\gamma=0}^{\gamma_{\max}} \Phi_{\lambda|(H,\rho)|(L,\gamma)} \times \text{PDF}(\rho, \gamma) \tag{14}$$

$$\Theta_{\text{RR}} = \sum_{\rho=0}^{\rho_{\max}} \sum_{\gamma=0}^{\gamma_{\max}} \Theta_{\text{RR}|(H,\rho)|(L,\gamma)} \times \text{PDF}(\rho, \gamma) \tag{15}$$

Ultimately, the PDR is depicted in the following manner:

$$\text{PDR} = \frac{\Theta_{\text{RR}}}{\Phi_\lambda} \tag{16}$$

## 4 Proposed Algorithm for Prioritizing Various DENM Types

Our study analyzed how our hybrid vehicular access protocol manages the prioritization and transmission of safety-critical messages, such as CAMs and DENMs, among ITS-S. We now shift our focus to examine the Edge Server's (ES) (Fig. 1) algorithm for prioritizing different DENM types. The ES assigns varying levels of significance to different DENM types. These are categorized into New, Update, Cancellation, and Negation types, as we detailed in [15]. Each DENM within these categories is given a specific Priority Level (PL). This ensures the timely transmission of critical information. Factors like urgency, importance, message age, and deadline are taken into account. The prioritization algorithm, formalized in Algorithm 1, functions at the Application level. It guides the sequence in which messages are evaluated.

---

**Algorithm 1.** Edge-DENM-Prioritization $(\mathcal{D}, \tau)$

---

**Require:** $\mathcal{D}$, set of DENM requests; $\tau$, current time
**Ensure:** $Q$, prioritized DENM queue
1: $Q \leftarrow \emptyset$; $F \leftarrow \emptyset$
2: **for** $d \in \mathcal{D}$ **do**
3:     $\Delta T \leftarrow d.T_0 - \tau$; $\Pi(d) \leftarrow d.PL$
4:     **if** $\Delta T > 0 \wedge \Pi(d) \neq \text{NULL}$ **then**
5:         $F \leftarrow F \cup \{(d, \Pi(d), \Delta T)\}$
6:     **end if**
7:     Sort $F$ by $\Pi$ in descending order; $F \leftarrow F \cup \text{Sort\_TTC}(F)$
8: **end for**
9: **for** $(d, \Pi, \Delta T)$ in $F$ **do**
10:     $T_0 \leftarrow (d.R_{\text{dur}} > 0 \wedge d.R_{\text{int}} > 0)?\text{Set\_T\_Repetition}(d.R_{\text{dur}}, d.R_{\text{int}}) : d.T_0$
11:     $Q \leftarrow Q \cup \{\text{Create\_DENM}(d, T_0, \Pi)\}$
12: **end for**
13: **return** $Q$

---

The ES systematically prioritizes DENM messages by their type. New DENMs, which signal imminent hazards, are prioritized over Update DENMs, reflecting traffic changes. Cancellation DENMs, indicating the end of hazards, rank lower than Updates, while Negation DENMs, which correct erroneous alerts, receive the lowest priority. For DENMs of the same category, the ES differentiates based on urgency, importance, time elapsed since the message was generated, and deadlines.

Moreover, upon receiving DENM requests (denoted as $\mathcal{D}$ in Algorithm 1), the ES categorizes each by type. It then calculates the remaining valid time $(\Delta T)$ for each message by subtracting the current time $(\tau)$ from the message's original expiry time $(T_0)$. Messages with expired timers are discarded, and the ES notifies the sender of the failure.

For valid messages, the ES assesses the traffic safety situation by considering the Time To Collision (TTC), a measure of how soon a collision is expected if the current trajectories and speeds are maintained. The TTC is a critical factor in assigning PLs to DENMs, ranging from PL-2 to PL-0, with PL-0 being the most urgent. This categorization is based on the severity of the traffic situation and the immediacy of response required. It is depicted in Fig. 4.

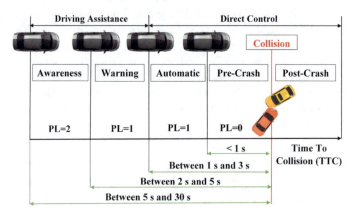

**Fig. 4.** Mapping of Priority Levels (PLs) to Time to Collision (TTC) in Vehicle Safety Response Scenarios.

The ES uses a mapping function, $\Pi(d)$, to assign PLs to DENMs, correlating them with their respective TTC values. DENMs with shorter TTCs, indicating more urgent scenarios, receive higher priority. The ES sorts the messages into a Filtering set $(F)$, arranging them in descending order of PL. The rate at which $(F)$ is updated corresponds directly to the frequency of receiving traffic information. Regularly refreshing $(F)$ is essential for reordering DENMs based on the most recent TTC values.

For DENMs that involve repetitive broadcasting, the ES calculates new repetition times using the Set_T_Repetition function, which considers repetition duration $(R_{\mathrm{dur}})$ and interval $(R_{\mathrm{int}})$. These parameters ensure the message is repeated within the valid timeframe without exceeding the original expiry time $(T_0)$. If no repetition is required, the original expiry time remains as the validity marker.

Finally, the ES compiles the prioritized queue $(Q)$, ready for the transmission of DENMs, ensuring that the most critical alerts are communicated first and thereby enhancing traffic safety.

# 5    Configuration and Analysis of the Simulation

## 5.1    Configuration of the Simulation Environment

Our simulation utilized the Artery framework to assess a vehicular network at a railway level crossing. The environment, a 2 km area with two intersections and a 2-lane road crossing train tracks, incorporated 200 CAVs traveling at an average speed of 50 km/h. The ES was centrally located with a 200-meter communication radius, while Road Side Units (RSUs), spaced 400 m apart, facilitated extensive network connectivity. The simulation used a 600-meter range for the 802.11p channel and a 500-meter range for V2X communications. CAMs were broadcasted at 100 ms intervals. Key communication parameters included a SINR-based PHY radio model, 550-meter carrier sense range, and 20 μs time slots. The contention window was set at 32, with AIFS at 76 μs. The network operated at 6 Mbps bandwidth with 16-QAM modulation and a 1/2 coding rate, simulating Constant Bit Rate (CBR) traffic over UDP. The sizes for DENM and CAM messages were set to 500 and 150 bytes, respectively.

## 5.2    Analysis and Outcomes of the Simulation

### 5.2.1    Comparative Analysis of Access Protocols

In our simulation study, the VDA/CSMA Hybrid Vehicular Access Protocol was benchmarked against BH-MAC [7], QCH-MAC [8], CS-TDMA [9], HER-MAC [10], and a hybrid TDMA/CSMA protocol [11], focusing on ENT and PDR metrics.

Figure 5 displays ENT variations among vehicular access protocols in response to increasing CAV flow at the Level-Crossing Zone (L-CZ). BH-MAC records the highest ENT with an average of 147.96 ms, escalating with more CAVs due to its demanding slot reservation and bitmap broadcasting. This is primarily caused by BH-MAC's incorporation of TDMA reserved slots and slot allocation Bitmap within BSM packets. Consequently, it amplifies the size of the packet payload, increasing $S_p$, and, in turn, $DTrns_p$. CS-TDMA and HER-MAC also show rising ENTs (averaging 147.22 ms and 116.12 ms), indicating scalability issues in high traffic. QCH-MAC's average ENT increment (113.49 ms) points to contention problems in dense traffic. The hybrid TDMA/CSMA protocol moderately improves ENT (with an average of 110.63 ms) but struggles at high densities. Conversely, our VDA/CSMA scheme maintains an average ENT of 50.44 ms, ensuring swift emergency message delivery.

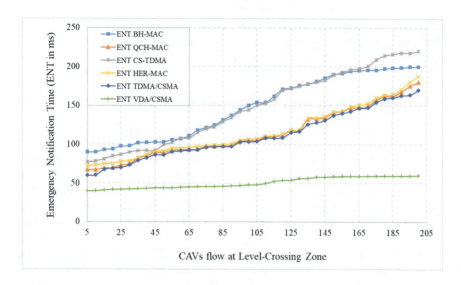

**Fig. 5.** Comparison of Emergency Notification Time (ENT) Across Various CAV Densities in Level-Crossing Zone (L-CZ)

The data on PDR depicted in Fig. 6 supports our observations regarding ENT, showcasing the VDA/CSMA protocol's stable performance across different traffic densities. The VDA/CSMA protocol achieves an average PDR of 0.932, outperforming the TDMA/CSMA by 0.032, QCH-MAC by 0.072, HER-MAC by 0.109, CS-TDMA by 0.137, and BH-MAC by 0.161. This superior performance stems from the VDA's systematic allocation of $T$, effectively reducing collisions. The application of Markov Chain Modeling for vehicle behavior prediction under varying loads is instrumental in finely tuning the distribution of vehicles across different message generation rates. As the sum of the expected number of messages per unit time in state H ($E[k_{x_H}]$) rises, $\Phi_{\lambda|(H,\rho)|(L,\gamma)}$ and consequently $\Theta_{RR|(H,\rho)|(L,\gamma)}$ increase, reflecting successful transmissions. The consistent calculation of both the overall message generation rate ($\Phi_\lambda$) and the overall message reception rate ($\Theta_{RR}$), validates the PDR behavior's reliability and accuracy.

Advancing from the performance analysis of access protocols, we now examine the impact of prioritization algorithms on the sequencing of various DENM types in the upper network layers.

### 5.2.2 Comparative Analysis of Message Prioritization Techniques

Extensive simulations validated our method's efficacy in prioritizing different DENM types. The performance, particularly in Lat and FAR was compared with established algorithms like PBS [12], DCC-gatekeeper [13], and Adaptive Message Prioritization (AMP) [14].

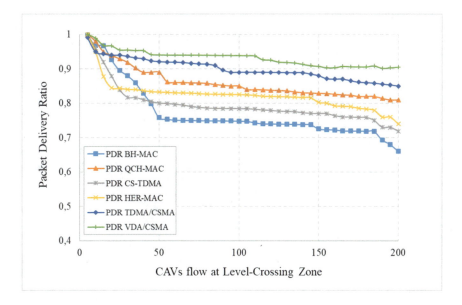

**Fig. 6.** Examining Packet Delivery Ratio (PDR) Fluctuations Relative to CAV Density in the Level-Crossing Zone (L-CZ)

Figure 7 examines Lat variations of our Edge-DENM Prioritization Algorithm (EDPA) against other algorithms under different CAV densities at the L-CZ. The PBS algorithm shows increased Lat, ranging from 34.11474 to 90.00002 ms with higher CAV flows, due to its limited adaptability in dynamic traffic, leading to delays in crucial message transmission in congested conditions. DCC-gatekeeper's Lat ranges from 27.25873 to 82.99954 ms, attributed to packet processing delays and extended queuing times, particularly in dense traffic. AMP algorithm, while showing lower Lat fluctuations (18.36974 to 74.00015 ms), still falls behind EDPA, which maintains a Lat range of 10.37775 to 69.33621 ms. AMP's strategy of aligning transmission priorities results in CAMs degradation and subsequent delays in vital message delivery during high traffic. In contrast, EDPA achieves the lowest Lat by prioritizing DENMs through a multi-faceted approach considering urgency, importance, message age, deadline, and TTC. It leverages the mapping function $\Pi(d)$ for sophisticated PL assignment, correlating PL with TTC for accurate message timing. This prioritization mechanism at the ES ensures prompt transmission of critical messages, significantly lowering Lat in high-density scenarios.

In tandem with Lat trends, Fig. 8 illustrates FAR evolution and underscores differences among the four algorithms as CAV density rises. Comparing these algorithms to the EDPA, we observe higher FAR values. For instance, with 10 CAVs at the L-CZ, the difference is 2.77918% greater than EDPA, increasing to 13.37809% with the PBS algorithm for 195 CAVs. PBS's high FAR in dense traffic results from RSU-based data scheduling, which inadequately distin-

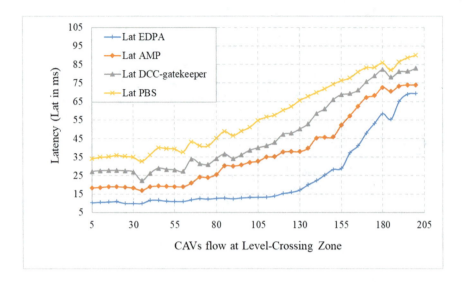

**Fig. 7.** Comparative Analysis of Latency (Lat) in Diverse CAV Densities at the Level-Crossing Zone (L-CZ) Using Multiple Algorithmic Strategies

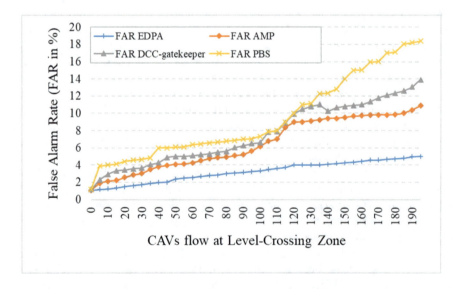

**Fig. 8.** Evaluating False Alarm Rate (FAR) Variations Across Varying CAV Densities in the Level-Crossing Zone (L-CZ) Using Various Algorithmic Approaches

guishes emergency messages, potentially causing delays or blockages, especially in high-density scenarios. DCC-gatekeeper's FAR compared to EDPA ranges from 1.69215% higher for 10 CAVs to 8.89607% higher for 195 CAVs. Its multi-queue system managing varying PLs contributes to high FAR in dense traffic, often causing delays or drops, particularly in overloaded queues and high-density situations, increasing false alarm likelihood. For the AMP algorithm, FAR is marginally higher than EDPA. It goes from 0.87996% more for 10 CAVs to 5.8864% more for 195 CAVs. This stems from prioritizing less frequent CAM transmission over critical safety messages in high-density scenarios, leading to an imbalance and increased FAR. In contrast, the EDPA consistently achieves the lowest FAR values (Fig. 8), owing to its advanced strategy in prioritizing DENMs. The algorithm's regular updates to the Filtering set ($F$), aligned with traffic data intervals, enable prompt restructuring of DENMs according to TTC values. This, along with the assignment of PLs based on traffic severity determined by $\Pi(d)$, ensures prioritization of the most critical DENMs, thus reducing false alarms in dense traffic areas.

## 6    Conclusions and Prospects for Future Research

This study introduces a novel hybrid vehicular access protocol, blending Vehicular Deterministic Access (VDA) with Carrier Sense Multiple Access with Collision Avoidance (CSMA/CA) and a message prioritization algorithm. It effectively addresses vehicular network challenges, optimizing the transmission of critical safety messages and ensuring high delivery rates with low collision risks under varied traffic conditions. The protocol's efficacy in real-time safety communication for Connected and Automated Vehicles (CAVs) is validated through analytical modeling and simulations in the Artery framework.

Future research will focus on adapting our protocol for complex urban vehicular networks, addressing intricacies such as dense traffic, varied road geometries, and diverse environmental conditions. We also intend to leverage machine learning for enhanced message prioritization. We believe that conducting field trials will assess real-world applicability and scalability. Additionally, investigating security measures and cyber threat resilience will further strengthen the protocol in evolving transportation systems.

## References

1. Deng, Z., Yang, K., Shen, W., Shi, Y.: Cooperative platoon formation of connected and autonomous vehicles: toward efficient merging coordination at unsignalized intersections. IEEE Trans. Intell. Transp. Syst. **24**(5), 5625–5639 (2023). https://doi.org/10.1109/TITS.2023.3235774
2. ETSI EN 302 637-2 (V1.4.1): Intelligent Transport Systems (ITS); Vehicular Communications; Basic Set of Applications; Part 2: Specification of Cooperative Awareness Basic Service, April 2019
3. ETSI EN 302 637-3 (V1.3.1): Intelligent Transport Systems (ITS); Vehicular Communications; Basic Set of Applications; Part 3: Specifications of Decentralized Environmental Notification Basic Service, April 2019

4. Huang, T., et al.: V2X cooperative perception for autonomous driving: recent advances and challenges. arXiv:2310.03525 (2023)
5. Rezgui, J., Cherkaoui, S., Chakroun, O.: Deterministic access for DSRC, 802.11p vehicular safety communication. In: 7th International Wireless Communications and Mobile Computing Conference, Istanbul, Turkey, pp. 595–600 (2011). https://doi.org/10.1109/IWCMC.2011.5982600
6. Gilani, M.H.S., Sarrafi, I., Abbaspour, M.: An adaptive CSMA/TDMA hybrid MAC for energy and throughput improvement of wireless sensor networks. Ad Hoc Netw. **11**(4), 1297–1304 (2013). ISSN 1570-8705, https://doi.org/10.1016/j.adhoc.2011.01.005
7. Kumar, S., Kim, H.: BH-MAC: an efficient hybrid MAC protocol for vehicular communication. In: 2020 International Conference on COMmunication Systems & NETworkS (COMSNETS), Bengaluru, India, pp. 362–367 (2020). https://doi.org/10.1109/COMSNETS48256.2020.9027322
8. Boulila, N., Hadded, M., Laouiti, A., Azouz Saidane, L.: QCH-MAC: a Qos-aware centralized hybrid MAC protocol for vehicular Ad Hoc NETworks. In: 2018 IEEE 32nd International Conference on Advanced Information Networking and Applications (AINA), Krakow, Poland, pp. 55–62 (2018). https://doi.org/10.1109/AINA.2018.00021
9. Zhang, L., Liu, Z., Zou, R., Guo, J., Liu, Y.: A scalable CSMA and self-organizing TDMA MAC for IEEE 802.11p/1609.x in VANETs. Wireless Pers. Commun. **74**, 1197–1212 (2014). https://doi.org/10.1007/s11277-013-1572-3
10. Dang, D.N.M., Dang, H.N., Nguyen, V., Htike, Z., Hong, C.S.: HER-MAC: a hybrid efficient and reliable MAC for vehicular Ad Hoc networks. In: 2014 IEEE 28th International Conference on Advanced Information Networking and Applications, Victoria, BC, Canada, pp. 186–193 (2014). https://doi.org/10.1109/AINA.2014.27
11. Nguyen, V., Oo, T.Z., Chuan, P., Hong, C.S.: An efficient time slot acquisition on the hybrid TDMA/CSMA multichannel MAC in VANETs. IEEE Commun. Lett. **20**(5), 970–973 (2016). https://doi.org/10.1109/LCOMM.2016.2536672
12. Kumar, V., Vaisla, K.S., Sudarsan, S.D.: Priority based data scheduling in VANETs. In: 2016 International Conference on Advances in Computing and Communication Engineering (ICACCE), Durban, South Africa, pp. 19–22 (2016). https://doi.org/10.1109/ICACCE.2016.8073717
13. Kuhlmorgen, S., Festag, A., Fettweis, G.: Evaluation of multi-hop packet prioritization for decentralized congestion control in VANETs. In: IEEE Wireless Communications and Networking Conference (WCNC), San Francisco, CA, USA, pp. 1–6 (2017). https://doi.org/10.1109/WCNC.2017.7925899
14. Kunibe, M., Yamazaki, R., Murakawa, T., Shigeno, H.: Adaptive message prioritization for vehicular cooperative perception at target intervals. J. Inf. Process. **31**, 57–65 (2023). https://doi.org/10.2197/ipsjjip.31.57
15. Dardour, M., Mosbah, M., Ahmed, T.: A messaging strategy based on ITS-G5 for a bus blockage emergency at a level crossing. In: 14th IFIP Wireless and Mobile Networking Conference (WMNC), Sousse, Tunisia, pp. 108–112 (2022). https://doi.org/10.23919/WMNC56391.2022.9954293

# Adaptive Cost-Reward Scheduling for Optimizing Radio Utilization and NR Numerology Efficiency in B5G New Radio Networking

Wei-Teng Chang[1] and Ben-Jye Chang[2(⊠)]

[1] Graduate School of Engineering Science and Technology, National Yunlin University of Science and Technology, Douliu, Yunlin, Taiwan, ROC

[2] Department of Computer Science and Information Engineering, National Yunlin University of Science and Technology, Douliu, Yunlin, Taiwan, ROC

changb@yuntech.edu.tw

**Abstract.** The importance of having fast and reliable communication that can move with you is clear in critical areas like emergency response, ASD, eV2X communication, and LEO connectivity. The emergence of B5G/5G technology represents a transformational leap within the 3GPP standard, providing a flexible NR SCS mode tailored to the diverse 5QI requirements, which include latency, data rate, jitter, loss, and reliability. Efficient flow scheduling in B5G radio networks can be a challenging task, especially in SDN paradigms. Existing research has proposed scheduling methods. However, they often prioritize traditional QoS requirements and overlook critical factors such as bearer costs and associated rewards. In this paper, we present a novel Cost-Based Flow Scheduling (eSCFS) framework that utilizes an extended Sigmoid function to dynamically prioritize flows, taking into account all relevant factors. Our main goal is to reduce latency while maximizing the use of radio RB and bringing maximization net benefits to B5G NR networks. The eSCFS method has been validated through numerical simulations, demonstrating superior performance across key metrics such as network latency, resource utilization, and overall profitability. This Paper therefore aims to achieve and contribute to several objectives: 1) analyze the QoS requirements of various services within limited radio resources, 2) proposing a vRB state-dependent dynamic flow scheduling scheme to maximize network performance, 3) proposing an adaptive scheduling approach based on cost-reward considerations.

## 1 Introduction

In this section, firstly describe current tendencies and critical technical specifications in B5G cellular networks. Next, studies several research topics, include flexible numerology, bandwidth part (BWP) technologies and flow scheduling in B5G New Radio (NR).

This research was supported in parts by the Ministry of Science and Technology of Taiwan, ROC, under Grants MOST-111-2221-E-224-044-MY3 and Qualcomm Technologies, Inc., USA, under Agreements NAT-414698, NAT-414699, NAT-514843, NAT-457592, NAT-487837.

L. Barolli (Ed.): AINA 2024, LNDECT 199, pp. 129–139, 2024.
https://doi.org/10.1007/978-3-031-57840-3_12

Finally, this paper examines key challenges concerning RB resources assignment in B5G NR, and succinctly outline the motivation, objectives, and contributions of our study.

## 1.1  B5G/6G Network Trends and Key Standards

According to B5G/6G QoS characteristics [11], B5G/6G networks are designed for diverse types of application flows, such as very high data rate, ultra-high reliability, low latency and jitter, and so on. Typical scenarios include eMBB, uRLLC, eMergency messaging and mMTC, etc. For assuring above QoS-based services in 5G/B5G radio communications [13], several mature techniques [1–4] are specified, including Network Slicing (NS) [17], Software Defined Network (SDN) [15, 16], Virtualizing Network Function (VNF) [14], flow steering and forwarding, and Cooperative-AI system, and so on.

B5G and 6G builds upon 5G and specify even more extreme performance requirements. They combine Artificial Intelligence (AI) mechanisms, utilizing machine learning (ML) [5, 10], with highly collaborative communications through Low Earth Orbit (LEO) satellites to enable Adaptive Infinite Connectivity (ALC) [6]. When combined with LEO, B5G/6G acts as a global radio access wireless network that enables diverse offloading and computing of B5G RAN traffic flow [7]. In addition, AI-based B5G/6G networks are well-suited for eV2X applications, owing to their stringent QoS requirements, such as ultra-high reliability and ultra-low latency [8, 9].

For flexibly and efficiently allocating radio RBs to different requirements of QoS KPIs, in 3GPP Rel. 17, the most key technology specified in 5G NR would be the NR Numerology of SCS modes [12] that defines 7 modes, $\mu \in \{0, 1, ..., 6\}$, of frequency SCS of $15 \cdot 2^{\mu} (Khz)$ and time cyclic prefix of $2^{-\mu} (ms)$. For instance, the SCS mode $\mu = 6$ offering the least access delay of $2^{-6} (ms)$ can provide much low delay for eMergency flowing. Conversely, for a mMTC application, the SCS mode $\mu = 0\ or\ 1$ provides the delay of $2^{-0}\ or\ 2^{-1} (ms)$ for mMTC.

## 1.2  Related Studies of BWP and Frequency Numerology Key Technologies in 5G/B5G NR

In 3GPP Rel. 17, the Bandwidth Part (BWP) [3] is introduced for the NR air interface as an essential component of the overall channel bandwidth configuration for implementing delay constraints. The UE can configure BWPs with highly flexible numerology parameters [13], as detailed in Table 1, along with various static fully partitioned BWP configurations. In the context of 5G BWP, [19] examines the collaboration between SCS mode selections and NR BWP configurations. NR has the flexibility to switch between FR1 and FR2, thereby achieving throughput according to the QoS metrics. Nevertheless, frequent transitions between different BWP sizes will lead to increased switching delays and signaling overhead [18], resulting in additional delays. The higher probability of switching failure of RB allocation is due to the fragmentation of available free-state RBs. Consequently, meeting the QoS requirements for uRLLC in B5G/6G networks will become challenging.

**Table 1.** Shows several NR numerology parameters that BWP can utilise. [12].

| $f$ | $\mu$ | SCS (kHz) | $N_{slot}^{sf}$ | Slot length (ms) | Max pRB |
|---|---|---|---|---|---|
| Sub-6G (FR1) | 0 | $15 \times 2^0$ (15) | 1 | $2^{-0}$ (1) | 270 |
| | 1 | $15 \times 2^1$ (30) | 2 | $2^{-1}$ (0.5) | 273 |
| | 2 | $15 \times 2^2$ (60) | 4 | $2^{-2}$ (0.25) | 135 |
| mmWave (FR2) | 2 | $15 \times 2^2$ (60) | 4 | $2^{-2}$ (0.25) | 264 |
| | 3 | $15 \times 2^3$ (120) | 8 | $2^{-3}$ (0.125) | 264 |
| | 4 | $15 \times 2^4$ (240) | 16 | $2^{-4}$ (0.0625) | 138 |

### 1.3 Research on Flow-Based Queueing Scheduling

The aims of B5G/6G radio flow scheduling are to prioritize processing based on various flow types for different services. Nevertheless, to ensure that flows meet diverse QoS requirements, such as loss, latency, and jitter, while minimizing carrying costs, maximizing system rewards and radio resource utilization, undoubtedly presents an important challenge.

Various criteria can be used to classify flow scheduling methodologies, including multi-QoS-based approaches like New Radio flexibility (NRflex) [17], weighted-based methods such as Weighted Fairness Queue (WFQ) [20] and Weighted RR (WRR) [23], priority-based methods like Priority Queue (PQ) [21] and Frame Level Scheduler (FLS) [22], and fair-based methods like Round-Robin (RR).

Although various scheduling methods can flexibly and elastically schedule packets according to singular or plural B5G QoS regulation, respectively, but they are affected by several critical impacts. Typical drawbacks include 1) arduous in addressing both supplementary and non-supplementary QoS requirements simultaneously, 2) neglecting manifold flowing service requirements in B5G/6G, 3) neglecting the net-profit, reward and cost of the provided network system, etc.

### 1.4 Critical Issues, Motivations and Objectives of This Paper

Some critical challenges and issues demand thorough resolution, particularly,

1) Requiring a dynamic scheduling method based on the RB resources state under various numerology parameters, while fulfilling the necessary QoS for diverse flow types,
2) Requiring studies on how to fully allocate the limited wireless RB resources under various Sub-carrier Spacing modes for various required 5QI services,
3) Requiring the analyses of UE's QoS flow requirements and the RB resource state under a cost-reward algorithm, and then maximizing the net profit and reward of the network system while minimizing carrying costs.

To achieve the mechanism of dynamic prioritization of flow scheduling based on the state of RB resources, we propose an extended Sigmoid-based cost-based flow scheduling (eSCFS) algorithm. The major contributions include: 1) proposing an eSCFS algorithm based on the RB resource state; 2) dynamically scheduling packets in 5G NR according to net-profit (priority is given to packets with the highest net-profit); 3) minimization of probability of packet loss and queue delay for the highest flow type.

The remainder of this paper is organized below. Section 2 describes the network model that includes various types of flowing queues and radio RB allocation within the 5G/B5G NR frequency numerology. In Sect. 3, we elaborate on the proposed eSCFS approach, followed by the presentation of numerical results in Sect. 4. Finally, in Sect. 5, we summarize the conclusions drawn and future studies.

## 2   Network Model

Figure 1 models a B5G/6G network based on flexible NR numerology. According to the 5G QoS Identifier, the B5G NR supports different network queues for various types (i.e., priorities) of flows, $k \in \{1, ..., K\}$, originating in a UE. We present an example of end-to-end SFC for eMergency (red) slicing, where eMergency flow packets arrive at flow queues to be scheduled in 5G/B5G gNB. The emergency slices require low latency, therefore the SFC for this flow is dynamically steered to the destination server by Mobile Edge Computing (MEC).

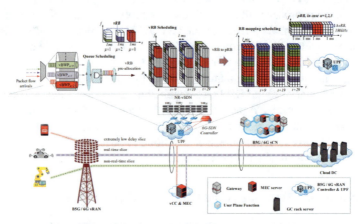

**Fig. 1.**  Network model of 5G/B5G based on flexible NR numerology with flow queues and RB.

## 3   The Proposed eSCFS Approach

Section 3 describes the proposed approach of the Extended Sigmoid-based Cost-based Flow Scheduling (eSCFS) in B5G/5G NR. For the static preconfigured RB allocation specified in 3GPP, the purpose of eSCFS is to optimize flow scheduling and wireless RB allocation in NR from several aspects: 1) minimizing loss probability and latency for the highest type of flow, and 2) maximizing RB utilization. Importantly, the eSCFS method dynamically adjusts flow scheduling priorities based on the state of radio RBs.

Figure 2 illustrates the approach model of eSCFS for implementing QoS requirements of various applications in NR numerology and flow scheduling. The proposed
eSCFS enables the SDN controller and NFV orchestrator to manage end-to-end NS of
QoS-based flow SFC. With eSCFS, the B5G NR UL incoming queue receives different types of flow packets from various UEs. Subsequently, NR schedules the incoming
packets based on flow ID and type. eSCFS prioritizes scheduling of flow packets with
the highest net-benefit.

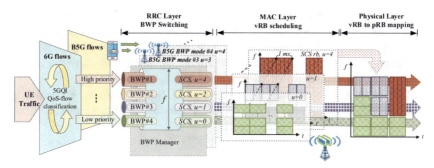

**Fig. 2.** The model of the eSCFS approach for the B5G NR.

## 3.1 Exponient Cost Function, $\mathbb{C}_\mu(\cdot)$

In this paper, an exponential cost formula is proposed that depends on two important
parameters: 1) the total number of vRB within the delay bound of each flow type $\mu$,
expressed as $Q_\mu$, , and 2) the state vRB of each flow type $\mu$, which refers to the amount
of blocked and allocated vRBs used, expressed as $U_\mu$.

The exponential-based cost formula, $\mathbb{C}_\mu(Q_\mu, U_\mu)$, is formulated in Eq. (1),

$$\mathbb{C}_\mu(Q_\mu, U_\mu) = 0.5 \cdot \left( \left( \frac{U_\mu}{Q_\mu} \right)^e + H_\mu \right) \tag{1}$$

where $H_\mu$ represents a fine-tune scaling factor of accessible RBs. $H_\mu$ intensify, while
the number of available RBs drops, and conversely. In this instance, the carrying cost
$\mathbb{C}_\mu(\cdot)$ will be significantly higher.

The formula for the cost of different flow types varies depending on the number of
used vRBs and NR SCS parameters. It is important to note that the carrying cost of
all types of flows grows exponentially with the number of vRBs used. eSCFS achieves
adaptive scheduling with RB state using the proposed exponential cost formula instead
of a linear function, focusing on RB state rather than performance metrics like delay,
loss, rate, and so on. If the number of used RBs attains the maximum number of available
RBs, the maximum cost is 1.0. On the other hand, if the number of used RBs is 0, the
cost will be 0.0.

As various flow types have distinct delay constraint bounds, for instance, 5 ms for $\mu_2$,
10 ms for $\mu_1$ and 50 ms for $\mu_0$, and so on, the extension of the maximum number of RBs

to 220 (44*5) for $\mu_2$, 480 (48*10) for $\mu_1$, and 2500 (50*52), respectively, accommodates various flow types within their specific delay constraints.

## 3.2 Extended Sigmoid-Based Reward Function, $\mathbb{R}_\mu(\cdot)$

Secondly, eSCFS proposes an extended Sigmoid reward formula to determine the reward for bringing in data packets from the arrival flow. The proposed scheduling is based on cost-reward and prioritizes the flow packet with the highest net-income. Two goals can be achieved. Firstly, it provides dynamic scheduling based on the RB state. Secondly, the scheduling scheme prioritizes the flow packet with the maximum net profit as the highest priority for scheduling and obtains several benefits: distinguishing the scheduling prioritization among various flow priorities, minimizing network costs, and maximizing the system's net profit and the reward.

eSCFS analyzes the Sigmoid formula given in Eq. (2) with an S-sharp impact parameter $t$ and the exponential or Euler number $e$, resembling the error function.

$$S(t) = \left(1 + e^{-t}\right)^{-1} \tag{2}$$

In Eq. (2), the Sigmoid formula exhibits an inflection point. Typically, the formula logically increases on one side (for example, the slope to the right of the inflection points or the result exceeds the value of the inflection point). However, the curve of the formula increases exponentially in the opposite direction (for example, the slope to the left of the inflection points or the result is less than the value of the inflection point).

As shown in Eq. (3), eSCFS utilizes the peak of the S-sharp sigmoid curve to introduce the extended sigmoid-based reward formula $\mathbb{R}_\mu(\cdot)$, denoted as R, based on several significant parameters outlined below,

1) $Q_\mu$ is the overall number of vRBs that are within the delay limit of the flow type $\mu$,
2) $U_\mu$ represents the number of vRB states that are blocked and allocated, and it belongs to flow type $\mu$,
3) $F_\mu^1$ represents the pivot point of the flow type $\mu$,
4) $F_\mu^2$ represents the slope rate of the flow type $\mu$, and
5) a fine-tuning factor for Inflation Rate, expressed as *IR*.

$$\mathbb{R}_\mu \leftarrow \left(1 + \left(e^{-1 \cdot F_\mu^2 \cdot \left(\left(Q_\mu - U_\mu\right) - F_\mu^1\right)}\right)\right)^{-1} + IR_\mu \tag{3}$$

These reward formulas reduce based on extended Sigmoid functions as number of vRBs used grows. The highest flow type, $\mu = 2$, gives the sharpest reward formula curve and the reward value of the majority of states of vRBs used (e.g. between 0 and 33) is 1 and the reward value reduces to very low value or 0 as the number of vRBs used reaches 44. As a result, the highest flow type brings the highest reward and should be scheduled first.

Conversely, when $\mu = 0$, it results in a smooth reward formula curve with the reward decreasing significantly as the number of vRBs in use decreases. The reward decreases significantly to low values when the number of vRBs used reaches 27. It nearly reaches

0 when the number of vRBs used grows to 30 and then to 52. This indicates that the lowest flow type is rewarded less and scheduled with the lowest priority.

Further discussion of the analysis of the defined impact parameters is undertaken to ascertain the optimal reward formula for different types of traffic. Firstly, concerning Type $\mu$ flows, the available vacant vRB quantity is determined by subtracting the utilized vRBs quantity ($U_\mu$) from the total vRBs quantity within the latency range ($Q_\mu$), denoted as $Q_\mu - U_\mu$.

Second, delve into the analysis of the adaptive pivot point, labeled $F_\mu^1$, within the reward formula related to type $\mu$ flows. The goal here is to determine the proximity of the pivot's reward to either the maximum or minimum utilization of vRBs.

When analyzing rewards, it's preferable for the pivot point $F_\mu^1$ to be near the maximum number of vRBs used for higher flow types, resulting in a greater reward. Conversely, for lower flow types, the pivot point should be closer to the minimum, leading to lower rewards. Notice that the pivot point is dynamically adjusted based on the current vRB status.

For example, consider flow type $\mu_0$ with a bandwidth of 10 MHz and a sub-frame time of 1 ms. Within a 1 ms sub-frame, there are 52 vRBs available, ranging from 0 to 51 in usage. Therefore, the central pivot point for flow type $\mu_0$ is determined as $F_{\mu_0}^1 = 52/2 \, or \, 26$, as illustrated by curve S1 in Fig. 3(a). To elaborate further, a higher value of $F_{\mu_0}^1 = 30$ corresponds to a greater reward for S7, while a lower value of $F_{\mu_0}^1 = 15$ results in a reduced reward for S5, and so on.

Thirdly, we explore the analysis of the slope rate within the reward formula, denoted as $F_\mu^2$, to understand the changing pace of reward delivery between the maximum (1.0) and minimum (0.0) rewards. As $F_\mu^2$ approaches 1.0, the reward function sharpens, while nearing 0.0, it becomes smoother. It's important to note that when $F_\mu^2 = 0$, the reward formula curve becomes horizontal. Consequently, for higher-flow types, a larger slope rate $F_\mu^2$ is preferred to encompass more vRB states with higher rewards. Conversely, for lower-flow types, a smaller slope rate $F_\mu^2$ is favored to cover fewer vRB states with higher rewards. By finely tuning the slope rate $F_\mu^2$ for different flow types, the reward formulas can be dynamically optimized.

Figure 3(b) illustrates the extended Sigmoid reward formulas corresponding to different SCS modes.

### 3.3 Scheduling the Flow Packet with the Highest Net-Profit

As shown in Fig. 3(b), based on the considered dynamical parameters, the reward formulas for various types of flows are obtained adaptively. After the analysis and obtaining the exponential cost formula and the augmented sigmoid reward formula for different flow types, the proposed scheduling method (eSCFS) chooses the flow packet in a flow queue with the largest positive net gain as the optimal highest scheduling priority (denoted by $\mathbb{N}_\mu^{OPT}$), as formulated in Eq. (4),

$$\mathbb{N}_\mu^{OPT} \leftarrow \arg\max_{\forall\mu} \{\mathbb{R}_\mu - \mathbb{C}_\mu\} \tag{4}$$

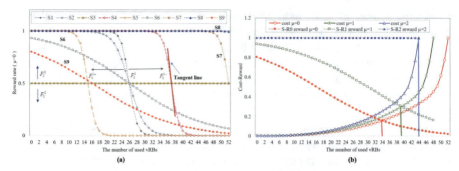

**Fig. 3.** Illustrates two components: (a) Extended Sigmoid reward formulae generated based on various parameter settings outlined in Table 1 ($\mu_0$). (b) The cost formula and the reward formula for various flow types.

Specifically, from Fig. 3(b) and Eq. (5), the proposed eSCFS scheduling algorithm has some superior features: first, the condition for becoming a scheduling candidate is a flow packet whose reward is greater than the carrying cost, that is, $\mathbb{R}_\mu > \mathbb{C}_\mu$. Second, when eSCFS scheduling maximizes network net profit, flow packets with the highest positive net profit are scheduled first, while ensuring various QoS regulation for various services. Finally, the eSCFS scheduling method considers multiple parameters of the states of the flow queue and the vRB, and can maximize the utilization of the Radio resources of the vRB according to the state of the flow queue.

## 4    Numerical Results

This section evaluates the proposed approach (eSCFS) and related studies: 3GPP 5G NR radio resource allocation specification (5G Std.) [2, 3], the WFQ [20], the RR scheduling [20], the PQ [21], the Nrflex [17], and FLS [22]. The performance metrics compared under different numbers of UEs' impact factors include average delay, net-profit, and finally vRB utilization.

In network model, the 5G NR is composed of two tiers of 7 gNBs for NR and one core network, where each gNB assumes a 10 MHz bandwidth for the NR SCS mode 0, namely, $\mu = 0$. That is, each 5G physical frame is made up of 10 sub-frames in 10 ms. The traffic model generates three types of flows: uRLLC/eMERGENCY in eV2X application, mMTC+ for B5G-IIoT, and eMBB+. These flows are initiated at UEs and follow a Pareto distribution with various parameters of $\alpha\_on$, $\alpha\_off$ and $\beta$, as shown in Table 2.

Figure 4 shows that for the highest priority (mode), both WFQ and PQ produce the lowest latency, while NRFlex produces the highest delay. Both eSCFS and the 5G Std. Methods contribute to competitively high latency. However, all latency values for the compared methods remain below the 5 ms latency limit. In eSCFS, the vRB with the lowest MX is pre-allocated to minimize delay, and scheduling is based on a cost-reward algorithm rather than queue priority.

Figure 5 evaluates and compares the mean vRB utilization of various SCS modes. The higher the mean vRB utilization, the higher the success rate and the lower the

**Table 2.** Simulation parameters

| Parameters and tools | Values |
| --- | --- |
| Length of simulation | 1.5 s |
| Simulator tools | GCC v4.9.2 and Dev C+ + v5.11 |
| gNBs count | 1 |
| Frequency | 10 MHz |
| SCS parameters for μ | $\{0, 1, 2\}$ |
| Slot parameters in each subframe | $\{0, 2, 4\}$ |
| Number of RBs in each sub-frame | $\mu_0 = 44, \mu_1 = 48, \mu_2 = 52$ |
| Number of UEs | $\{500, 1000, 2000, 3000, 4000, 5000, 6000\}$ |
| Max delay bound | $\mu_0 = 50$ ms, $\mu_1 = 10$ ms, $\mu_2 = 5$ ms, |
| Packet length | 512, 1024, 1500 Bytes |
| Arrival rate | 2 Mb/s |

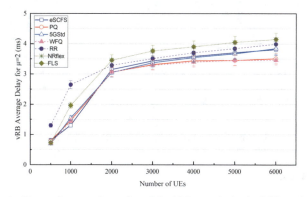

**Fig. 4.** Shows the mean latencies of the highest priority in SCS mode $\mu_2$.

numerology MX in vRB scheduling. For the highest priorities of SCS mode $\mu_2$, Fig. 5 illustrates that eSCFS attains the highest vRB utilization, leveraging its cost-reward scheduling, vRB free-state related cost formula, and proactive allocation of vRB with minimal MX. In contrast, RR demonstrates the lowest vRB utilization, attributed to its fair scheduling strategy. Additionally, 5G Std., PQ, and FLS result in highly competitive vRB usage, which is significantly lower than that of eSCFS. WFQ and NRFlex, due to fairness weighting and delay binding, exhibit lower vRB usage.

Figure 6 compares the overall carrying network cost, the overall revenue from UEs and the overall net profit of all approaches. It is clear that approaches with increased vRB utilization result in higher carrying costs. Additionally, various SCS modes yield different rewards. Therefore, maximizing net profit is a significant challenge. Figure 6 compares the net profit, showing that the proposed eSCFS yields the highest net profit, while the others generate competitive but lesser net profits.

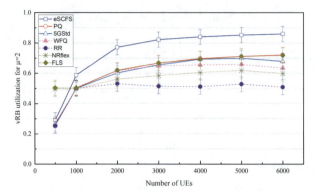

**Fig. 5.** Mean utilization of the vRB in SCS mode $\mu_2$

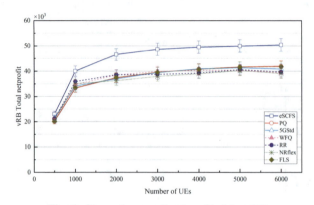

**Fig. 6.** Shows the overall net profit of the vRB.

## 5    Conclusions

The advanced technology of B5G NR flexible numerology ensures QoS requirements for diverse types of flows in eV2X and LEO. However, most of the related studies focus on QoS metrics for flow scheduling, and B5G NR specifies radio RB allocation of static preconfigured resources, which causes the static RB allocation to be affected by dynamic traffic. Therefore, this paper proposes using the cost formula associated with RB status as a scheduling method to prioritize flows differentially on a dynamic basis. This method is called extended Sigmoid-based cost-based flow scheduling (eSCFS). Numerical results show that the proposed eSCFS method outperforms related works in terms of average access latency, vRB utilization, and total net profit.

## References

1. 3GPP TS 23.501 (2022) System Architecture for the 5G System-Stage 2 (Release 17) Version 17.4.0
2. 3GPP TS 38.211 (2022) NR: Physical channels and modulation (Release 17) Version 17.1.0

3. 3GPP TS 38.214 (2022) NR: Physical layer procedures for data (Release 17) Version 17.1.0
4. Chang, B.-J., Chang, W.-T.: Cost-reward-based carrier aggregation with differentiating network slicing for optimizing radio RB allocation in 5G new radio network. In: IEEE IEMCON 2019, pp. 814–820 (2019)
5. Wild, T., Braun, V., Viswanathan, H.: Joint design of communication and sensing for beyond 5G and 6G systems. IEEE Access $9(1)$, 30845–30857 (2021)
6. Ma, T., et al.: UAV-LEO integrated backbone: a ubiquitous data collection approach for B5G internet of remote things networks. IEEE Sel. Areas Commune. $39(11)$, 3491–3505 (2021)
7. Israel, L.-M., Beatriz, S., et al.: Inter-plane inter-satellite connectivity in dense LEO constellations. IEEE Trans. Wireless Commun. $20(6)$, 3430–3443 (2021)
8. Mario, H., Castaneda, G., et al.: A tutorial on 5G NR V2X communications. IEEE Commun. Surv. Tutorials $23(3)$, 1972–2026 (2021)
9. Chang, B.-J., Hung, W., Lin, Y., Chang, W.-T.: Dynamic keeping reserved resource probability with slicing flow steering in 5G sidelink SPS for platooning ADAS and autonomous self driving. In: 2020 CACS (2020)
10. Santos, J., et al.: Towards low-latency service delivery in a continuum of virtual resources: state-of-the-art and research directions. IEEE Commun. Surv. Tutorials $23(4)$, 2557–2589 (2021)
11. Diego, G., Soares, P.: A detailed relevance analysis of enabling technologies for 6G architectures. IEEE Access **11**, 89644–89684 (2023)
12. 3GPP TR 38.101-3 (2022) NR, User Equipment (UE) radio transmission and reception (Rel. 17) Version 17.7.0
13. 3GPP, TS 38.300 (2022) NR and NG-RAN Overall Description, Stage 2 (Release 17) Version 17.2.0
14. Caio, L.S., et al.: Modelling and analysis of 5G networks based on MEC-NFV for URLLC services. IEEE Latin Am. Trans. $19(10)$, 1745–1753 (2021)
15. Mosahebfard, M., et al.: Modelling the admission ratio in NFV-Based converged optical-wireless 5G networks. IEEE Trans. Veh. Technol. $70(11)$, 12024–12038 (2021)
16. Ibrahim, A.A.Z.: Heuristic resource allocation algorithm for controller placement in multi-control 5G based on SDN/NFV architecture. IEEE Access **9**, 2602–2617 (2020)
17. Boutiba, K., Ksentini, A., Brik, B.: NRflex: enforcing network slicing in 5G new radio. Comput. Commun. **181**, 284–291 (2022)
18. Abinader, F., et al.: Impact of bandwidth Part (BWP) switching on 5G NR system performance. In: 2019 IEEE 5GWF, pp. 161–166 (2019)
19. Jeon, J.: NR wide bandwidth operation. IEEE Comms. Mag. $36(3)$, 42–46 (2018)
20. Seah, W.K., et al.: Combined communication and computing resource scheduling in sliced 5G multi-access edge computing systems. IEEE Trans. Veh. Technol. $71(3)$, 3144–3154 (2022)
21. Dike, J.N., Ani, C.I.: Performance evaluation of an optimized hybrid packet scheduler for bandwidth constrained voice over internet protocol networks. Uniport J. Eng. Sci. Res. (UJESR) **5**, 128–138 (2020)
22. Ang, E.M., et al.: Two-level scheduling framework with frame level scheduling and exponential rule in wireless network. In: 2014 ICISA, pp. 1–4 (2014)
23. Dighriri, M., et al.: Comparison data traffic scheduling techniques for classifying QoS over 5G mobile networks. In: 2017 WAINA (2017)

# A New Redundant Intelligent Architecture to Improve the Operational Safety of Autonomous Vehicles

Hajer Boujezza[✉] and Anis Boubakri

Private Higher Institute of Digitalization Technologies (ISPTed), Tabarka, Tunisia
{hboujezza,aboubakri}@ted-university.com

**Abstract.** Considering that connected autonomous vehicles are the future of transportation, it is essential to investigate their dependability and accessibility in order to enhance their market acceptance and make them commercially viable. One of the primary challenges faced by autonomous connected vehicles is the frequent occurrence of dropped packages, which significantly undermines their reliability. To ensure a high level of reliability in autonomous vehicles equipped with a level 5 autonomous driving system, we propose a novel architecture that incorporates redundancy. This architecture aims to bolster reliability while mitigating the risk of failures. Additionally, we introduce a neuro-fuzzy communication controller that enables the reduction of dropped message rates by seamlessly switching between the various connectivity modules present within the vehicle.

**Keywords:** V2V connectivity · Platoon · Autonomous vehicles · neuro-fuzzy controller · Redundant

## 1 Introduction

With the continuous advancement in computer capacity and embedded computing [1], accompanied by the progress in artificial intelligence [2], a significant paradigm shift is occurring. Simultaneously, there has been a notable rise in the number of accidents, where over 95% of these incidents can be attributed to human drivers [3,4]. In light of this, autonomous vehicles [5] emerge as a promising solution to minimize human driver involvement, or even eliminate the need for a human driver altogether, as the electronic control unit (ECU) assumes complete control over vehicle management. This represents the pinnacle of autonomous driving, known as the 5th level [6].

Fully autonomous vehicles [5] are real-time systems that operate with utmost sensitivity, leaving no room for error. Their decision-making process, including trajectory planning [7], control execution, and speed management, must occur within extremely short timeframes. A single erroneous decision can potentially result in an accident [8].

L. Barolli (Ed.): AINA 2024, LNDECT 199, pp. 140–152, 2024.
https://doi.org/10.1007/978-3-031-57840-3_13

To make informed decisions, these vehicles heavily rely on onboard sensors such as cameras, lidar, and IMU [9,10] to perceive their surrounding environment. However, the information gathered from these sensors alone is insufficient to obtain a comprehensive view of the overall traffic scenario. This is where autonomous connected vehicles (CAVs) [11] come into play by facilitating the exchange of kinematic data, such as position, speed, acceleration, and direction, through vehicle-to-vehicle (V2V) communication [12]. This communication can be accomplished using technologies like Dedicated Short-Range Communication (DSRC) [13,14] or Cellular Vehicle-to-Everything (C-V2X) [15] leveraging 5G cellular networks.

The concept of traffic in the form of platoons has emerged as a promising technology for managing road traffic effectively. In this context, two types of communication are relevant: intra-platoon communication [16], which involves the allocation of radio resources among vehicles within a platoon, and inter-platoon communication, which focuses on resource sharing between platoon leaders [16,17]. However, the main challenge that hinders the development of a robust and reliable cooperative driving system [18] lies in communication issues [16]. These problems can be summarized as follows: limited range of communication in urban environments, typically around 300 m [19]; a high rate of dropped or abandoned messages; and network unavailability. For instance, in a platoon-based traffic scenario [16], if the data transmitted by the leading vehicle, which dictates decisions to its followers, becomes outdated, the reliability and accuracy of the decisions implemented within the platoon are compromised.

To ensure the safe operation of autonomous vehicles (AVs) on the road, it is crucial to prioritize the following primary objective: enhancing the availability and reliability of the AVs' control system, an aspect that has not been thoroughly investigated previously. However, achieving this objective necessitates improvements in the performance of the communication system. In this research paper, we propose and evaluate an intelligent strategy that utilizes a neuro-fuzzy controller to effectively minimize the occurrence of dropped messages. Our solution is built upon a redundant architecture, and the core concept of our controller involves assessing the power of each communication module. Based on the operating time and power levels, our controller intelligently selects the most suitable communication module to ensure seamless and reliable communication.

The main research contributions are summarized as follows:

- Improved availability of inter-vehicle communication services, which is a critical component for VA traffic. Since if data is collected with a delay, then dangerous situations may occur. Better communication quality is achieved by switching from DSRC to C-V2X, when the power level deteriorates.
- Software redundancy to improve the availability of data processing to make good decisions even if the vehicle ECU fails.
- Improved accuracy of the decisions produced.

This research paper is structured into three main sections. The first part introduces our proposed architecture, which incorporates both hardware and software redundancy, thereby enhancing the availability of autonomous vehicle

(AV) functionalities. In the second part, we present the neuro-fuzzy controller, which enables the automatic management of connectivity availability. To validate the effectiveness of our proposed solution, the third part presents a comprehensive performance evaluation. Finally, the paper concludes with a summary of findings and implications.

## 2   System Model

Autonomous vehicles are comprised of two key components: hardware and software. The software part plays a critical role in enabling the hardware to function safely. Given the complexity of autonomous vehicles and their reliance on software management, ensuring operational safety [20,21] is of utmost importance. Operational safety encompasses key factors such as reliability and availability, as it has been observed that approximately 80% of failures in autonomous vehicles can be attributed to issues within the software component [22].

### 2.1   Description of the Current AV Architecture

Presently, autonomous vehicles (AVs) typically consist of a set of sensors such as Lidar and cameras. These sensors collect local information and transmit it to the Electronic Control Unit (ECU) for processing. The ECU is responsible for processing the data received from sensors, managing the communication system, and making decisions for the AV. Furthermore, AVs are equipped with a transmitter/receiver that supports either DSRC communication [13] or C-V2X cellular communication. This communication module is essential for gathering information from nearby vehicles, enabling the AV to operate autonomously.

However, challenges arise as the rate of abandoned messages within the communication system remains high, thus compromising the quality of the decisions made by the AV. Consequently, the operational safety of AVs is called into question. To address this issue, it becomes imperative to adopt an alternative architecture that prioritizes reliability and availability. Such an architecture would ensure the operational safety of AVs by mitigating the impact of high message abandonment rates and promoting dependable communication for optimal decision-making.

### 2.2   Proposed Architecture

The achievement of level 5 autonomous vehicles (AVs) has not yet been realized, despite advancements in in-car systems technology. This can be attributed to various challenges, including the following:

- The sensors: From the sensors that the AVs can discover the traffic environment. In situations, these sensors become blind and question the reliability and availability of AVS functionality;

– Communication: for the computer to process the correct information on time, connectivity must be ensured. The AVs based on DSRC communication sulfur very high rate of dropped packets. The quality of connectivity deteriorates in dense scenario;
– Availability: in level 5 the driver cannot intervene on decisions. The vehicle is taken over by the ECU. For this, we must have a powerful and highly available calculator;
– The ringer effects: given the enormous amount of noise in the road, we cannot differentiate whether it is a police stop sign or a passenger in the road;

With the increasing level of automation in various processes, the focus on achieving level 5 automation in autonomous vehicles (AVs) has become paramount. Level 5 automation implies that vehicle management can be performed without any human intervention. However, this reliance on automation systems introduces a significant challenge. In the event of failures, such as ECU malfunctions or errors in the driving program, there is a heightened risk of incorrect decision-making or inadequate availability of necessary data within the required timeframe. This scenario renders the vehicle blind and increases the likelihood of accidents and associated damages. Moreover, in the context of platooning, where vehicles travel in a closely coordinated manner, an error in one vehicle can propagate throughout the platoon and potentially impact other follower platoons.

To address these challenges, each vehicle within a platoon is equipped with two communication modules, namely DSRC and C-V2X. This redundancy ensures that each vehicle can potentially function as a leader vehicle within the platoon.

In order to enhance the reliability and availability of autonomous driving systems in AVs, our proposed solution entails the implementation of a novel architecture (see Fig. 1). This architecture incorporates hardware redundancy at the ECU level, as well as redundancy in communication systems, alongside software redundancy.

The master controller processes the information received from the I/O unit (Sensors, actuator) via the local network (Modbus) and traffic environment information via one of the DSRC, C-V2X, NDN networks. The processed data is saved locally on the storage medium. As traffic data has a large capacity, we keep only the information necessary for instant circulation, and the other data is saved in the cloud. The processed data is displayed on the IMH to inform the driver of the condition of his vehicle and the tasks to be performed.

**Hardware Redundancy.** The concept of redundancy is duplication, triplication, etc. one or more components of a system that perform the same function. There are two types of redundancy: hardware redundancy and software redundancy. Regarding hardware redundancy, this can involve several components such as ECU, power supply, communication bus, input/output modules (sensors and actuators), communication module, etc. From a material point of view, the major risk is the unavailability of the computer. In circulation in the form of

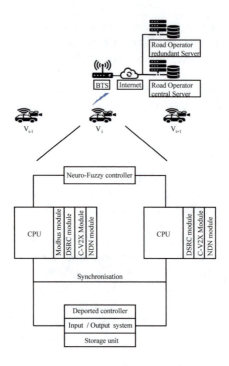

**Fig. 1.** Proposed redundant architecture based on redundancy

centralized Platoons, the leader manages all the Platoon member vehicles. If there is a fault with the leader computer, accidents can occur, since the inter-vehicle distance is very short. You can only react to brake when you detect that the leader computer is damaged, thereafter there is a risk of demining the flow of road traffic. Therefore, we propose to have two ECUs: One master ECU and the other in reserve. If an anomaly has been detected, the switchover takes place automatically to the reserve ECU. We can also make a redundancy at the level of the sensors, but since the vehicles are connected so if there is an anomaly in a sensor information can be obtained from the network by interpolating the values received from the vehicles which precede it and the one which follows it. A redundant power supply must also be provided, because generally autonomous vehicles are of the electric type, so that the vehicle does not run out of fuel (depleted energy), there must be another source of energy allowing circulation to the less until the next charging point, which is beyond the scope of this article's consideration.

As autonomous vehicles are connected and connectivity is paramount to ensure smooth operation. Therefore, there are two transmitters/receivers. A transmitter/receiver compatible with DSRC technology, it works primarily for V2V communication between Platoon vehicles if you do not have DSRC coverage, the switchover is automatic to the other transceiver which is compatible with the 5G network, the failover procedure is shown in Fig. 1. If DSRC coverage returns, the switchover from 5G to DSRC is automatic to minimize the additional overhead costs of 5G communication. If DSRC coverage returns, the switchover from 5G to DSRC is automatic to minimize the additional overhead costs of 5G communication.

We can make even days and odd days, but since we do not have, any indication of the driving time is what it is equal between even and odd days, another solution is to share the charges from one ECU to the other trip but the trips are not the same duration. Therefore, the best solution is to do the changeover during drive time.

**Software Redundancy.** A priori, failures in on-board computer systems such as autonomous vehicles autonomous can either be of material origin (computer, communication system, sensors, etc.), either of software origin (autonomous driving system, perception, planning, etc.). In practice, more than 80% are of software origin. To increase the operational reliability of AVs, it is proposed to make software redundancy in addition to hardware redundancy.

The switchover takes place automatically from the master controller to the reserve controller. This leads us to manage the software part. Since the ECU switching to the other is automatic is in real time. For this switchover to take place instantly, the data to be processed must be shared between the two ECUs. Our solution is software redundancy. The high availability part of the program is loaded both in the master ECU and in the reserve ECU. When it is being processed in the master ECU computer, it is not processed in the reserve ECU computer. The corresponding jump in the reserve ECU avoids possible discrepancies between the two parts of the program caused e.g. security messages. The program is thus ready on the reserve station for further processing. The flowchart below in Fig. 2 shows the operating principle of software redundancy from the aspect of the master and backup ECU.

In order for the driving system to become highly available, it is not necessary to start all over again in the event of a master ECU failure; the latter continuously transfers the treatment data to the reserve station.

# 3   Communication System

To ensure proper functioning of autonomous vehicles it must be ensured that all the functionalities are present and functioning correctly. The functions which must be performed by the AVs and which are related to communication are:

- Perception: the stain of perception [23] uses the sensors installed in the vehicle, allowing the analysis and monitoring of the traffic environment at all times. Which is then used to share the condition of each vehicle with these neighbors;
- Scheduling: scheduling [24] consists of determining possible and safe routes for the vehicle in question, depending on perception and HD card;
- Decision-making: the decision making task [24] is to choose the optimal route based on all the possible routes of the previous task.

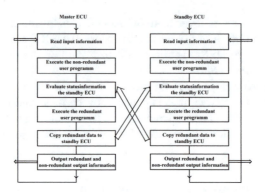

**Fig. 2.** Structure of software redundancy

The types of networks for AVs are:

- DSRC: DSRC technology has a low latency time, which is equal to 10 ms. But, this technology has a short range that does not exceed 300 m in urban areas. As the AVs circulate in the form of Platoons so the inter-vehicle distance is low, so even with these limits we will use this technology as a priority for V2V communication, for sending CAM messages and alarm messages. If a communication break is detected, which can cause traffic anomalies i.e. a communication break greater than 500 ms, we switch to the C-V2X communication and we activate a test function, which makes it possible to test whether the DSRC communication is resumed;
- C-V2X: C-V2X communication has a low latency time, which is equal to 15 ms. Also has a high range. Nevertheless, these characteristics deteriorate exponentially in heavy traffic. For this, we will reserve the C-V2X communication for the inter-Platoon communication;
- Named Data Network (NDN) [25]: It is a data base network where information is independent of its producer [26]. We can get the information from the nearest neighbor who used this information. Usually vehicles do not know where to look for information, but they do know the information they are looking for. Therefore, the NDN network is suitable for the vehicular network [25]. This network is characterized by discontinuous connectivity, does not require the transmitter and receiver to be connected at the same time. This justifies that the latency time is variable and is of the order of a few seconds, which does not meet the requirements of AVs. On the other hand, in view of these important characteristics of this network we will benefit from them in the spots where time is not a primary factor. As in the localization which is done off line. Another point gained by the use of NDN makes it easier to update HD maps [5]. Since the frequency of changing lanes is low, and to make the update it is necessary that the vehicles cross this road. Using NDN the update is even done by a smartphone that has used the information, therefore, the detection of change in the road becomes easy and the update is done quickly;

As shown in Table 1, the use of NDN network minimizes traffic on the DSRC and C-V2X network, subsequently increases the rate of non-abandoned messages also solves the problem of overload at the level of cellular network base stations.

**Table 1.** Assigns communication type

| Type of network | Application | Switch to |
|---|---|---|
| DSRC | Periodic message/Safety message | C-V2X |
| C-V2X | Communication between leader | To DSRC if applicable |
| NDN | scheduling (map HD) | Does not require |

The neuro-fuzzy network, allowing the management of the right communication system choice, has four layers as shown in Fig. 3. The first layer receives its data from the vehicle in question, communicates the input data (power of each of the communication modules (DSRC and C-V2x), operating time and history) to the second layer (fuzzification) which, in turn, calculates the degrees to which the input variables belong to the various classes of situations. Each input has, either three classes. The fuzzy set of connectivity module power level allowing communication management PP, MP, GP, which correspond to small connectivity power, medium connectivity power, large connectivity power.

The fuzzy set of operating times PT, MT, GT, which correspond to small operating time, medium operating time, large operating time. The output class corresponding to the communication management is given by two actions either maintain the communication system or switch to the other module.

The third layer calculates the degree of membership of the premise part, by the realization of the logical AND operator, by taking the minimum of the memberships having activated the $k^{th}$ rule.

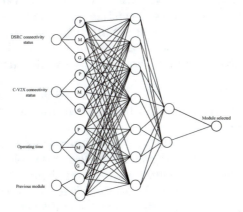

**Fig. 3.** Switching curve between master ECU and backup ECU

Let $I_k$ be the set of indices of the membership functions that have activated the $k^{th}$ rule, is given by:

$$z_k^{(3)} = \min_{I \in I_k} \left( z_l^{(2)} \right) \tag{1}$$

The weights linking the third layer to the fourth layer are equal to one. The fourth layer calculates the degrees of membership of the conclusion parts whose classes are defined by the activation functions of the neurons of the layer in question. It performs the logical OR function. The $j^{th}$ variable of the fourth layer is given by:

$$y_j^{(4)} = \max_{k \in I_j} \left( z_k^{(3)} \right) \tag{2}$$

with: $I_j$ the set of indices of the variables of the third layer, allowing to determine the $j^{th}$ variable of the fourth layer.

**Table 2.** Output classes

| Abbreviation | Description |
|---|---|
| M | Maintain module |
| S | Switch module |

The fifth layer of the neural network performs the defuzzification of the values provided by the activated rule set. The output classes are shown in Table 2.

## 4   Simulation and Performance

To validate our architecture, we performed a simulation in an intersection of two roads as shown in Fig. 4; each road is composed of two voices to go and two voices to return. The vehicles circulating form a chain of Platoons on each voice, the leader vehicle communicates with each other via C-V2X technology as a priority. The vehicles of each Platoon communicate with each other using DSRC communication as a priority. Each vehicle is equipped with our redundant architecture.

We assume that a hardware failure has occurred in a Vi vehicle of a Platoon Pi, the changeover to the reserve ECU is studied from the switching time and its influence on the circulation.

Figure 5 shows that the use of the DSRC communication module has an abandoned message rate of about 30%. While the use of the cellular communication module presents a rate of abandoned messages of about 12%, this performance degrades if the number of vehicles increases, this is due to the saturation of the base station. On the other hand, the automatic switching between the modules gives a better performance with a rate of abandoned messages is about 5%.

**Fig. 4.** Test Environment

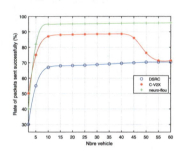

**Fig. 5.** Rate of messages sent successfully according to the number of vehicles in circulation

If we vary the number of vehicles, we can see that the use of the cellular communication module is invariant to the number of RSUs, as shown in Figs. 6 and 7. On the other hand, the DSRC communication module improves significantly and exceeds the performance of the cellular communication module from 15 RSUs. While the performance of our proposed solution improves slightly and becomes indifferent beyond 2 RSUs as shown in Figs. 6 and 7.

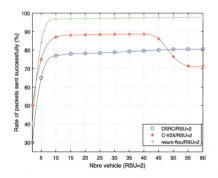

**Fig. 6.** Rate of messages sent successfully according to the number of vehicles in circulation, with 2 RSUs implemented

As the implementation of RSUs is very expensive and sometimes impossible, this shows the performance evolution of our automatic neuro-fuzzy controller compared to the use of separate communication modules. Our controller converges in a short time of about 600 ms, as shown in Fig. 8, which further proves the performance improvement of our solution.

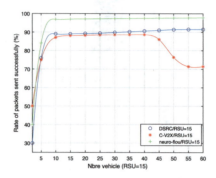

**Fig. 7.** Rate of messages sent successfully according to the number of vehicles in circulation, with 15 RSUs implemented

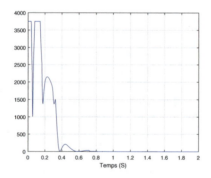

**Fig. 8.** Criteria for optimizing decisions produced as a function of time

## 5   Conclusion

This article introduces a redundant architecture designed to enhance the operational safety of AVs. Our research demonstrates that the proposed architecture effectively increases availability, which is a crucial yet previously understudied metric in the field of AVs. Additionally, we showcase how our solution significantly reduces the rate of abandoned messages, aiming for near-zero occurrences. Leveraging the NDN network further facilitates the updating of HD maps, addressing a significant challenge in the field. Future publications will explore avenues for further improvement and refinement of this work.

# References

1. Lee, E.A., Seshia, S.A.: Introduction to Embedded Systems: A Cyber-Physical Systems Approach. MIT Press (2017)
2. Bahrammirzaee, A.: A comparative survey of artificial intelligence applications in finance: artificial neural networks, expert system and hybrid intelligent systems. Neural Comput. Appl. **19**(8), 1165–1195 (2010)
3. Szénási, S., Kertész, G., Felde, I., Nádai, L.: Statistical accident analysis supporting the control of autonomous vehicles. J. Comput. Meth. Sci. Eng. **21**(1), 85–97 (2021)
4. Kyriakidis, M., et al.: A human factors perspective on automated driving. Theor. Issues Ergon. Sci. **20**(3), 223–249 (2019)
5. Van Brummelen, J., O'Brien, M., Gruyer, D., Najjaran, H.: Autonomous vehicle perception: the technology of today and tomorrow. Trans. Res. Part C Emerg. Technol. **89**, 384–406 (2018)
6. Okuda, R., Kajiwara, Y., Terashima, K.: A survey of technical trend of ADAS and autonomous driving. In: Technical Papers of: International Symposium on VLSI Design, Automation and Test, vol. 2014, pp. 1–4 (2014)
7. Van Brummelen, J., O'Brien, M., Gruyer, D., Najjaran, H.: Autonomous vehicle perception: the technology of today and tomorrow. Transp. Res. Part C Emerg. Technol, **89**, 384–406 (2018)
8. Davies, A.: Google's self-driving car caused its first crash (2016). http://www. wired.com/2016/02/googles-selfdrivingcar-may-caused-first-crash/
9. Nunen, E., Koch, R., Elshof, L., Krosse, B.: Sensor safety for the European Truck Platooning Challenge, November 2016
10. Vanholme, B., Gruyer, D., Lusetti, B., Glaser, S., Mammar, S.: Highly automated driving on highways based on legal safety. IEEE Trans. Intell. Transp. Syst. **14**(1), 333–347 (2013)
11. Crane, D.A., Logue, K.D., Pilz, B.C.: A survey of legal issues arising from the deployment of autonomous and connected vehicles. Mich. Telecomm. Tech. L. Rev. **23**, 191 (2016)
12. Boubakri, A., Mettali Gammar, S.: Intra-platoon communication in autonomous vehicle: a survey. In: 2020 9th IFIP International Conference on Performance Evaluation and Modeling in Wireless Networks (PEMWN), pp. 1–6 (2020)
13. IEEE 802.11p part11: Wireless LAN medium access control (MAC) and physical layer (PHY) specifications: Amendment 7: Wireless access in vehicular environment, July 2010
14. Ucar, S., Ergen, S.C., Ozkasap, O.: Security vulnerabilities of IEEE 802.11p and visible light communication based platoon. In: 2016 IEEE Vehicular Networking Conference (VNC), pp. 1–4 (2016)
15. Gonzalez-Martín, M., Sepulcre, M., Molina-Masegosa, R., Gozalvez, J.: Analytical models of the performance of C-V2X mode 4 vehicular communications. IEEE Trans. Veh. Technol. **68**(2), 1155–1166 (2018)
16. Boubakri, A., Gammar, S.M.: Intra-platoon communication in autonomous vehicle: a survey. In: 2020 9th IFIP International Conference on Performance Evaluation and Modeling in Wireless Networks (PEMWN), pp. 1–6. IEEE (2020)

17. Abualhoul, M.Y., Marouf, M., Shag, O., Nashashibi, F.: Enhancing the field of view limitation of visible light communication-based platoon. In: 2014 IEEE 6th International Symposium on Wireless Vehicular Communications, WiVeC 2014, pp. 1–5 (2014)

18. Tsugawa, S., Kato, S., Tokuda, K., Matsui, T., Fujii, H.: A cooperative driving system with automated vehicles and inter-vehicle communications in Demo 2000. In: 2001 IEEE Intelligent Transportation Systems. Proceedings (Cat. No. 01TH8585), ITSC 2001, pp. 918–923 (2001)

19. Yin, J., et al.: Performance evaluation of safety applications over DSRC vehicular ad hoc networks, pp. 1–9, January 2004

20. Koopman, P., Ferrell, U., Fratrik, F., Wagner, M.: A safety standard approach for fully autonomous vehicles. In: Romanovsky, A., Troubitsyna, E., Gashi, I., Schoitsch, E., Bitsch, F. (eds.) SAFECOMP 2019. LNCS, vol. 11699, pp. 326–332. Springer, Cham (2019). https://doi.org/10.1007/978-3-030-26250-1_26

21. Van Brummelen, J., O'Brien, M., Gruyer, D., Najjaran, H.: Autonomous vehicle perception: the technology of today and tomorrow. Transp. Res. Part C Emerg. Technol. **89**, 384–406 (2018)

22. Pham, H.: Software Reliability. Springer, Heidelberg (2000)

23. Fraedrich, E., Lenz, B.: Societal and individual acceptance of autonomous driving. In: Maurer, M., Gerdes, J.C., Lenz, B., Winner, H. (eds.) Autonomous Driving, pp. 621–640. Springer, Heidelberg (2016). https://doi.org/10.1007/978-3-662-48847-8_29

24. Cheng, H.: Autonomous Intelligent Vehicles: Theory, Algorithms, and Implementation. Springer, London (2011). https://doi.org/10.1007/978-1-4471-2280-7

25. Khelifi, H., et al.: Named data networking in vehicular ad hoc networks: state-of-the-art and challenges. IEEE Commun. Surv. Tut. **22**(1), 320–351 (2019)

26. Amadeo, M., Campolo, C., Molinaro, A.: Named data networking for priority-based content dissemination in VANETs. In: IEEE 27th Annual International Symposium on Personal, Indoor, and Mobile Radio Communications (PIMRC), vol. 2016, pp. 1–6. IEEE (2016)

# A Comparison Study of a Fuzzy-Based Simulation System and Testbed for Selection of Radio Access Technologies in 5G Wireless Networks

Phudit Ampririt[1](✉), Shunya Higashi[1], Paboth Kraikritayakul[1], Ermioni Qafzezi[2], Keita Matsuo[2], and Leonard Barolli[2]

[1] Graduate School of Engineering, Fukuoka Institute of Technology, 3-30-1 Wajiro-Higashi, Higashi-Ku, Fukuoka 811-0295, Japan
{bd21201,mgm23108,s23y1004}@bene.fit.ac.jp
[2] Department of Information and Communication Engineering, Fukuoka Institute of Technology, 3-30-1 Wajiro-Higashi, Higashi-Ku, Fukuoka 811-0295, Japan
qafzezi@bene.fit.ac.jp, {kt-matsuo,barolli}@fit.ac.jp

**Abstract.** The advent of 5th Generation (5G) wireless networks marks a significant leap forward in communications technology, supporting a vast array of services and applications that will profoundly impact various facets of life, including entertainment and transportation. 5G networks promise major improvements in reliability, data throughput, latency reduction, and enhanced mobility. They will feature a wide array of Radio Access Technologies (RATs), supported by a range of Base Stations (BSs) that span technologies from GSM to Wi-Fi. This technological diversity is key to fulfilling the promise of 5G in providing ubiquitous, high-quality wireless service across different environments and applications. In order to address this challenge, this paper presents a comparison study between a Fuzzy-based simulation system and a testbed for RAT selection in 5G networks. The system and testbed take into account three input parameters: Coverage (CV), User Priority (UP), and Spectral Efficiency (SE), which influence the Radio Access Technology Decision Value (RDV). The evaluation results show that experimental results were very close and have the same tendency with simulated results.

## 1 Introduction

The emergence of 5th Generation (5G) wireless networks is characterized by an exponential increase in connected devices, each with unique traffic demands, posing potential challenges to Internet bandwidth and Quality of Service (QoS). 5G networks are engineered to significantly improve reliability, throughput, latency, and mobility. They encompass a range of Radio Access Technologies (RATs) and Base Stations (BSs), covering a spectrum from GSM to Wi-Fi, and providing diverse radio coverage options from macrocells to picocells. This variety, coupled with varying transmission powers, aims to enhance the user's Quality of Experience (QoE), ensure energy efficiency, and bolster network redundancy

L. Barolli (Ed.): AINA 2024, LNDECT 199, pp. 153–165, 2024.
https://doi.org/10.1007/978-3-031-57840-3_14

and reliability [1–3]. This diversity and adaptability are key in addressing the burgeoning demands of modern wireless users and applications in the 5G era.

5G networks are poised to provide extensive support for a diverse range of services and applications, profoundly impacting various facets of our lives, including entertainment and transportation. Three fundamental application scenarios: Enhanced Mobile Broadband (eMBB), Massive Machine Type Communication (mMTC), and Ultra-Reliable & Low Latency Communications (URLLC) form the foundation of 5G technology's architectural framework for achieving this goal.

eMBB focuses on fulfilling human-centric usage scenarios, ensuring access to a diverse array of multimedia content, services, and data. The persistent growth in demand for mobile broadband necessitates the evolution of enhanced mobile broadband solutions. This expanded usage scenario encompasses not only existing Mobile Broadband applications but also introduces new application areas and criteria, all aimed at enhancing efficiency and delivering a more seamless user experience.

While, URLLC will be characterized by stringent demands on throughput, latency, and availability. Notable examples encompass wireless control in industrial manufacturing and production processes, remote medical surgeries, smart grid distribution automation, and transportation security, among others. Lastly, mMTC is a use scenario defined by a multitude of connected devices transmitting small, non-time-critical data. This scenario necessitates the use of cost-effective devices with extended battery life [4–7].

The primary objective of ongoing research in 5G wireless networks is the development of systems finely tuned to meet the heightened capacities and demands of these advanced networks. One prevalent trend is the integration of Software-Defined Networking (SDN) and Network Function Virtualization (NFV), which streamlines the efficient management of both administrative and technical networks, particularly those endowed with substantial processing power [8,9].

To mitigate latency issues, there's significant research focused on novel mobile handover strategies utilizing Software-Defined Networking (SDN). These approaches are designed to significantly reduce processing delays [10–12]. Furthermore, integrating diverse technologies enhances overall network management, enabling dynamic responses to evolving network conditions and user demands. This holistic approach is key in developing more efficient and adaptable network infrastructures, especially pertinent in the realm of advanced wireless networks.

Our research conducts a comparative analysis between a Fuzzy Logic (FL)-based simulation system and a practical testbed, focusing on Radio Access Technology (RAT) selection within 5G wireless networks. To assess the Radio Access Technology Decision Value (RDV), we consider three key parameters: Coverage (CV), User Priority (UP), and Spectral Efficiency (SE). The evaluation results show that experimental results were very close and have the same tendency with simulated results.

The structure of this paper is as follows: Sect. 2 provides an overview of Software-Defined Networking (SDN). The concept of 5G Network Slicing is introduced in Sect. 3. In Sect. 4, we detail our proposed Fuzzy Logic (FL)-based system and its testbed implementation. The analysis of both simulation and experimental results is discussed in Sect. 5. The paper concludes with insights and directions for future research in Sect. 6.

## 2  Software-Defined Networking (SDN)

The Software-Defined Networking (SDN) is revolutionizing the way networks are managed by enhancing their programmability and virtualization capabilities. This transformation primarily stems from a fundamental SDN architectural feature: the decoupling of the control plane from the data plane. This architectural separation fosters a more agile and dynamic approach to network management. By allowing independent operation and optimization of each plane, networks can achieve greater flexibility and efficiency in handling data traffic and managing network resources. It allows network administrators to adjust network operations through software, eliminating the need for physical modifications to network infrastructure. This adaptability is crucial to the efficiency of SDN, facilitating more nimble and responsive network handling, a vital aspect in modern network operations [13,14]. This innovation is particularly pertinent in the context of evolving 5G networks, where flexibility and responsiveness are key factors.

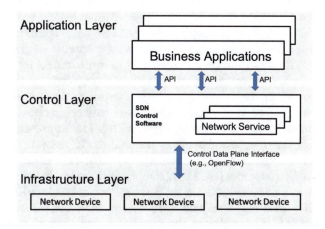

Fig. 1. The architecture of SDN.

The architecture of SDN is presented in Fig. 1, which has three distinct layers explained in following.

- **The Application Layer:** Essential for decision-making, the Application Layer collects data from the controller and interfaces with Northbound Interfaces for network programming. It guides the Control Layer, ensuring efficient Software-Defined Networking (SDN) configurations for optimal service quality and security, a key in 5G networks.
- **The Control Layer:** The Control Layer manages the Data Plane by implementing rules from the Application Layer to the Infrastructure Layer via Southbound Interfaces. This ensures efficient network connectivity and collaboration, vital in complex 5G environments.
- **The Infrastructure Layer:** This layer includes essential hardware like load balancers, switches, and routers. Controlled by the SDN controller, it ensures seamless operation and security of the network, adapting to the various requirements.

## 3   5G Network Slicing

Network Slicing (NS) is widely regarded as a pivotal technical strategy for empowering the future of communication networks. Built upon the foundations of SDN and NFV, network slicing empowers infrastructure providers (InPs) to maximize the utilization of their physical resources by concurrently accommodating diverse services with varying requirements on a shared infrastructure. In essence, network slicing establishes multiple distinct logical virtual networks, referred to as network slices or simply "slices". These slices are mutually isolated and can be dynamically created on-demand to meet specific service needs a key benefit of NS is its contribution to enhancing network reliability and security. Due to the independent operation of each slice, disruptions or complexities in one slice do not affect the others. Such logical segmentation facilitates the efficient and secure execution of various specialized network services, a significant advancement in modern network technology [15–18]. This approach is instrumental in addressing the diverse and evolving demands in 5G network environments.

The framework of 5G Network Slicing (5G NS), as defined by The Next Generation Mobile Networks (NGMN), comprises three key layers: the Resource Layer, the Network Slice Instance Layer, and the Service Instance Layer, depicted in Fig. 2 [19,20]. This organizational structure is fundamental in realizing the full potential of 5G NS, catering to the specific requirements of various network applications and services.

- **The Resource Layer:** This layer is encompassing a diverse range of services targeted at end-users. These services are provided by various entities such as application providers and mobile network operators. Within this layer, each service instance is distinct, designed to meet the specific needs of individual

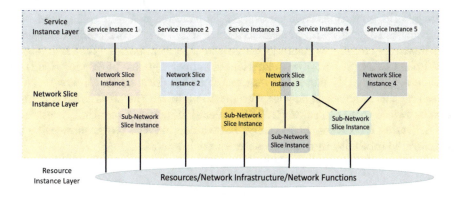

**Fig. 2.** The framework of 5G NS.

users. This customization is vital in 5G networks, as it allows for the personalization of services, ensuring that user requirements are met efficiently and effectively.

- **The Network Slice Instance Layer:** The Network Slice Instance Layer is a crucial component in 5G network structures. It comprises various network slice instances, each consisting of a set of resources specifically configured to meet distinct service demands. This layer is capable of supporting multiple sub-network instances, which can either function autonomously or share resources. This flexibility is determined by the network's design and its specific requirements. The ability to tailor resource allocation in this manner is fundamental to the adaptability and efficiency of 5G network architectures, enabling them to support a wide range of applications and services with varying performance needs.
- **The Service Instance Layer:** Combining physical and logical resources, this layer is key for creating and managing network slices. It ensures each service instance is supported by a dedicated network segment, optimizing efficiency and resource use, vital for the effectiveness and adaptability of the Network Slice (NS) paradigm in advanced networks.

## 4   Fuzzy-Based System and Implemented Testbed

Figure 3 provides an overview of the proposed system. The SDN controller plays a pivotal role in directing evolved Base Stations (eBS) for efficient data communication with User Equipment (UE). Each eBS is equipped to handle multiple network slices, allowing for specialized allocation to distinct services and applications. This capability is crucial for enhancing performance and efficiency in 5G network operations, demonstrating the integrated system's ability to adapt and optimize resource allocation for diverse network demands.

The proposed Fuzzy-based Radio Access Technology Selection System (FRSS) is integrated within the SDN controller, enhancing its capability to manage eBS operations and various RATs effectively. A key role of the SDN controller is to gather and analyze extensive data on network traffic conditions. For instance, consider a scenario where the User Equipment (UE) is linked to a Wireless LAN (WLAN) but experiences subpar Quality of Service (QoS). In such cases, the SDN controller assesses network data across different RATs. Based on this analysis, it decides whether to maintain the UE's WLAN connection or switch to an alternate RAT. This decision, made in real-time, is essential for delivering consistent, high-quality network services to users, showcasing the system's adaptability in dynamic network environments.

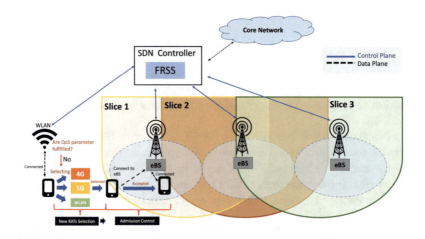

**Fig. 3.** Proposed system overview.

The structure of the FRSS is depicted in Fig. 4. It incorporates three input parameters: Coverage (CV), User Priority (UP), and Spectral Efficiency (SE). Based on these input parameters, the system calculates the output value called the Radio Access Technology Decision Value (RDV).

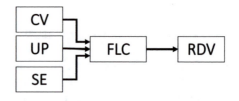

**Fig. 4.** Proposed system structure.

Our testbed, illustrated in Fig. 5, is built around a Raspberry Pi Model B (Pi4B) setup. One Pi4B functions as the SDN controller, while another serves as the User Equipment (UE). Three additional Pi4Bs act as base stations (eBSs)

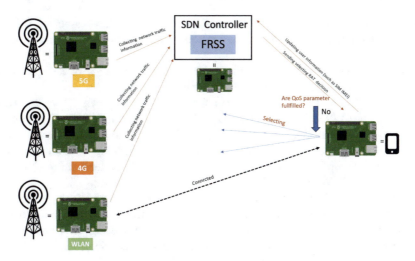

**Fig. 5.** Testbed structure.

for 5G, 4G, and Wi-Fi, each equipped with specific hats for respective RAT connectivity. This setup enables the collection of key network parameters like RSRP and RSSI, as well as Data Rate. The SDN controller also gathers user-specific information from the UE Pi4B, including SIM ID and IMEI. Leveraging this diverse data, our Flexible Radio Selection System (FRSS) efficiently identifies the most appropriate RAT for any given scenario.

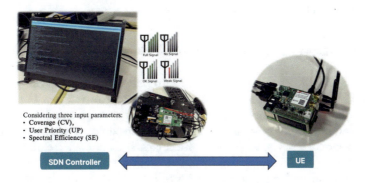

**Fig. 6.** Devices used for testbed implementation.

Figure 6 displays the devices set up in our testbed. WiFi data is captured by establishing a connection with a router. For 4G data collection, we employ a 4G antenna and a 4G wireless module, both integrated with a Raspberry Pi. Meanwhile, the generation of 5G data is achieved using the Network Simulator 3 (NS-3), enabling us to simulate and analyze 5G network scenarios effectively.

This diverse setup allows for a thorough and realistic assessment of different network technologies, crucial for our study in advanced wireless network systems [21,22].

In the following section, we will delve into the specifics of the input and output parameters pertinent to our research.

**Coverage (CV):** The CV is pivotal for gauging the reach of Radio Access Technology (RAT) cells, like 5G, 4G, and Wi-Fi. It's determined by analyzing signal robustness alongside RAT-specific performance metrics. Distinct RAT types exhibit varied CV due to their inherent characteristics in diverse settings. In our experimental framework, CV for 5G and 4G is gauged using the RSRP parameter, while Wi-Fi coverage employs the RSSI measure.

**User Priority (UP):** In our network architecture, UP denotes the importance assigned to users or devices, with a higher UP leading to preferential allocation of RAT for enhanced Service Quality (SQ). In our testing environment, UP assessment considers aspects like the user's current applications and ongoing data tariff structures, aiming to optimize RAT allocation.

**Spectral Efficiency (SE):** The SE is a crucial metric indicating the efficacy of spectrum utilization, expressed as the data transmission rate (bits per second) per unit of spectrum bandwidth (Hz). It essentially evaluates the capacity of a network to transmit data within its allocated spectral bounds. This metric is indispensable in quantifying network performance, particularly in scenarios demanding high bandwidth, such as 5G networks.

**Radio Access Technology Decision Value (RDV):** This output parameter is providing a numerical value to aid in choosing the optimal Radio Access Technology (RAT) for specific situations. This metric quantifies the suitability of various RATs, ensuring the most effective technology is utilized for each unique scenario in network deployment.

## 5   Simulation and Experimental Results

In Fig. 7 and Fig. 8, we show the relationship between RDV and SE considering different UP and CV values. We consider three scenarios when the CV value is 10%, 50% and 90%. In experimental results the average trend lines, depicted in red, green and blue, symbolize varying UP thresholds. Each individual data point corresponds to a specific test case for each examined technology (WiFi, 4G, and 5G).

In simulation and experimental results for CV 10%, we can see a similar trend with a noticeable spread of data points. The actual measured RDV values at various points of SE, particularly at higher SE percentages, are slightly higher than the simulation results. For example, at SE=70% and UP=90%, the value of RDV for simulations is approximately 0.6, whereas the testbed results show data points extending beyond 0.65. Also, considering the impact of UP on RDV, we changed the UP value to 10%, 50% and 90%. In the testbed results (see Fig. 8(a)), when we changed the UP 10% to 50% and 50% to 90% at SE 50%, the RDV value is increased by 10% and 10%, respectively. But, in the case of

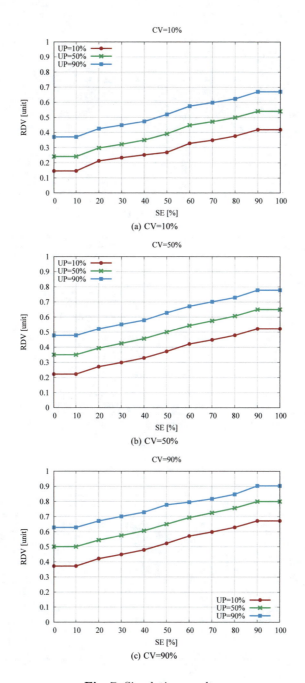

**Fig. 7.** Simulation results.

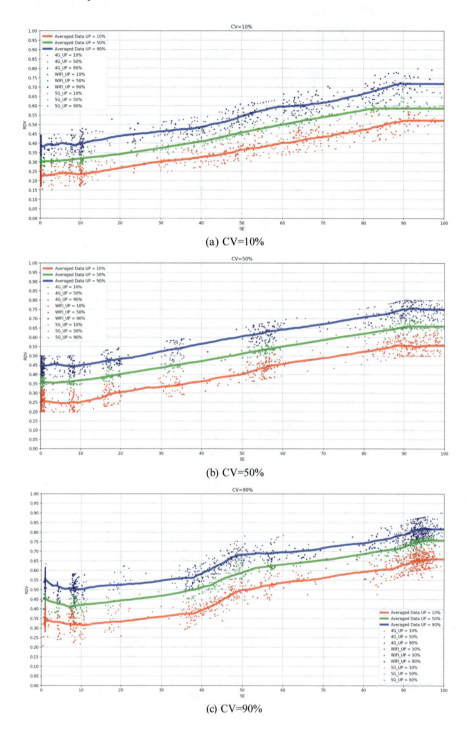

(a) CV=10%

(b) CV=50%

(c) CV=90%

**Fig. 8.** Testbed results.

simulation results (see Fig. 7(a)), the RDV value is increased by 13% and 12%, respectively. So, the difference between simulations and experiments results is very small.

In case when CV is 50%, the testbed results (see Fig. 8(b)) are concentrated is some Se values. By comparing with the case when CV value is 10% (change from 10% to 50%), the RDV value increased by 6% for users with UP value of 90% and SE value of 50%. For simulation results (see Fig. 7(b)), the RDV value increased by 10%. In this scenario, the difference of simulation results with experimental is a little bit higher compared with the case when CV is 10%, but still is small and the trend is the same.

In case when CV is 90%, the simulation results (see Fig. 7(c)) indicate that the RDV values are increased compared with the case of CV 10% and 50%. When UP is 90% and SE is 90%, the RDV value is 0.9. In the testbed results (see Fig. 8(c)), we see a concentration of data when SE is between 80% and 100%. In the simulation results, we consider the range of SE values between 90% and 100%, and UP is 90%, the RDV value is 0.9. In the testbed results, the RDV (on the average trend line) is approximately 0.8, but we can see that the scattered data points are close to 0.9. This shows that the simulation results and experimental results are very close.

## 6    Conclusions and Future Work

In this paper, we presented the implementation of a simulation system and a testbed using FL for selection of RATs in 5G wireless networks. We considered three input parameters: CV, UP and SE, while the output parameter is RDV. We compared the simulation results with experimental results. The comparison revealed that the trend of simulation results and experimental results is the same. Also, the simulation and experimental data are very close.

In our future research, we plan to conduct additional experiments using the testbed by considering more input parameters.

## References

1. Navarro-Ortiz, J., Romero-Diaz, P., Sendra, S., Ameigeiras, P., Ramos-Munoz, J.J., Lopez-Soler, J.M.: A survey on 5G usage scenarios and traffic models. IEEE Communications Surveys Tutorials **22**(2), 905–929 (2020). https://doi.org/10.1109/COMST.2020.2971781
2. Pham, Q.V., Fang, F., Ha, V.N., Piran, M.J., Le, M., Le, L.B., Hwang, W.J., Ding, Z.: A survey of multi-access edge computing in 5g and beyond: Fundamentals, technology integration, and state-of-the-art. IEEE Access 8, 116,974–117,017 (2020). https://doi.org/10.1109/ACCESS.2020.3001277
3. Orsino, A., Araniti, G., Molinaro, A., Iera, A.: Effective rat selection approach for 5g dense wireless networks. In: 2015 IEEE 81st Vehicular Technology Conference (VTC Spring), pp. 1–5 (2015). https://doi.org/10.1109/VTCSpring.2015.7145798
4. Akpakwu, G.A., Silva, B.J., Hancke, G.P., Abu-Mahfouz, A.M.: A survey on 5G networks for the internet of things: Communication technologies and challenges. IEEE Access **6**, 3619–3647 (2018)

5. Palmieri, F.: A reliability and latency-aware routing framework for 5g transport infrastructures. Computer Networks 179 (9), Article 107365 (2020). https://doi.org/10.1016/j.comnet.2020.107365

6. Kamil, I.A., Ogundoyin, S.O.: Lightweight privacy-preserving power injection and communication over vehicular networks and 5G smart grid slice with provable security. Internet of Things 8(100116), 100–116 (2019). https://doi.org/10.1016/j.iot.2019.100116

7. Series, M.: Imt vision-framework and overall objectives of the future development of imt for 2020 and beyond. Recommendation ITU 2083(0) (2015)

8. Hossain, E., Hasan, M.: 5G cellular: key enabling technologies and research challenges. IEEE Instrumentation Measurement Magazine 18, no. 3(3), 11–21 (2015). https://doi.org/10.1109/MIM.2015.7108393

9. Vagionas, C., Maximidis, R., Stratakos, I., Margaris, A., Mesodiakaki, A., Gatzianas, M., Kanta, K., Toumasis, P., Giannoulis, G., Apostolopoulos, D., Papatheofanous, E.A., Lentaris, G., Reisis, D., Soudris, D., Tsagkaris, K., Argyris, N., Syrivelis, D., Bakopoulos, P., Oldenbeuving, R.M., Roeloffzen, C.G.H., van Dijk, P.W.L., Dimogiannis, I., Kontogiannis, A., Avramopoulos, H., Miliou, A., Pleros, N., Kalfas, G.: End-to-end real-time service provisioning over a sdn-controllable analog mmwave fiber-wireless 5g x-haul network. Journal of Lightwave Technology pp. 1–10 (2023). https://doi.org/10.1109/JLT.2023.3234365

10. Yao, D., Su, X., Liu, B., Zeng, J.: A mobile handover mechanism based on fuzzy logic and mptcp protocol under sdn architecture*. In: 18th International Symposium on Communications and Information Technologies (ISCIT-2018), pp. 141–146 (2018). https://doi.org/10.1109/ISCIT.2018.8587956

11. Lee, J., Yoo, Y.: Handover cell selection using user mobility information in a 5G sdn-based network. In: 2017 Ninth International Conference on Ubiquitous and Future Networks (ICUFN-2017), pp. 697–702 (2017). https://doi.org/10.1109/ICUFN.2017.7993880

12. Moravejosharieh, A., Ahmadi, K., Ahmad, S.: A fuzzy logic approach to increase quality of service in software defined networking. In: 2018 International Conference on Advances in Computing, Communication Control and Networking (ICACCCN-2018), pp. 68–73 (2018). https://doi.org/10.1109/ICACCCN.2018.8748678

13. Li, L.E., Mao, Z.M., Rexford, J.: Toward software-defined cellular networks. In: 2012 European Workshop on Software Defined Networking, pp. 7–12 (2012). https://doi.org/10.1109/EWSDN.2012.28

14. Mousa, M., Bahaa-Eldin, A.M., Sobh, M.: Software defined networking concepts and challenges. In: 2016 11th International Conference on Computer Engineering & Systems (ICCES-2016), pp. 79–90. IEEE (2016)

15. An, N., Kim, Y., Park, J., Kwon, D.H., Lim, H.: Slice management for quality of service differentiation in wireless network slicing. Sensors 19, 2745 (2019). https://doi.org/10.3390/s19122745

16. Jiang, M., Condoluci, M., Mahmoodi, T.: Network slicing management & prioritization in 5G mobile systems. In: European Wireless 2016; 22th European Wireless Conference, pp. 1–6. VDE (2016)

17. Chen, J., Tsai, M., Zhao, L., Chang, W., Lin, Y., Zhou, Q., Lu, Y., Tsai, J., Cai, Y.: Realizing dynamic network slice resource management based on sdn networks. In: 2019 International Conference on Intelligent Computing and its Emerging Applications (ICEA), pp. 120–125 (2019)

18. Li, X., Samaka, M., Chan, H.A., Bhamare, D., Gupta, L., Guo, C., Jain, R.: Network slicing for 5G: Challenges and opportunities. IEEE Internet Comput. 21(5), 20–27 (2017)

19. Afolabi, I., Taleb, T., Samdanis, K., Ksentini, A., Flinck, H.: Network slicing and softwarization: A survey on principles, enabling technologies, and solutions. IEEE Communications Surveys Tutorials **20**(3), 2429–2453 (2018). https://doi.org/10.1109/COMST.2018.2815638

20. Alliance, N.: Description of network slicing concept. NGMN 5G P 1(1), 7 Pages (2016). https://ngmn.org/wp-content/uploads/160113_NGMN_Network_Slicing_v1_0.pdf

21. Patriciello, N., Lagen, S., Bojovic, B., Giupponi, L.: An e2e simulator for 5g nr networks. Simulation Modelling Practice and Theory 96, 101,933 (2019)

22. Koutlia, K., Bojovic, B., Ali, Z., Lagén, S.: Calibration of the 5g-lena system level simulator in 3gpp reference scenarios. Simulation Modelling Practice and Theory 119, 102,580 (2022)

# A Distributed Approach for Autonomous Landmine Detection Using Multi-UAVs

Amar Nath[1(✉)] and Rajdeep Niyogi[2]

[1] SLIET Longowal, Punjab 148106, India
amarnath@sliet.ac.in
[2] IIT Roorkee, Roorkee 247667, India
rajdeep.niyogi@cs.iitr.ac.in

**Abstract.** This paper emphasizes the advantages of multi-UAV systems for land-mine detection, such as their quick coverage of large areas and adaptability to hazardous or inaccessible environments. A single robot or UAV equipped with a landmine-detecting sensor has been used in most research on robotic landmine detection. However, using multiple UAVs or robots can significantly improve landmine detection speed, quality, and accuracy. This paper suggests a distributed algorithm for landmine detection using multi-UAVs to increase the effectiveness, safety, and accuracy of landmine detection. The proposed approach guarantees fast and accurate landmine detection by calculating the lowest number of points needed to cover the target region without overlap. We used the Eclipse platform for the implementation. The simulation results show that using multiple threads or drones can enhance the efficiency of the landmine detection process.

## 1 Introduction

A landmine, in essence, is an explosive device, a pressure-actuated explosive device surreptitiously placed beneath the ground surface. Its primary function is the automatic detonation and subsequent neutralization of enemy personnel or materiel traversing or approaching its vicinity. Each year, landmines cause thousands of casualties and fatalities. They can also keep people from getting to vital resources like water and land. Traditional landmine detection methods are often slow, expensive, and dangerous. However, using unmanned aerial vehicles (UAVs) can significantly improve the efficiency and effectiveness of landmine detection. Multi-UAV systems can be used to cover a large area quickly and efficiently. They can also be equipped with various sensors, such as ground-penetrating radar, to detect landmines in various environments. Figure 1 illustrates a situation involving landmines.

Massoud Hassani has developed the Mine Kafon Drone (MKD), a device capable of autonomously mapping, detecting, and neutralizing landmines[1]. The MKD, also known as the Minefield Clearance Drone, is engineered to offer a simpler, cost-effective, and safer approach to clearing landmines. It is outfitted with three distinct attachments and six rotors for diverse functionalities. Upon identification of the mines, the drone returns

---

[1] https://www.kickstarter.com/projects/massoudhassani/mine-kafon-drone.

L. Barolli (Ed.): AINA 2024, LNDECT 199, pp. 166–177, 2024.
https://doi.org/10.1007/978-3-031-57840-3_15

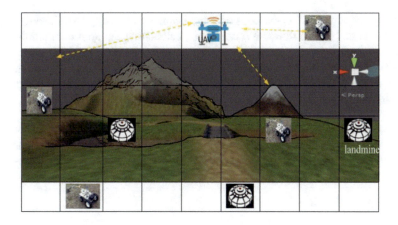

**Fig. 1.** Landmine site scenario

to the operator to swap the metal detector with a robotic arm. This arm is then used to precisely place small detonators, approximately the size of tennis balls, on top of the mines' locations. Once the detonators are positioned, the mines are remotely set off, and the drone withdraws to a secure distance.

However, MKD is a significant advancement in landmine clearance. It encounters several challenges, such as (i) detecting landmines that have been underground for many years, (ii) dealing with landmines that require the coordinated effort of multiple drones for safe detonation, and (iii) optimizing the entire process of mapping, detecting, and detonating landmines to be quicker and more automated. To overcome these hurdles, two critical factors must be considered: a) the capability to explore and analyze unknown and potentially hazardous terrains swiftly, and b) the development and implementation of a multi-robot task allocation algorithm. This algorithm would enable the effective distribution and collaborative execution of tasks among multiple Agents/robots/UAVs, ensuring efficient and coordinated operations.

Detecting landmines can be significantly enhanced by leveraging the capabilities of commercially available autonomous robots or UAVs. Key objectives in landmine detection and detonation include maximizing efficiency and safety while ensuring thorough and minimally overlapping coverage of the target area. The key objective of this study is to create a system that can reliably and independently detect landmines, lowering the risks that human operators must face. To efficiently detect and detonate landmines, the proposed approach uses various UAVs, each equipped with various sensors, such as thermal and visual cameras and ground-penetrating radars. The study has twofold contributions.

1. To suggest a multi-UAV-based model for landmine detection to efficiently explore the target area with minimum sensors, i.e., calculating the minimum points needed to be visited.
2. To develop a distributed algorithm for multi-UAV coordination to explore the target area so that none of the areas is left unexplored and there is no overlapping of area exploration.

The rest of the paper is organized as follows. Related works are listed in Sect. 2. The suggested methodology is explained in Sect. 3. Section 4 describes the implementation and results. Conclusions are drawn in Sect. 5.

## 2    Related Work

Over the past decades, autonomous landmine detection with robotic devices has been a focus of ongoing study. Several studies have been conducted in landmine detection using UAVs and multi-robot systems. This section provides an overview of the relevant research that has contributed to developing the proposed landmine detection approach using multi-UAVs.

The works [1–3] provide insight into the most recent development of landmine detection using robot/s. In work [1], significant lessons on data fusion and signal processing features are briefly covered. The outputs of the sensors, which are thought of as knowledgeable specialists, are fused in a fusion operation during the detection process, which is viewed as a global process. The study then briefly discusses the issue of using remote sensing to reduce area. The study [2] evaluates the current mine clearance technologies and discusses their development efforts and limits to automate the demining process. Furthermore, it presents the technological characteristics and designs potential of a mobile platform that are required to expedite the demining procedure and guarantee safety through economical means. In work [3], the importance of integrating robotics, communications, and information investigation in land mine identification is highlighted with a focus on real-time human detection and tracking in chaotic and dynamic situations.

Recent research on autonomous landmine detection has greatly emphasized the development of robotic systems for landmine detection[4]; most of these systems consist of a single robot outfitted with the necessary landmine detection sensors. Some robots, for example, are equipped with tools for removing vegetation and a device for stomping the ground and setting off explosives [5]. The paper [12] suggests combining detection data from many sources, such as multi-spectral, hyper-spectral, and infrared sensors installed on UAVs, to detect landmines optically. This paper uses a quad-copter drone platform in the air equipped with two optical sensors: a thermal infrared (TIR) matrix array camera and a visible/near-infrared (VNIR) multi-spectral frame camera. Extracting landmine features, comparing them to reference data from the features database, and making suitable judgments based on the results were the primary goals of the image analysis jobs for both sensor types. In their study, they used one drone.

A recent work [8] presents a comprehensive framework for the autonomous detection of landmines using UAVs, focusing on establishing an efficient coverage route to survey the target area. The process involves delineating the area of interest through deep learning-based segmentation and formulating an optimal coverage route plan for the aerial survey. An ideal UAV route is identified using multiple coverage path patterns, ensuring a thorough and systematic scan of the designated region. However, the work [8] did not use multi-UAVs for landmine detection.

# 3 Proposed Approach for Landmine Detection

Over recent years, the robotic exploration of unknown environments has gained significant attention from the robotic research community. A critical task in robotic exploration is the coverage of terrain or areas. It involves multiple robots using their sensors or tools to ensure complete coverage of every part of an environment. This technique of area coverage finds application in several areas of robotic systems, including aerial reconnaissance for military purposes, robotic exploration in unmanned search and rescue missions, autonomous exploration of extraterrestrial surfaces for space-related endeavors, and even domestic uses like automated crop harvesting. The innovative method developed for our system combines a swarming-based emergent system with a utility-based multi-agent model. This approach is employed to control teams of robots, enhancing the efficiency of the area coverage task [9]. The challenge of landmine handling can be broadly categorized into two stages: the detection and the subsequent detonation of the landmines. Here, in this study, we focus on the landmine detection problem.

We now discuss the proposed methodology for landmine detection, which is designed to be fast, secure, and reliable.

## 3.1 Landmine Detection

In our approach to landmine detection, we consider an environment whose overall boundaries are predefined but whose internal landscape is initially unknown. This strategy is particularly applicable for comprehensive landmine searches in dynamic terrains. We divide the area into smaller, more manageable subsections to manage the search operation effectively. This division is conceptualized as a grid, where each cell represents a distinct subsection of the larger area. The primary tool for mapping and navigating this environment is a grid matrix. This matrix is a dynamic map, continuously updated and shared among all participating agents (drones) in the search operation. These agents, including autonomous drones or other robotic systems, utilize this shared matrix to coordinate their efforts, systematically covering each grid subsection. The drones' positions and any detected landmines are recorded and illustrated visually at specific intervals, as shown in Fig. 2.

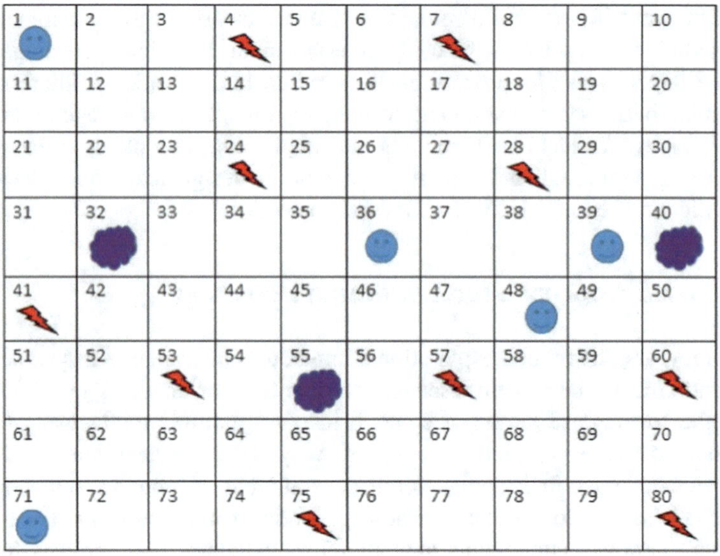

**Fig. 2.** The landmine-suspected area where large and small landmines are depicted in violet and red colors, respectively, while blue smiley icons represent drones

To detect landmines in unknown areas fast, securely, and reliably with multi-UAVs, the objective is to conduct a thorough landmine scan in a target area, ensuring there are no regions left without scanning or the same area is not explored multiple times, i.e., overlaps. The choice of algorithm for this task largely depends on the detection range of the landmine sensor equipped on the drone. The drone must navigate to certain points within a grid layout to guarantee exhaustive area coverage. We aim to reduce the number of these navigational points, thereby speeding up the scanning process. This challenge is known as the "coverage problem", which essentially measures the efficiency with which the sensors survey the target area. Considering the drone is equipped with a single sensor, determining the least number of required navigational points is tantamount to establishing the minimum number of sensors necessary to fully cover the area ($A$).

Let's consider a collection of points, denoted as $P = \{p_1, \ldots, p_n\}$, within the target area that needs to be covered by sensors (embedded with drones), all possessing an identical sensing range $r$. Although the target area $A$ to be covered could take on various shapes, for the sake of simplicity, let's assume it's a rectangle $A$ with a width of $w$ and a length of $h$. In this scenario, if each point within the area is within the range of at least $k$ sensors, the area is described as being $k$-covered. Here, $k$ represents the coverage's degree, indicating the minimum number of sensors that must encompass any given point in the area[2].

---

[2] https://personal.utdallas.edu/~dzdu/cs7301c/main.pdf.

**Definition 1.** *A point* $(x, y)$ *within the target area A is considered to be covered by a sensor* $s_i$ *positioned at* $(x_i, y_i)$ *if the distance between the point and the sensor, calculated as* $\sqrt{(x_i - x)^2 + (y_i - y)^2}$, *is less than or equal to the sensing range r. The target area A is called k-coverage when every point* $(x, y)$ *in A falls within the range of at least k distinct sensors [10].*

The requirement for $k$-coverage necessitates a corresponding increase in the number of sensors $n$, proportional to the value of $k$. Hence, higher $k$ values demand greater sensors for achieving $k$-coverage. Up to this point, the underlying assumption has been that since sensors are relatively inexpensive devices, deploying many is not a significant concern. Therefore, it has been presumed that enough sensors are available to ensure $k$ coverage across the designated target area [10].

## 3.2 Minimal Number of Sensors

To achieve $k$-coverage of a rectangular area having size $a \times b$, where all sensors share an identical coverage radius, determine the minimal number of sensors $N_k(a, b)$ required for full area coverage. For this calculation, it is assumed that each sensor provides a coverage radius of 1 unit. By convention, $N_k(a, b) = 0$ if either $a \leq 0$ or $b \leq 0$. A $k$-coverage utilizing $n$ sensors, satisfies Eq. 1 (Fig. 3).

$$N_k(a, b) \leq n \tag{1}$$

**Proposition 1.** $N_1(\sqrt{2} \cdot m, \sqrt{2} \cdot n) \leq m \cdot n$, *where m, n are elements of* $\mathcal{N}$ *(the set of positive integers) [10].*

In this scenario, the minimum number of points a drone must visit to cover the target area with minimal overlap fully is $m \times n$, which equals $80 \times 80 = 6400$ for the area under consideration (as depicted in Fig. 2). The primary goal of the robots/drones is to detect landmines using their specialized sensors. Multiple Unmanned Aerial Vehicles (UAVs) are deployed to accelerate the area coverage process.

## 3.3 Distributed Landmine Detection Algorithm

The distributed algorithm for effectively covering the area using multiple drones is outlined in Algorithm 1 (given in Fig. 4). The target area is set up like a grid of many identical cells. Each cell has an important point at the intersection of its diagonals. This point needs to be reached by a drone for effective landmine detection. The connections in the grid, shown as edges in a graph, represent the routes to neighboring points. Deploying multiple drones speeds up searching and scanning the target area, making it easier and quicker to find and accurately locate landmines. At the beginning, all drones take off from the base station. Each drone has its list, maintained in a queue Q, where it keeps track of the neighboring points or vertices it needs to visit. There's also an array, A, that records which points have been visited and which have not. Any landmines found are marked in a two-dimensional array called *LandmineMap*. A visual representation of how things look when the algorithm gets started is shown in Fig. 5.

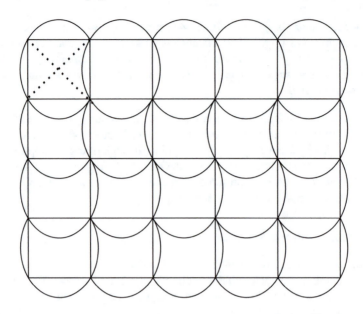

**Fig. 3.** 1-Coverage of the rectangular target area where the width $w = \sqrt{2} \cdot m$ and the height $h = \sqrt{2} \cdot n$.

Initially, a drone's state is set to IDLE, which transitions to EXPLORE once it begins surveying the target area for landmines. In the EXPLORE state, the drone navigates to the center of the grid cell, corresponding to a vertex in the graph. Upon arrival at a vertex, the drone adds this vertex to its queue Q and marks it as visited in the array A. The drone then switches its state from the EXPLORE to the DETECT. While in the DETECT state, the drone utilizes its sensors to search for landmines within that cell. Any detected landmines are recorded in the *LandmineMap*. After logging the land-mine's vertex/position, the drone changes to the ANALYZE state. In the ANALYZE state, the drone checks for Visited messages from neighboring drones (as outlined in Algorithm 1, lines 14–19). It updates its visited array A, removes the first vertex from its queue Q, and adds all unvisited neighboring vertices to Q. The drone then changes its state to the EXPLORE state and proceeds to the next vertex. The drone returns to the base station once Q is empty, indicating all vertices have been visited. Upon their return, all drones transmit their *LandmineMap* matrices to the base station, where these maps are consolidated into a single *FinalLandmineMap*. This process marks the algorithm's termination, culminating in the identification of all landmine positions.

---

**Algorithm 1:** Distributed landmine detection algorithm

---

    **Initialization**   : State of all drones Set to IDLE and all the drones has a map
                      (LandmineMap) which is initialized with 0.

1  The drones start for exploring the target area.
2  State of the drones changes to EXPLORE
   **Send Function**
4  Drone$_i$ insert the visiting vertex into its queue Q$_i$
5  Drone$_i$ mark the starting vertex as visited in its array A$_i$
6  **if** *(State$_i$* = EXPLORE *AND* Q$_i \neq \emptyset$ ) **then**
7     Drone$_i$ mark the vertex as visited in A$_i$
8     Drone$_i$ broadcast the array A$_i$ within its range by Visited message
9     State$_i$:=DETECT ▷ Drone uses its sensor to detect the
           landmine in that cell
10    **if** *(State$_i$* = DETECT*)* **then**
           update LandmineMap$_i$
           State$_i$:=ANALYZE

   **Receive Function**
14  **if** *(State$_i$* = ANALYZE *AND received* Visited *message)* **then**
15     Update its visited array A$_i$
16     Removes the first vertex of the queue Q$_i$
17     Inserts all the unvisited neighbors of the vertex into the queue Q$_i$;
18     State$_i$:=EXPLORE
19     Drone$_i$ moves to the next points, i.e., vertex
20  **if** *(Q$_i$ = $\emptyset$)* **then**
       return LandmineMap$_i$
22  Merge all the LandmineMaps received from all the drones
23  All the drones return to the base station

**Fig. 4.** Distributed landmine detection algorithm

## 3.4   Implementation Strategy for Landmine Detonation

The process of landmine neutralization commences upon receiving the map of the land-mine site. Drones designated for excavation and detonation head to the specific land-mine locations as indicated in the *FinalLandmineMap*. Certain landmines can be managed by a pair of drones, one for digging and another for detonation. However, in cases where a landmine is particularly heavy or deeply embedded in the ground, two more drones may be necessary for the detonation task, as illustrated in Fig. 6. The coordination of these multiple drones and the formation of teams can be effectively managed using a distributed algorithm, as detailed in our previous work [11].

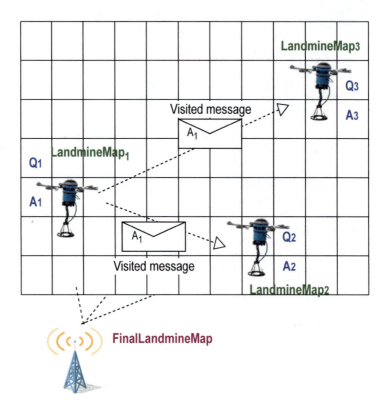

**Fig. 5.** The initial setup at the beginning of the distributed algorithm

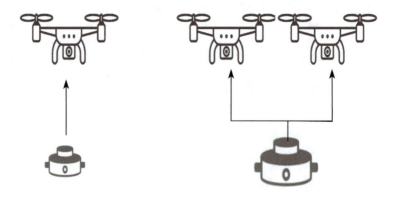

**Fig. 6.** Landmine detonation by single and multiple drones

# 4 Implementation

To evaluate the efficacy of the distributed method for landmine detection, we used the Eclipse platform. In particular, we used Eclipse version *2022-12 (4.26.0)* on a system with an Intel® Core™ i3-2200 CPU@3.40 GHz×8, 4 GB RAM, and 64-bit Window operating system. Several threads were used in the distributed approach's implementation to improve efficiency and performance. We ignored the message overhead to streamline the experimental process and concentrate only on assessing the fundamental features and capabilities of the distributed approach.

We conducted rigorous testing of the proposed system to know its potential and limitations of the Java threading concept. We designed and tested some Java classes, i.e., *Graph.java, Marven.java, ThreadSearch.java*, and *MultithreadSearch.java*. The details of all the Java classes are given below.

- Graph.java: This is created to code the algorithm as abstract data structure type (ADT). This class makes working efficiently with graph data possible, which lets us control and modify graph-related operations.
- Marven.java: We introduced this class to create a more complex graph. We use this class to build a large Marvel graph, allowing us to test the distributed algorithm on a larger, more realistic dataset. Ensuring our method can manage enormous amounts of data and real-world events is crucial.
- ThreadSearch.java: We developed this class to analyze and access the performance and functionality of our algorithm. The algorithm should only be tested on one thread with this class. We may learn more about the behavior and efficacy of the algorithm under usual settings by assessing its performance in a single-threaded context.
- MultiThreadSearch.java: This Java class is created for evaluation and comparison. Several threads were created using this class to test the multi-UAVs concept of the algorithm. We learn more about the behavior and efficacy of the algorithm under usual settings by assessing its performance in a single-threaded context with this class. With parallelism, we spread the processing load across several threads, potentially increasing output and speeding up execution times.

We found that the multi-threaded approach outperforms the single-threaded version after testing the single and multi-threaded versions. This significant result demonstrates how the distributed approach might perform better when parallel processing is used, especially when dealing with larger and more complex networks or datasets. Combining these courses provides a comprehensive and robust framework for implementing and evaluating the decentralized approach.

We employ two threads in most of our testing and occasionally three under specific conditions. These multi-threaded tests differ from single-threaded tests primarily because the former uses multiple threads while the latter does not. Interestingly, using many threads significantly shortens the time required to complete the search. In Fig. 7, we illustrate the relationship, considering varying thread counts, between the number of locations to be visited inside a target landmine area and the time needed. The figure unequivocally demonstrates that the efficiency gains attained by parallel processing are attained by raising the number of threads, leading to a significant reduction in the traversal time of the graph.

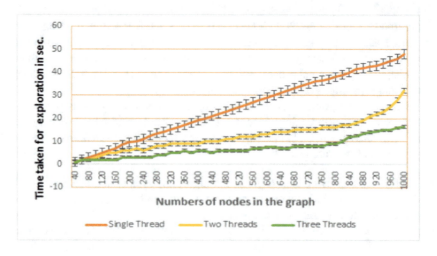

**Fig. 7.** The amount of time taken by an agent to go across a target area's points while changing the number of threads

## 5   Conclusion

Landmines must be detected and detonated as soon as possible from several important suspected areas, like along the military convoy's route. Our proposed approach uses UAVs to map, detect, and detonate landmines more quickly and efficiently. We can achieve rapid, safe, and reliable landmine detection by adopting a distributed approach, where the minimum number of sensing points in a target area is calculated. Then, multi-UAVs are used for landmine detection and detonation. This approach holds the potential to significantly improve the overall speed and safety of landmine detection and subsequent detonation, crucially protecting military personnel and ensuring mission success. The simulation results demonstrate that utilizing multiple threads or drones can enhance the efficiency of the landmine detection process. Our goal for the future is to implement our suggested approach in a real-world setting and use multi-autonomous drones to accomplish landmine detection detonation tasks safely and quickly.

**Acknowledgement.** The second author was in part supported by a research grant from Google.

# References

1. Acheroy, M.: Mine action: status of sensor technology for close-in and remote detection of antipersonnel mines. In: 3rd International Workshop on Advanced Ground Penetrating Radar (IWAGPR 2005), pp. 3–13 (2005)
2. Habib, M.K.: Humanitarian demining: reality and the challenge of technology-the state of the arts. Int. J. Adv. Rob. Syst. **4**(2), 151–172 (2007)
3. Patil, D., Ansari, M., Tendulkar, D., Bhatlekar, R., Pawar, V.N., Aswale, S.: A survey on autonomous military service robot. In: International Conference on Emerging Trends in Information Technology and Engineering (ic-ETITE), pp. 1-7-. IEEE (2020)
4. Dasgupta, P., Baca, J., Guruprasad, K.R., Munoz-Melendez, A., Jumadinova, J.: The COM-RADE system for multirobot autonomous landmine detection in postconflict regions. J. Robot. Hindawi Publ. Corp. **2015**, 921370 (2015)
5. Stefan, H.: Some robotic approaches and technologies for humanitarian demining. In: Maki, K.H. (ed.) Humanitarian Demining. In Tech Open Access Publishing, Vienna (2008)
6. Havlik, S.: Land robotic vehicles for demining. In: Habib, M.K. (ed.) Humanitarian Demining. InTechOpen (2008)
7. Gustavo, S.C.A., Pereira, G.A.S., Pimenta, L.C.A., Iscold, P.: Multi-UAV routing for area coverage and remote sensing with minimum time. Sensors **15**(11), 27783–27803 (2015)
8. Ahmed, B., Kumar, K., Kumar, N., Thakur, N., Alzahrani, B., Almansour, A.: Unmanned ariel vehicle (UAV) path planning for area segmentation in intelligent landmine detection systems. Sensors **23**(16), 7264 (2023)
9. Sze, K.C., Peng, N.A., Rekleitis, I.: Distributed coverage with multi-robot system. In: ICRA, pp. 2423–2429 (2006)
10. Tabirca, T., Tabirca, S., Yang, L.T.: Smallest number of sensors for k-covering. Int. J. Comput. Commun. 1–8 (2013). ISSN 1841-9836
11. Nath, A., Niyogi, R.: Distributed framework for task execution with quantitative skills. In: 21st International Conference on Computational Science and Its Applications (ICCSA 2021), pp. 413–426 (2021)
12. Popov, M., Stankevich, S.A., Mosov, S.P., Titarenko, O.V., Topolnytskyi, M.V., Dugin, S.S.: Landmine detection with UAV-based optical data fusion. In: IEEE EUROCON 19th International Conference on Smart Technologies, pp. 175–178. IEEE (2021)

# A Cuckoo Search Based Simulation System for Node Placement Problem in WMNs: Performance Evaluation for Normal, Exponential and Weibull Distributions of Mesh Clients

Shinji Sakamoto[1]([✉]), Leonard Barolli[2], Yi Liu[3], and Makoto Takizawa[4]

[1] Department of Information and Computer Science, Kanazawa Institute of Technology, 7-1 Ohgigaoka, Nonoichi, Ishikawa 921-8501, Japan
shinji.sakamoto@ieee.org

[2] Department of Information and Communication Engineering, Fukuoka Institute of Technology, 3-30-1 Wajiro-Higashi, Higashi-Ku, Fukuoka 811-0295, Japan
barolli@fit.ac.jp

[3] Department of Computer Science, National Institute of Technology, Oita College, 1666, Maki, Oita 870-0152, Japan
y-liu@oita-ct.ac.jp

[4] Department of Advanced Sciences, Faculty of Science and Engineering, Hosei University, Kajino-Machi, Koganei-Shi, Tokyo 184-8584, Japan
makoto.takizawa@computer.org

**Abstract.** Wireless Mesh Networks (WMNs) are cost-effective and highly robust networks. However, WMNs have various challenges associated with wireless communication. To deal with these issues, it is needed to find appropriate locations of mesh routers. However, the problem is known as NP-hard problem. In order to solve this problem, we consider Cuckoo Search (CS) algorithm and implement an intelligent simulation system called WMN-CS. In this work, we evaluate the performance of WMN-CS considering different distributions of mesh clients. Simulation results show that for Normal distribution of mesh clients, the WMN-CS can find suitable mesh router locations for 30 phases. However, 200 phases are insufficient for WMN-CS convergence in case of Exponential and Weibull distributions.

## 1 Introduction

Wireless Mesh Networks (WMNs) are cost-effective and they can be quickly deployable with high robustness. Nevertheless, they face several issues when the location of mesh routers is not appropriate resulting in congestion, interference, reduced data transfer rate, packet loss, and increased latency [12]. To deal with these issues, the location of mesh routers is very important. However, the allocating mesh routers at optimal position in WMNs is known to be an NP-hard problem [2,5,8].

L. Barolli (Ed.): AINA 2024, LNDECT 199, pp. 178–186, 2024.
https://doi.org/10.1007/978-3-031-57840-3_16

In our previous work, we proposed and implemented a Cuckoo Search (CS) algorithm based intelligent system to find better locations of mesh routers [3, 15]. The proposed system is called WMN-CS. In this study, we evaluate the performance of proposed WMN-CS considering different distributions of mesh clients: Normal, Exponential and Weibull distributions.

The remainder of this paper is structured in the following manner. Section 2 presents a short decription of related work. The node placement problem in WMNs is defined in Sect. 3. In Sect. 4, we present the CS algorithm. In Sect. 5, we introduce the WMN-CS simulation system. The simulation results are discussed in Sect. 6. We conclude the paper in Sect. 7.

## 2    Related Work

Resource allocation, node selection, and routing are the most attractive topics in WMNs [1, 16]. Many research works are proposed to deal with these topics. Some of them examine optimisation models using Mixed-Integer Linear Programming (MILP) and meta-heuristic techniques for solving node placement problems in WMNs [4, 6, 7, 9–11, 13, 14]. Other works consider Hill Climbing (HC), Simulated Annealing (SA), Genetic Algorithm (GA) and Particle Swarm Optimisation (PSO). Heuristic approaches are often used to solve NP-hard problems because they are more time-efficient than Mixed Integer Linear Programming (MILP) techniques [17].

The performance of WMNs depends on the environment and/or locations of mesh nodes. So, the performance can be evaluated by computer simulation with some assumptions. In particular, the heuristic-based solutions show potential solutions to solve the node placement problem in WMNs and other related issues [18].

## 3    Node Placement Problem in WMNs

Several issues related to wireless communication must be considered in order to use the full potential performance of WMNs. In WMNs, the signal strength decreases with distance and is obstructed by physical barriers such as walls that lead to inhomogeneous coverage [16]. Thus, there are some dead zones that may be out of covering signal. Moreover, the scalability of the network is constrained by the finite number of available channels. The signal interference may decrease the network performance, especially in high-density network areas where multiple mesh nodes requiring the network connection.

By optimizing router location can be solved several issues related to the nature of wireless communication in WMNs. The goal is to optimize the placement of mesh router nodes to balance coverage, connectivity, and network capacity. However, finding the optimal location of mesh routers is known to be an NP-hard problem [2].

For the node placement problem in WMNs, we consider a continuous 2D area with specified width and height. The mesh client nodes and mesh router nodes are allocated at arbitrary points in the considered area. The problem is to identify optimal locations of mesh router nodes that maximise connectivity and provide extensive coverage for users. To assess the network connectivity of WMNs, we consider a graph representing the mesh routers and employ the Size of Giant Component (SGC) as a metric. While, for user coverage, we use the Number of Covered Mesh Clients (NCMC) metric. The NCMC is determined by counting the number of mesh clients within the radio coverage of at least one mesh router node. The network performance is directly impacted by the connectivity of the network and the user coverage. Hence, the SGC and NCMC measures have a considerable significance.

We consider an adjacency matrix of mesh nodes for establishing a framework for the problem. Each node has location information and communication distance information, representing a vector $v = < x, y, r >$. Every node is located in the considered 2D area, and the adjacency matrix is built using the vector information. That is, if node $v$ is located within the communication distance of node $u$ and node $u$ is located within the communication distance of node $v$, then these nodes have the edge in the WMNs graph.

## 4 Cuckoo Search Algorithm

### 4.1 Overview of Cuckoo Search

In this work, we consider a meta-heuristic approach named Cuckoo Search (CS), which imitates the cuckoo behaviour about the phenomena of brood parasitism.

The CS algorithm is based on three assumptions [19].

1. A cuckoo can lay a single egg at a time.
2. The high-quality eggs can survive for the next generations.
3. The host bird may detect the egg laid by a cuckoo with a probability $p_a$, where $p_a$ is a value between 0 and 1. In case of detecting the laid egg by cuckoo, the host bird will remove the egg or move away from the nest.

The CS algorithm pseudo-code is shown in Algorithm 1. The CS algorithm involves three key hyperparameters: the number of nests, the host bird recognition rate, and the scale parameter of the Lévy distribution.

Firstly, the CS algorithm starts with the initial solution and parameters. Then, the fitness value for each nest is calculated and the most efficient nest is found. By repeatedly iterating the above process, the CS identify appropriate solutions. This iterative process continues until a predetermined stopping criterion is met.

The most optimal nest is determined from the set of nests and suitable solutions are indentified by iteratively repeating the process. During each iteration, the CS algorithm generates new solutions by following Lévy flight, which is a type of random walk. Then, the fitness value of the new solution is calculated. If the fitness value of the new solution is higher than the previous one (the quality and effectiveness of new solution compared to the existing solution), the previous solution is substituted with new solution. If the current solution is worse than the previous solution, it can be eliminated using a stochastic process.

After calculating all fitness values, the optimal solutions are retained while the remaining cuckoos that have not laid their eggs continue to exist. The new solution replaces the retained solution if the new solution is better than the retained solution. Then, the fitness value of each nest is re-evaluated. This process continues until the termination criterion is met.

**Algorithm 1.** Pseudo code of CS algorithm.

```
1:  Initialize Parameters:
2:      Computation Time t = 0 and T_max
3:      Number of Nests n(n > 0)
4:      Host Bird Recognition Rate p_a(0 < p_a < 1)
5:      Lévy Distribution Scale Parameter γ(γ > 0)
6:  Fitness Function to get Fitness Value as f
7:  Generate Initial n Solutions S_0
8:  while t < T_max do
9:      while i < n do
10:         j := i % len(S_t) // j is the remainder of dividing i by number of solutions.
11:         Generate a new solution S_{t+1}^i from S_t^j by Lévy Flights.
12:         if (f(S_{t+1}^i) < f(S_t^j)) and (rand() < p_a) then
13:             Discard Solution S_{t+1}^i
14:         end if
15:         i = i + 1
16:     end while
17:     t = t + 1
18: end while
19: return Best solution.
```

## 4.2 Lévy Flight

The Lévy flight is a kind of random walk, which is used in the CS algorithm to generate new solutions through a stochastic process. The Lévy flight is recognised for its inclination to display long-tailed characteristics. This attribute allows the extensive exploration of the search space by CS, which has the potential to find optimal solutions.

The Lévy flight generates short movements with a high probability. It can generate long movements with a small likelihood. This form of motion is not restricted only to theoretical approaches but also is observed in the natural world. For instance, honeybees utilise the Lévy flight pattern in order to find flower gardens. They typically undertake brief flights to locate a suitable blossom in a flower garden. When they are unable to find a suitable bloom in the vicinity, they consider extensive journeys to locate another flower garden. This phenomenon is not exclusive to insects, humans also exhibit behaviours that align with the concepts of the Lévy flight.

The Lévy distribution is used in probability theory and statistics for the purpose of representing non-negative random variables. It is a specific case of the inverse-gamma distribution.

The probability density function of Lévy distribution can be expressed as following:

$$P(x; \mu, \gamma) = \sqrt{\frac{\gamma}{2\pi} \frac{e^{-\gamma/2(x-\mu)}}{(x-\mu)^{3/2}}}, \tag{1}$$

where $\mu$ is the local parameter ($\mu \leq x$) and $\gamma$ is the scale parameter ($\gamma > 0$).

While, the Lévy distribution cumulative distribution function can be expressed by the following equation:

$$F(x; \mu, \gamma) = \mathrm{erfc}\left(\sqrt{\frac{\gamma}{2(x - \mu)}}\right), \tag{2}$$

where erfc is the complementary error function.

By using the inverse transformation method, we can generate the values which follows Lévy distribution.

$$F^{-1}(x; \mu, \gamma) = \frac{\gamma}{2(\mathrm{erfc}^{-1}(x))^2} + \mu \tag{3}$$

## 5   WMN-CS Simulation System

In the following, we describe the startup process, the nesting and egg-laying behaviour and the fitness function of the WMN-CS intelligent simulation system.

In the WMN-CS, solutions are called nests and selected solutions are called eggs. Firstly, a random process creates the initial solution. Cuckoos try to lay their eggs on the most appropriate nest so that they move following Lévy's flight to find a better nest to lay eggs. Once the cuckoo identifies a suitable nest for laying the egg, it replaces the host bird's eggs with its own. This intelligent system imitates the behaviour of the cuckoo bird, which uses a Lévy flight pattern to find a more suitable nest for laying eggs. WMN-CS evaluates the newly found solutions using the fitness function, which quantitatively measures the quality of nests found by cuckoo search. We show the flowchart of WMN-CS in Fig. 1.

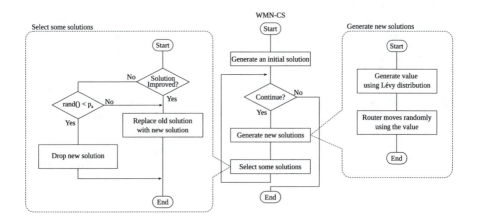

**Fig. 1.** WMN-CS flowchart.

The WMN-CS can evaluate the founded solution by using fitness value calculated as follows:

$$\text{Fitness} = \alpha \times \text{SGC}(x, y, r) + \beta \times \text{NCMC}(x, y, r).$$

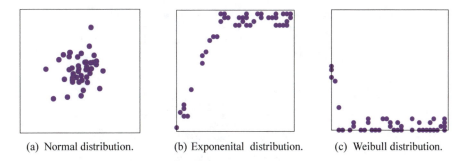

(a) Normal distribution.    (b) Exponenital distribution.    (c) Weibull distribution.

**Fig. 2.** Examples for distribution of mesh clients.

For the fitness function, we consider SGC and NCMC metrics. The SGC measures the connectivity of WMN, while the NCMC measures the client coverage. The weight coefficients $\alpha$ and $\beta$ are used in order to balance the effects of SGC and NCMC.

By WMN-CS, we can generate various distributions of mesh clients. For this research work, as shown in Fig. 2, we consider Normal, Exponential and Weibull distributions of mesh clients.

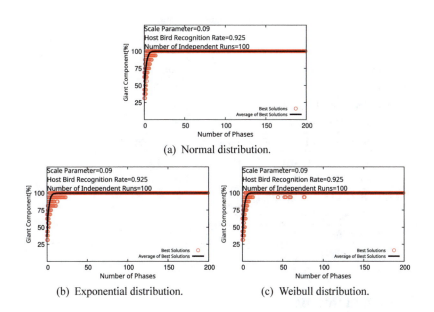

(a) Normal distribution.

(b) Exponential distribution.    (c) Weibull distribution.

**Fig. 3.** Simulation results of SGC for different distributions of mesh clients.

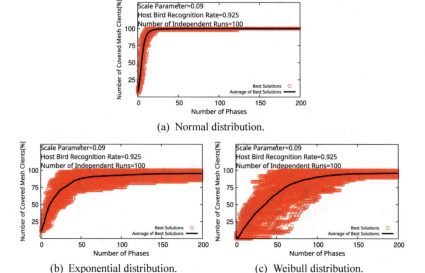

(a) Normal distribution.

(b) Exponential distribution.　　　(c) Weibull distribution.

**Fig. 4.** Simulation results of NCMC for different distributions of mesh clients.

**Table 1.** Parameter settings.

| Parameters | Values |
| --- | --- |
| Clients Distribution | Normal, Exponenital, Weibull |
| Area Size | $32 \times 32$ |
| Number of Mesh Routers | 16 |
| Number of Mesh Clients | 48 |
| Communication distance of a Mesh Router | From 2.0 to 3.0 |
| Fitness Function Weight-coefficients $(\alpha, \beta)$ | 0.7, 0.3 |
| Number of Nests | 70 |
| Host Bird Recognition Rate $(p_a)$ | 0.925 |
| Scale Parameter $(\gamma)$ | 0.09 |
| Total Iterations | 2000 |
| Iteration per Phase | 10 |

## 6　Simulation Results

We show the simulation parameters in Table 1. For simulations, we consider Normal, Exponential and Weibull distribution of mesh clients. We assume a small scale WMN so that the area size is 32 units by 32 units, and the number of mesh routers and mesh clients is 16 and 48, respectively.

　　We show the performance of SGC in Fig. 3. The SGC reaches the maximum value for all distributions of mesh clients. However, as shown in Fig. 4, for NCMC metric the

WMN-CS can reach the maximum value only for Normal distribution of mesh clients. So, the Normal distribution has the best performance compared with Exponential and Weibull distributions. Comparing Exponential and Weibull distributions, the convergence of Weibull distribution is slower than Exponential distribution, but the performance at 200 phase for Weibull distribution is better than Exponential distribution.

As shown in Fig. 3(a) and Fig. 4(a), the WMN-CS can find suitable mesh router locations for 30 phases in case of Normal distribution of mesh clients. However, as shown in Fig. 4(b) and Fig. 4(c), 200 phases are insufficient for WMN-CS convergence in case of Exponential and Weibull distributions.

## 7    Conclusions

In this reserach work, we evaluated the performance of WMN-CS simulation system considering Normal, Exponential and Weibull distributions of mesh clients. The simulation results show that the WMN-CS can find suitable mesh router locations for 30 phases in case of Normal distribution of mesh clients. However, 200 phases are insufficient for WMN-CS convergence in case of Exponential and Weibull distributions.

In our future work, we will consider other mesh client distributions and parameters. Also, we would like to compare the performance of the proposed system with other intelligent systems.

## References

1. Ahmed, A.M., Hashim, A.H.A.: Metaheuristic approaches for gateway placement optimization in wireless mesh networks: a survey. Int. J. Comput. Sci. Netw. Secur. (IJCSNS) **14**(12), 1 (2014)
2. Amaldi, E., Capone, A., Cesana, M., Filippini, I., Malucelli, F.: Optimization models and methods for planning wireless mesh networks. Comput. Netw. **52**(11), 2159–2171 (2008)
3. Asakura, K., Sakamoto, S.: A cuckoo search based simulation system for node placement problem in wireless mesh networks. In: Barolli, L. (ed.) CISIS 2023. LNCS, vol. 176, pp. 179–187. Springer, Heidelberg (2023). https://doi.org/10.1007/978-3-031-35734-3_18
4. Barolli, A., Bylykbashi, K., Qafzezi, E., Sakamoto, S., Barolli, L., Takizawa, M.: A comparison study of chi-square and uniform distributions of mesh clients for different router replacement methods using wmn-psodga hybrid intelligent simulation system. J. High Speed Netw. **27**(4), 319–334 (2021)
5. Basirati, M., Akbari Jokar, M.R., Hassannayebi, E.: Bi-objective optimization approaches to many-to-many hub location routing with distance balancing and hard time window. Neural Comput. Appl. **32**, 13,267-13,288 (2020)
6. Coelho, P.H.G., do Amaral, J.F., Guimaraes, K., Bentes, M.C.: Layout of routers in mesh networks with evolutionary techniques. In: The 21st International Conference on Enterprise Information System (ICEIS-2019), pp. 438–445 (2019)
7. Elmazi, D., Oda, T., Sakamoto, S., Spaho, E., Barolli, L., Xhafa, F.: Friedman test for analysing WMNS: a comparison study for genetic algorithms and simulated annealing. In: 2015 9th International Conference on Innovative Mobile and Internet Services in Ubiquitous Computing, pp. 171–178. IEEE (2015)
8. Gharehchopogh, F.S., Shayanfar, H., Gholizadeh, H.: A comprehensive survey on symbiotic organisms search algorithms. Artif. Intell. Rev. **53**, 2265–2312 (2020)

9. Lee, S.C., Tan, S.W., Wong, E., Lee, K.L., Lim, C.: Survivability evaluation of optimum network node placement in a hybrid fiber-wireless access network. In: IEEE Photonic Society 24th Annual Meeting, pp 298–299. IEEE (2011)

10. Lin, C.C.: Dynamic router node placement in wireless mesh networks: a PSO approach with constriction coefficient and its convergence analysis. Inf. Sci. **232**, 294–308 (2013)

11. Oda, T., Elmazi, D., Barolli, A., Sakamoto, S., Barolli, L., Xhafa, F.: A genetic algorithm-based system for wireless mesh networks: analysis of system data considering different routing protocols and architectures. Soft. Comput. **20**, 2627–2640 (2016)

12. Qiu, L., Bahl, P., Rao, A., Zhou, L.: Troubleshooting wireless mesh networks. ACM SIG-COMM Comput. Commun. Rev. **36**(5), 17–28 (2006)

13. Sakamoto, S.: A hybrid intelligent system for wireless mesh networks: assessment of implemented system for two instances and three router replacement methods using v max parameter. Int. J. Web Grid Serv. **19**(3), 389–400 (2023)

14. Sakamoto, S., Obukata, R., Oda, T., Barolli, L., Ikeda, M.: Implementation of an intelligent hybrid simulation system for node placement problem in wmns considering particle swarm optimization and simulated annealing. In: The 31st IEEE International Conference on Advanced Information Networking and Applications (AINA-2017), pp. 697–703. IEEE (2017)

15. Sakamoto, S., Asakura, K., Barolli, L., Takizawa, M.: An intelligent system based on cuckoo search for node placement problem in WMNs: tuning of scale and host bird recognition rate hyperparameters. In: Barolli, L. (ed.) BWCCA 2023. LNCS, vol. 186, pp. 168–177. Springer, Heidelberg (2023). https://doi.org/10.1007/978-3-031-46784-4_15

16. Sanni, M.L., Hashim, A.H.A., Anwar, F., Naji, A.W., Ahmed, G.S.: Gateway placement optimisation problem for mobile multicast design in wireless mesh networks. In: 2012 International Conference on Computer and Communication Engineering (ICCCE), pp 446–451. IEEE (2012)

17. Seetha, S., Anand John Francis, S., Grace Mary Kanaga, E.: Optimal placement techniques of mesh router nodes in wireless mesh networks. In: Haldorai, A., Ramu, A., Mohanram, S., Chen, M.-Y. (eds.) 2nd EAI International Conference on Big Data Innovation for Sustainable Cognitive Computing. EICC, pp. 217–226. Springer, Cham (2021). https://doi.org/10.1007/978-3-030-47560-4_17

18. Taleb, S.M., Meraihi, Y., Gabis, A.B., Mirjalili, S., Ramdane-Cherif, A.: Nodes placement in wireless mesh networks using optimization approaches: a survey. Neural Comput. Appl. **34**(7), 5283–5319 (2022)

19. Yang, X.S.: Nature-Inspired Metaheuristic Algorithms. Luniver press (2010)

# Performance Evaluation of FC-RDVM and LDIWM Router Replacement Methods for Small and Middle Scale WMNs Considering UNDX-M Crossover Method and Stadium Distribution of Mesh Clients

Admir Barolli[1], Shinji Sakamoto[2], Leonard Barolli[3(✉)], and Makoto Takizawa[4]

[1] Department of Information Technology, Aleksander Moisiu University of Durres, L.1, Rruga e Currilave, Durres, Albania
`admirbarolli@uamd.edu.al`

[2] Department of Information and Computer Science, Kanazawa Institute of Technology, 7-1 Ohgigaoka Nonoichi, Ishikawa 921-8501, Japan
`shinji.sakamoto@ieee.org`

[3] Department of Information and Communication Engineering, Fukuoka Institute of Technology, 3-30-1 Wajiro-Higashi, Higashi-Ku, Fukuoka 811-0295, Japan
`barolli@fit.ac.jp`

[4] Department of Advanced Sciences, Faculty of Science and Engineering, Hosei University, 3-7-2, Kajino-machi, Koganei-shi, Tokyo 184-8584, Japan
`makoto.takizawa@computer.org`

**Abstract.** In this paper, we present a comparison study of Linearly Decreasing Inertia Weight Method (LDIWM) and Fast Convergence Rational Decrement of Vmax Method (FC-RDVM) for optimization of mesh routers in Wireless Mesh Networks (WMNs). We consider UNDX-m crossover method and Stadium distribution of mesh clients. Then, we perform simulations for small and middle scale WMNs. In case of both scales of WMNs, for both methods (LDIWM and FC-RDVM), all mesh routers are connected. Considering NCMC, for small scale WMNs, in case of LDIWM all mesh clients are covered, while for FC-RDVM four mesh clients are not covered. However, for middle scale WMNs and both methods, all mesh clients are covered. While, considering load balancing, for both scales of WMNs, LDIWM has a better loading balancing than FC-RDVM.

# 1 Introduction

Networks of today are going through rapid evolution and they are becoming very complex. Therefore, the optmization of these networks is very important. However, for optimization process are needed different parameters which make the problem NP-Hard. Therefore, it is needed a trade-off among these parameters.

The optimization process finds optimal values for specific parameters of a given system in order to fulfill all design requirements. Optimization problems can be found in all fields of science. The conventional optimization algorithms such as single-based solution can converge to local optimum and unknown search space issues. In order to deal with these limitations, many researchers have developed several metaheuristics to address complex/unsolved optimization problems.

During last years are appearing many applications of Wireless Mesh Networks (WMNs). They have many advantages such as scalability and cost-effectiveness. However, they have some issues which are related with the placement of mesh routers in a considered area. In order to deal with this problem are needed many parameters for optimization, which makes the problem NP-Hard. Therefore, many research works consider intelligent and meta-heuristic approaches to find solutions for NP-Hard problems.

In some previous research works, some researchers deal with the mesh node placement in WMNs [2,4–6,12] and some intelligent algorithms are investigated [1,3,7,8]. In our previous work [9,10], we implemented intelligtent simulation systems by considering single intelligent algorithms based on Particle Swarm Optimization (PSO), Hill Climbing (HC), Genetic Algorithms (GAs), Simulated Annealing (SA) and so on. Then, we designed and implemented a hybrid simulation system based on PSO and Distributed GA (DGA) called WMN-PSODGA.

In this paper, we present the implementation of WMN-PSOHCDGA hybrid simulation system by integrating three intelligent algorithms: PSO, HC and DGA. In WMN-PSOHCDGA, we implemented Linearly Decreasing Inertia Weight Method (LDIWM) and Fast Convergence Rational Decrement of Vmax Method (FC-RDVM) replacement methods. For the optimization of mesh routers in WMNs, we take into account three parameters: SGC (Size of Giant Component), NCMC (Number of Covered Mesh Clients), and NCMCpR (Number of Covered Mesh Clients per Router). We compare the results of LDIWM and FC-RDVM considering Stadium distribution of mesh clients and UNDX-m crossover method for small and middle scales WMNs. In case of both scales of WMNs, for both methods (LDIWM and FC-RDVM), all mesh routers are connected. Considering NCMC, for small scale WMNs, in case of LDIWM all mesh clients are covered, while for FC-RDVM four mesh clients are not covered. However, for middle scale WMNs and both methods, all mesh clients are covered. While, considering load balancing, for both scales of WMNs, LDIWM has a better loading balancing than FC-RDVM.

The rest of the paper is organized as follows. After the introduction, in Sect. 2, we brifely decribe PSO, HC and DGA algorithms. In Sect. 3 is presented WMN-PSOHCDGA hybrid simulation system. The simulation results are given in Sect. 4. Finally, we give conclusions and future work in Sect. 5.

# 2   PSO, HC and DGA Intelligent Algorithms

In this section, we give a short description of of PSO, HC and DGA algorithms, which are combined to implement our WMN-PSOHCDGA hybrid intelligent system.

## 2.1   PSO Algorithm

In PSO, a flock of fishes or birds that moves in a group can profit from collaborative behaviour. For instance, when a bird flying and searching randomly for food, all birds in the flock can share their discovery and help the entire flock get the best hunt.

We can simulate the movement of a flock of birds in order to find the optimal solution in a high-dimensional space. This is a heuristic solution, which may be is not the global optimal solution. But, the solution found by PSO is very close to the global optimal solution.

The PSO is a powerful meta-heuristic optimization algorithm and can be simulated as a simplified social system by graphically simulating the graceful but unpredictable choreography of a bird flock. We mathematically can model this behaviour by making the swarm finds the global minimum of the fitness function.

The PSO algorithm has these advantages. It is insensitive to scaling of design variables and is derivative free. The algoritm uses few parameters and can be easily parallelized for concurrent processing. However, the algorith has a weak local search ability.

## 2.2   HC Algorithm

Th HC algorithm is a simple local search algorithm. At the beginning, the algorithm starts with an initial solution and then iteratively makes small changes to improve the solution. These changes continue until a local maximum is achieved or there are no further improvement.

There are several variations of HC algorithm and it can be used in different optimization problems. The HC algoritm is simple and has an easy implementation. Also, it is efficient and can be easily modified by including additional heuristics or constraints. However, it can get stuck in local optima and is sensitive to the choice of initial solution. Also, it cannot explore the search space efficiently, which can limit its ability to find better solutions.

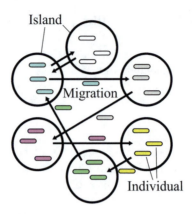

**Fig. 1.** Model of Migration in DGA.

### 2.3    GA and Distributed GA

The GAs are adaptive heuristic search algorithms, which are part of evolutionary algorithms. In GAs is used the concept of natural selection and genetics. For optimization and search problems, they can generate high-quality solutions. In natural selection process, individuals adapting to changes in their environment are able to survive and reproduce offsprings. Each individual represents a point in search space can be represented by an individual, which is a possible solution. The individuals can be represented by a string of bits, which are analogous to chromosomes.

In our hybrid simulation system, we considered DGA, which consider multiple islands to escape from local minimum. As shown in Fig. 1, each island computes its own solution and migrates to other islands.

In this paper, we consider UNDX crossover method, which is shown in Fig. 2. We implement in our simulation system UNDX-m, where m+2 parents are selected, then the first m+1 parents are used to span the m-dimensional subspace. For distribution of children is used normal distribution. Then, in order to give perturbation to the children to remainin in the sub-space a third parent is used. When $m$ is one (UNDX-1), the method is the same with original UNDX.

## 3    Structure of WMN-PSOHCDGA Simulation System

In this section, we present the WMN-PSOHCDGA hybrid simulation system, which combines PSO, HC and DGA algorithms by considering their advantages. The flowchart of our system is shown in Fig. 3.

In WMN-PSOHCDGA system, the initial solution are generated randomly by *ad hoc* methods [13]. Then, it carries out the optimization process for each island. After that the solutions are migrated (swapped) between islands to find a better solution. If the stopping criteria is not acchieved the optimization process is continuing, otherwise the best solution is found.

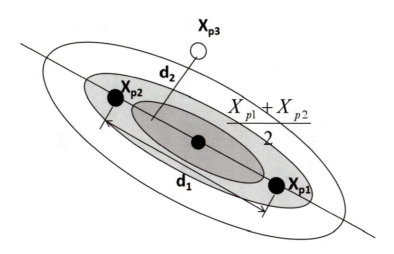

**Fig. 2.** UNDX

In the evaluation of each islands, we use DGA algorithms. In the case when the solution from DGA part is found, then it goes to PSO part, which updates the velocities and positions of particles by both router replacement methods. In order to improve the PSO algorithm convergence is utilized HC algorithm.

We show in Fig. 4 the solution for each particle-pattern. A WMN is represented by a gene and each individual in the population is a combination of mesh routers.

The fitness function has the following form.

$$Fitness = \alpha \times SGC(\boldsymbol{x}_{ij}, \boldsymbol{y}_{ij}) + \beta \times NCMC(\boldsymbol{x}_{ij}, \boldsymbol{y}_{ij}) \tag{1}$$
$$+\gamma \times NCMCpR(\boldsymbol{x}_{ij}, \boldsymbol{y}_{ij}).$$

In this work, we consider Stadium distribution of mesh clients, which considers users located as shown in Fig. 5 and Fig. 6.

There are many mesh router replacing methods for WMNs. In this paper, we consider LDIWM and FC-RDVM.

In LDIWM, the values of $C_1$ and $C_2$ are set the same as RIWM. While, $\omega$ parameter values are changed linearly from 0.9 to 0.4 (from unstable region to stable region).

In FC-RDVM [11], the $V_{max}$ value decreases with the increase of $T$, $k$ and $\delta$ values as shown in Eq. (2).

$$V_{max}(k) = \sqrt{W^2 + H^2} \times \frac{T - k}{T + \delta k} \tag{2}$$

In Eq. (2), the variables are the same with other methods, while $\delta$ is a curvature parameter.

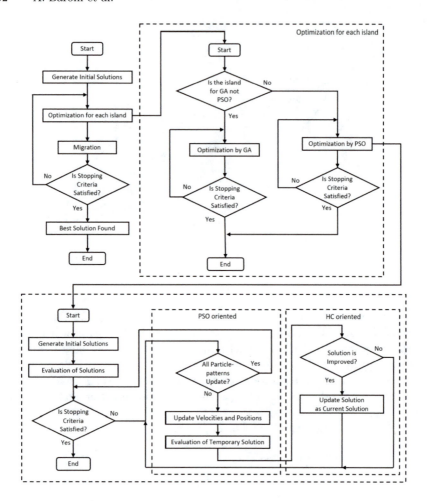

**Fig. 3.** Flowchart of WMN-PSOHCDGA hybrid simulation system.

## 4    Simulation Results

In this section, we show the simulation results.The simulation parameters are presented in Table 1.

We show the visualization results for small scale WMNs in Fig. 7. In Fig. 7(a) and Fig. 7(b) are shown the SGC results for LDIWM and FC-RDVM, respectively. For both methods (LDIWM and FC-RDVM), all mesh routers are connected, so the SGC is 100%. In Fig. 8 are shown the simulation results of NCMC for each mesh router. Considering Fig. 7 and Fig. 8, for LDIWM all mesh clients are covered, while for FC-RDVM four mesh clients are not covered.

The load balancing indicator for small scale WMNs is shown in Fig. 9. The parameter $r$ is the correlation coefficient. When the standard deviation

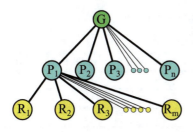

G: Global Solution
P: Particle-pattern
R: Mesh Router
n: Number of Particle-patterns
m: Number of Mesh Routers

**Fig. 4.** Relationship among global solution, particle-patterns, and mesh routers in PSO part.

**Fig. 5.** Three examples of Stadium distribution of mesh clients.

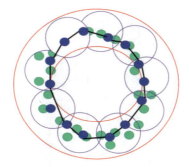

**Fig. 6.** A case of Stadium distribution of mesh clients.

is decreased line, the load balancing among mesh routers is considered better. Thus, the load balancing of LDIWM is better than FC-RDWM.

We show the visualization results for middle scale WMNs in Fig. 10. Also, for middle scale WMNs, for both methods all mesh routers are conneted by maximizing SGC. We show in Fig. 11 the NCMC simulation results for middle scale WMNs. Considering Fig. 10 and Fig. 11, we can see that both methods cover all mesh clients. We show the indicator of load balancing for middle scale WMNs in Fig. 12. Also for middle scale WMNs, the load balancing of LDIWM is better than FC-RDWM.

**Table 1.** Parameters used for simulations.

| Parameters | Values | |
|---|---|---|
| | Small Scale WMN | Middle Scale WMN |
| $\alpha : \beta : \gamma$ | 8 : 1 : 1 | |
| Number of GA Islands | 16 | |
| Number of Evolution Steps | 10 | |
| Number of Migrations | 200 | |
| Number of Mesh Routers | 16 | 32 |
| Number of Mesh Clients | 48 | 96 |
| Mesh Client Distribution | Stadium Distribution | |
| Selection Method | Rulette Selection Method | |
| Corssover Method | UNDX-m | |
| Mutation Method | Uniform Mutation | |

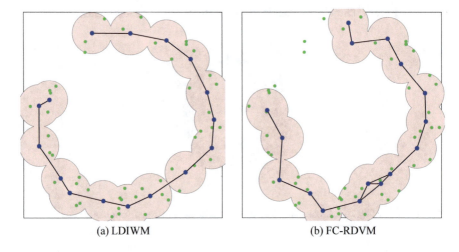

(a) LDIWM                    (b) FC-RDVM

**Fig. 7.** Visualization results after optimization for small scale WMNs.

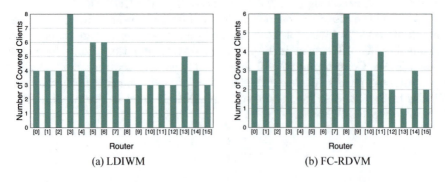

(a) LDIWM                    (b) FC-RDVM

**Fig. 8.** Number of covered mesh clients for small scale WMNs.

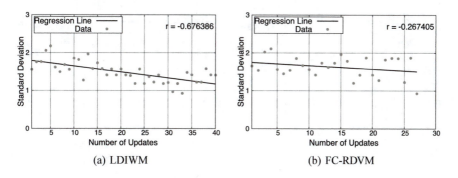

**Fig. 9.** Standard deviation for small scale WMNs.

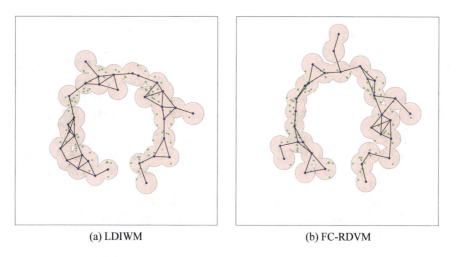

**Fig. 10.** Visualization results after optimization for middle scale WMNs.

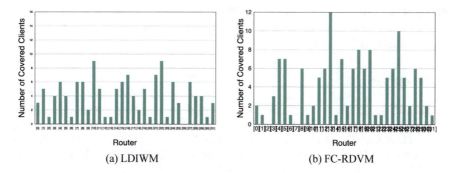

**Fig. 11.** Number of covered mesh clients for middle scale WMNs.

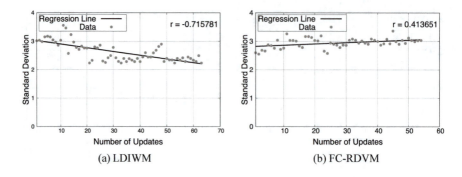

**Fig. 12.** Standard deviation for middle scale WMNs.

## 5  Conclusions

In this work, we evaluated the performance of LDIWM and FC-RDVM for small and middle scale WMNs. We considered Stadium distribution of mesh clients and UNDX-M crossover method, and evaluated both methods by computer simulations. Our conclusions from simulation results are summarized as follows.

- In case of small scale WMNs, for both methods the SGC is maximized. Considering NCMC, for LDIWM all mesh clients are covered, while for FC-RDVM four mesh clients are not covered. The load balancing of LDIWM is better than FC-RDWM.
- In case of middle scale WMNs, for both methods all mesh routers are conneted and all mesh clients covered. Also, the load balacing of LDIWM is better than FC-RDWM.

In future work, we will consider the implementation of different crossover, mutation and other router replacement methods. We will evaluate the performance of the implemented system for different number of phases and migrations. Furthermore, we plan to implement a testbed and compare the simulation results with experimental results.

## References

1. Barolli, A., Sakamoto, S., Ozera, K., Barolli, L., Kulla, E., Takizawa, M.: Design and implementation of a hybrid intelligent system based on particle swarm optimization and distributed genetic algorithm. In: Barolli, L., Xhafa, F., Javaid, N., Spaho, E., Kolici, V. (eds.) Advances in Internet, Data & Web Technologies, pp. 79–93. Springer, Cham (2018). https://doi.org/10.1007/978-3-319-75928-9_7
2. Franklin, A.A., Murthy, C.S.R.: Node placement algorithm for deployment of two-tier wireless mesh networks. In: Proceedings of Global Telecommunications Conference, pp. 4823–4827 (2007)
3. Girgis, M.R., Mahmoud, T.M., Abdullatif, B.A., Rabie, A.M.: Solving the wireless mesh network design problem using genetic algorithm and simulated annealing optimization methods. Int. J. Comput. Appl. **96**(11), 1–10 (2014)

4. Lim, A., Rodrigues, B., Wang, F., Xu, Z.: k-Center problems with minimum coverage. Theoret. Comput. Sci. **332**(1–3), 1–17 (2005)
5. Maolin, T., et al.: Gateways placement in bckbone wireless mesh networks. Int. J. Commun. Netw. Syst. Sci. **2**(1), 44–50 (2009)
6. Muthaiah, S.N., Rosenberg, C.P.: Single gateway placement in wireless mesh networks. In: Proceedings of 8th International IEEE Symposium on Computer Networks, pp. 4754–4759 (2008)
7. Naka, S., Genji, T., Yura, T., Fukuyama, Y.: A hybrid particle swarm optimization for distribution state estimation. IEEE Trans. Power Syst. **18**(1), 60–68 (2003)
8. Sakamoto, S., Kulla, E., Oda, T., Ikeda, M., Barolli, L., Xhafa, F.: A comparison study of simulated annealing and genetic algorithm for node placement problem in wireless mesh networks. J. Mob. Multim. **9**(1–2), 101–110 (2013)
9. Sakamoto, S., Kulla, E., Oda, T., Ikeda, M., Barolli, L., Xhafa, F.: A comparison study of hill climbing, simulated annealing and genetic algorithm for node placement problem in WMNs. J. High Speed Netw. **20**(1), 55–66 (2014)
10. Sakamoto, S., Oda, T., Ikeda, M., Barolli, L., Xhafa, F.: Implementation and evaluation of a simulation system based on particle swarm optimisation for node placement problem in wireless mesh networks. Int. J. Commun. Netw. Distrib. Syst. **17**(1), 1–13 (2016)
11. Sakamoto, S., Barolli, A., Liu, Y., Kulla, E., Barolli, L., Takizawa, M.: A fast convergence RDVM for router placement in WMNs: performance comparison of FC-RDVM with RDVM by WMN-PSOHC hybrid intelligent system. In: International Conference on Computational Intelligence in Security for Information Systems Conference, pp 17–25. Springer, Cham (2022). https://doi.org/10.1007/978-3-031-08812-4_3
12. Wang, J., Xie, B., Cai, K., Agrawal, D.P.: Efficient mesh router placement in wireless mesh networks. In: Proceedings of IEEE International Conference on Mobile Adhoc and Sensor Systems (MASS-2007), pp. 1–9 (2007)
13. Xhafa, F., Sanchez, C., Barolli, L.: Ad hoc and neighborhood search methods for placement of mesh routers in wireless mesh networks. In: Proceedings of 29th IEEE International Conference on Distributed Computing Systems Workshops (ICDCS-2009), pp. 400–405 (2009)

# Performance Evaluation of V2X Communication Based on Channel Bonding Wireless Link Method for Autonomous Driving

Yoshitaka Sibata[1,3]([✉]), Yasushi Banshy[2], and Shoichi Noguchi[3]

[1] Sendai Foundation for Applied Science and Iwate Prefectural University, 152-89 Sugo, Takizawa, Iwate, Japan
shibata@iwate-pu.ac.jp
[2] Holonic Systems Ltd., Kami-Hirasawa 7-3, Shiwa, Iwate, Japan
bansyo@holonic-systems.com
[3] Sendai Foundation for Applied Science, 1-5-1 Nisiki-Machi, Aoba-Ku, Sendai, Miyagi, Japan
Noguchi@sfais.or.jp

**Abstract.** In this paper, in order to realize autonomous driving for challenged communication environment such as mountain and snowy areas, a new vehicle to everything (V2X) communication which is based on multi-vendors channel bonding method is introduced. Mobile router using multi-link by various communication vendors with different frequencies are used for data transmission between the mobile routers and between the vehicle and road side unit, eventually stable and high throughput communication states can be attained. In this paper, system design and configuration of V2X Communication system based on multi-link bonding method are introduced and implemented. Prototyping and performance evaluation of the proposed V2X system is made to clarify its functional and performance effects.

## 1 Introduction

As progress of recent mobility communication technology, autonomous driving ability has been progressed from level 2 toward level 3 and 4. In level 1 and 2, the autonomous driving cars usually control their speed and steering angle by on-board sensors such as camera, Li-DAR, millimeter wave radar. In level 3, autonomous driving car is autonomously driven based on on-board sensors but controlled from outside location such as at remote control center where the running states and video images of driving car are always monitored by multiple attached cameras in addition to various sensors in realtime through external communication network such as V2X and mobile network. When the there are any dangerous objects and locations on road, then autopilot system immediately recognizes and reacts to the events and quickly to avoid the incidents and accident. Also the autonomous driving car is manually operated by the operator at remote control center by considering the various camera images and sensor data through V2X communication.

In level 4, there is no operators on board and the car is completely autonomously controlled by AI based computer although the autonomous car runs on the limited road

L. Barolli (Ed.): AINA 2024, LNDECT 199, pp. 198–206, 2024.
https://doi.org/10.1007/978-3-031-57840-3_18

or predetermined line. Furthermore, in level 5, autonomous driving car is completely controlled at free road without any road restrictions. Thus, autonomous driving cars are monitored and controlled through V2X communication.

Thus, in order to realize the autonomous driving in more than revel 3 in wide area of roads basedon various sensors, the V2X performs very important role for safe and reliable and stable driving and requires higher speed, low latency and long distant communication capability.

However, so far, the autonomous driving in mountain and rural areas are not considered because communication environment in those areas is supported by only single communication link such as 3G or LTE. Especially in Japan, since more than 60 percent of land is consisted of mountains and forests, the strength of the mobile communication links in many areas are damped and often disconnected. On the other hand, 5G and local 5G networks can provide higher, low latency. However, it is very difficult to widely cover the wide mountain and rural because the network characteristics and communication distance in those area are interrupted its communication deployment and very limited. Furthermore, since population decline and aging society of those areas are getting serious and worse, autonomous driving platform are indispensable as public transportation for those old residents.

In order to realize long distance V2X communication environment for autonomous driving environment in those areas, a new communication system based on bonging wireless link method is expected. In this method, as V2X communication, various cellular vendors including LTEs, Wi-MAX links are multiplexed. By different carrier links, multiple wireless links are multiplexed. As advantages of this method are as follows;

- Since multiple carrier's links are multiplexed, the total communication bandwidth can increase as the number of carrier link is increased.
- Since each carrier's LTEs widely cover the wide area, long distance communication for V2X can be easily realized.
- Since VPN network can be easily constructed, more secure V2X communication can be realized.
- Since different vendor's links are applied multiplexed, traffic validation on different locations can be averaged, eventually, more stable and high throughput can be provided.
- Depending on the development of each mobile cellular network vendors, different category of links including LTE, 5G, Wi-MAX with different frequency cannels can be multiplexed.
- By pre-measuring the carrier's link characteristics, the optimal combination of links including uplink and downlink can be selected.

On the other hand, as disadvantages of proposed method, the decision of the optimal combination of multiple links is difficult. In fact, the throughput, latency of the total links depends on the locations of base station. The latency of multiple links is larger than that of single link.

In this paper, system design and configuration of V2X communication system based on multi-link bonding method are introduced and implemented. Performance evaluation of the proposed V2X system is made to clarify its superiority of its functional and performance effects.

In the followings, the related works with V2X communication is introduced in Sect. 2. Target autonomous driving in challenged communication and road environment for snow mountain area is explained in Sect. 3. The proposed V2X communication methods to extend the communication distance for autonomous driving and prototype system of V2X communication of proposed system is shown in Sect. 4. In final, discussion and concluding remarks are summarized in Sect. 5.

## 2  V2X Communications

There are a number of articles concerned with research and standard with V2X communications. Initially Dedicated Short Range Communication (DSRC) is designed for relatively short distance range based on IEEE802.11p, 5.9 GHz in U.S.A and Europe. Article [1–3] show that IEEE802.11p can satisfy the moderate traffic service requirement such as message broadcasting with relatively low latency between vehicle-to-roadside unit in ITS. However, higher traffic services such as video senser data transmission are hard to realize because of highspeed throughput and low latency requirement for long distant range.

In order to overcome the drawback of IEEE802.11p, The IEEE802.11bd standard is designed to extend transmission reliability, throughput, and transmission range with IEEE802.11p. Particularly, C-V2X technology based on cellular network is well defined as direct and network communication models and classified as connectivity options such as Vehicle-to-Vehicle (V2V), Vehicle-to-Infrastructure (V2I), Vehicle-to-Vehicle (V2V) and Vehicle-to- Vehicle-to-Pedestrian (V2P), and standardization status in [4, 5]. The direct communication model provides direct exchange of realtime information between individual devices using PC5 interface as shown in Fig. 1. This mode performs V2X without network assistance as adhoc network and SIM-less communication. On the other hand, in network mode, all of the network nodes including vehicles, roadside units, network infrastructure, pedestrian are communicated each other through carrier network interface as shown in Fig. 2. Using carrier networks such as 3G, LTE and 5G, long range communication over wide area network can be realized. As actual realization of V2X, there are a few implementation cases for above standards. Long distance communication with high throughput bit rate is difficult to support ITS or autonomous driving. Only the cases of V2R and V2I are prototyped for performance evaluation.

In our previous research [6–8], in order to overcome the problems of conventional V2X communication, N-wavelength wireless cognitive network is introduced to support long distance communication for small control message by long wavelength wireless LAN and short distance communication for large data by short wavelength wireless LAN. Using this method, communication distance can be attained more than 2 km and total throughput is 250 Mbps between vehicle and roadside at 50 km/h. It is desired to extend the communication distance to cover whole typical city and town.

As extended the communication distance, hand-over methods are proposed to support long range of V2R between the vehicle and road side unit or remote control center [9, 10]. In those researches, communication distance can be improved but the connectivity is tend to blockages because of its inability to penetrate through the network nodes.

In order to resolve those problem, a new communication system based on multi-link bonding method is introduced. In this method, as V2X communication, various cellular carrier's SIMs including LTE, Wi-MAX and 5G links are multiplexed. By different links. Multiple mobile links are multiplexed. The details of the proposed multi-link bonding method is described in Sect. 4.

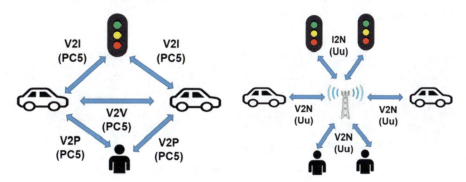

**Fig. 1.** Overview of C-V2X PC5 links      **Fig. 2.** Overview of C-V2X Uu links

## 3   Objective Autonomous Driving System

Figure 3 shows a typical EV driving vehicle and its basic control functions. The vehicle runs by own battery energy charged on Li-Ion battery on the electric magnet lines which were installed under several ten cm depth from the ground. Thus, the EV car can keep driving on this road by detecting the magnetic force line by various guide sensors. The edge computer in EV control system analyzes the position of guide line and drive on the preset collect route. When the EV drives on the magnet sensor, voltage deviations are generated on magnet sensor, the edge computer can analyze this voltage and control the action of EV. It is also prepared override function by which the EV can automatically change from automatic operation to manual and manually switch back to automatic operation to keep safety of autonomous driving. Furthermore, the obstacle detection function is also existed. By memorizing the driving space state of the own vehicle beforehand using 3D stereo vision camera, the objects on the running road can be identified as obstacle, otherwise the objects are not regarded as obstacle such as the stone on the pavement.

Various sensors including dynamic accelerator, gyro sensor, infrared temperature sensor, humidity sensor, quasi electrical static sensor, camera and GPS measure the time series physical sensor data. Then, those sensor data are processed by the road surface decision unit (Machine Learning) and the current road state can be identified in realtime. Next, those road state data are input to the ECU to calculate the amount of breaking and steering, and sent to the braking and steering components to optimally control the speed and direction of the EV. This close loop of the measuring EV speed and direction, sensing road data, deciding road state, computing and controlling braking/steering processes is repeated within a several msec.

On the other hand, those road state data also transmitted to the road state server in cloud computing system through the edge computing by V2Xcommunication protocol and processed to organized wide the road state GIS platform. Those data are distributed to all of the running EVs to know the head road state of the current location. From the received the head state of the current location, the EV can look ahead road state and predict proper target set values of speed and direction of the EV. Thus, by combining the control of both the current and feature speed and direction of EV, more correct and safer automotive driving can be attained (Fig. 4).

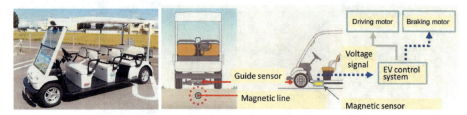

**Fig. 3.** EV Autonomous Driving system

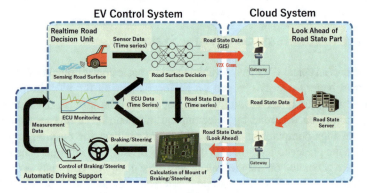

**Fig. 4.** Automatic EV Control System

## 4   V2X Communication and Prototype System

Figure 5 shows system configuration of V2X communication based on multi-link bonding network toward the level 4. Currently level 4 functions are not implemented but is expected to extend after confirming those V2X communication functions have correctly performed during recent test phase. In this system, the multi-link router at the autonomous driving vehicle performs a wireless multi-link function by bonding multiple carrier links with SIMs. The LTEs and Wi-MAX by various communication carriers including docomo, au, softbank and UQ Wi-Max are multiplexed to make a VPN between the multi-link router at the autonomous EV and the bonding router at remote control center through Internet environment.

By introducing this multi-link bonding with different carriers, the total communication bandwidth can increase as the number of SIM is increased. Since each carrier's LTEs widely cover a wide area, long distance communication for V2X can be easily realized.

Also, since VPN network can be easily constructed, more secure V2X communication can be realized. By applying multiplexed different vendors links, traffic validation on different locations can be averaged and as result, more stable throughput can be provided, and by pre-measuring the link characteristics, the optimal combination of links including uplink and downlink can be selected.

As prototype system, total 8 cameras are attached to front, rear, both sides and internal room to seamlessly monitor the surround states of autonomous EV for maintaining safety and reliable driving. HD quality (1280 × 720, 2.4~4.5 Mbps) sized images from 8camera are captured by 30 f/s and encoded to H.264 video format. The coded 8 video streams are aggregated at the multi-link router as a VPN and transmitted to the objective bonding router in the remote control center through Internet. On VPN based multi-link router, the 8 video streams are packetized and sent through each LTEs and Wi-MAX links. Thus, the packetized 8 video stream are distributed to the available links to average the traffic on all of links. The received packets at the destination router are sorted on each packet video stream and finally displayed on the monitor. Those video streams are also stored in video storage and control server.

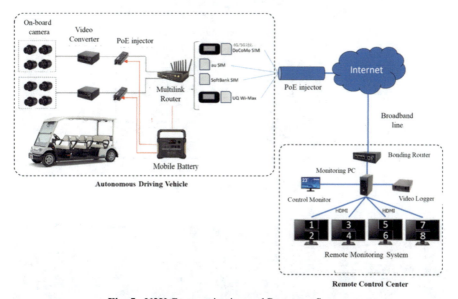

**Fig. 5.** V2X Communication and Prototype System

The autonomous vehicle driving area, Kami-Koami-Mura which is located on a rural mountain area in Northern part of Japan as shown in Fig. 6. There is a large amount of snow and icy in winter and communication environment is challenging. Currently the autonomous vehicle is taken as one of public transportation for elder residents every day. Now autonomous vehicle runs by Level 2 but expected to extend to level 4. For this reason, V2x communication facility is introduced and evaluated its functional and performance. Figure 7 shows 8 video stream and control monitor image. Each video images are individually displayed for operator to easily identify the states of autonomous vehicle and road surface. The each throughput of the carrier LTE and Wi-MAX varies depending on the location at Received signal strength intensity. In this measurement as shown in Fig. 8, the throughput is that au is the largest, softbank is the second, and docomo is the third. However, the total aggregated throughput is more than 50 Mbps and can satisfy the requirement with the around 50 Mbps. The frame rates of those 8 video streams are constantly 30 f/s throughput even though each link varies and unstable. This is due to averaging the total throughput by bonging function of multi-link router. Thus, by bonding multiple carrier's links, a larger and stable throughput can gain to satisfy the required throughput.

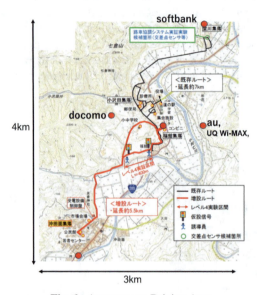

**Fig. 6.** Autonomous Driving Area

**Fig. 7.** Monitored Video Streams Surrounding Autonomous Vehicle

**Fig. 8.** Performance Indication of Multi-link Bonging Router

## 5   Conclusions

In this paper, in order to realize autonomous driving for challenged communication environment such as mountain and snowy areas, a new vehicle to everything (V2X) communication which is based on multi-vendors channel bonding method is introduced. Mobile router using multi-channel by various communication vendor with different frequencies are used for data transmission between the mobile routers and between the vehicle and road side unit, eventually stable communication states can be attained. In this paper, system design and configuration of V2X Communication system based on channel bonding method are introduced and implemented. Prototyping and performance evaluation of the proposed V2X system is made to clarify its functional and performance effects. Currently more detailed functional and performance in this prototype are analyzed by comparing the single LTE, Wi-Fi network by changing the number of LTE links and selecting the best combination of multiple-links. The latency performance of the proposed method is also investigated to realize the autonomous driving in level 4 in the near fure.

**Acknowledgments.** The research was supported by Sendai Foundation for Applied Science, by Japan Keiba Association Grant Numbers 2021M-198, JSPS KAKENHI Grant Numbers JP 20K11773, Strategic Information and Communications R&D Promotion Program Grant Number 181502003 by Ministry of Affairs.

## References

1. Hassan, M.I., Vu, H.L., Sakurai, T.: Performance analysis of the IEEE802.11 MAC protocol DSRC safety application. IEEE Trans. Veh. Technol. Veh. Technol. **60**(8), 3882–3896 (2011)

2. Hameed Mir, Z., Filali, F.: LTE and IEEE 802.11p for vehicular networking: a performance evaluation. EURASIP J. Wireless Commun. Netw. **2014**(1), 1–15 (2014). https://doi.org/10.1186/1687-1499-2014-89

3. Specification for telecommunication and Information Exchange Between Roadside and Vehicle Systems – 5 GHz Band Dedicated Short-Range Communication (DSRC), Medium Access Control (MAC) and Physical Layer (PHY) specification, ASTM Standard E2213–03 (2018)

4. Weber, R., Misener, J., Park, V.: "C-V2X – A Communication Technology for Cooperative, Connected and Automated Mobility", Mobile Communication - Technologies and Applications; 24, pp. 111–116. ITG-Symposium, VDE VERLAG GMBH (2019)

5. Torgunakov, V., Loginov, V., Khorov, E.: A study of channel bonding in IEEE 802.11bd networks. IEEE Access **10**, 25514–25533 (2022)

6. Chang, Y.H., Liu, H.H., Wei, H.Y.: Group-based sidelink communication for seamless vehicular handover. IEEE Access **7**, 56431–56442 (2019)

7. Hussain, S.M., Yusof, K.M., Hussain, S.A., Asuncion, R.: Performance evaluation of vertical handover in internet of vehicles. Int. J. Smart Sens. Intell. Syst. **14**(1), 1–16 (2021)

8. Sakuraba, A., Shibata, Y., Tamura, T.: Evaluation of performance on lpwa network realizes for multi-wavelength cognitive V2X wireless system. In: The 10th IEEE International Conference on Awareness Science and Technology, (iCAST2019), pp. 434–440, Morioka, Japan, Nov. (2019)

9. Sakuraba, A., Shibata, Y., Sato, G., Uchida, N.: Performance evaluation of 2-wavelength cognitive wireless network for V2R and V2V communication. Int. J. Mob. Comput. Multimedia Commun. **11**(4), 84–101 (2020)

10. Shibata, Y., Sakuraba, A., Sato, G., Uchida, N.: Performance evaluation of V2V and V2R communication based on 2-wavelength Cognitive Wireless Network on Road State Information GIS Platform. In: The 11th International Conference on Intelligent Networking and Collaborative Systems, (INCoS-2019), pp. 212–222, Sept. (2019)

# New Strategy to Reduce Communication Interference of Platoon

Anis Boubakri[✉] and Hajer Boujezza

ISP Ted University, Tunis, Tunisia
{aboubakri,hboujezza}@ted-university.com

**Abstract.** The internet of things allows to have the comfort of these users. The number of connected objects is increasing exponentially. There is therefore a risk of degrading the quality of comfort by the communication phenomena that are due to interference phenomena. Autonomous connected vehicles are a special case of internet of thing. As autonomous vehicles are sensitive to the time parameter. Therefore, interference degrades the performance of autonomous driving systems. In this paper, we present a Platoons architecture to reduce the interference throughout the traffic.

**Keywords:** Autonomous vehicles · Platoon · Interference · Managment

## 1 Introduction

Autonomous vehicles (AV) [1], have become a reality to ensure the comfort and safety of road users. In order to ensure autonomous driving [2, 3], it is necessary that each AV has an idea about its traffic environment (neighbor, obstacles, signs and so on) [4, 5]. Subsequently, each driver must localize with high accuracy [6–8] in his traffic environment, in the form of a kinematic model [9]. In order to improve the performance of autonomous driving systems, the model will be shared with neighboring AVs. Connected vehicles [1] are a good candidate for sharing data between vehicles. The communication between vehicles is done using V2V connectivity [10], using either Dedicated short-range communications (DSRC) technology [11] based on the IEEE802.11p standard [12] or by using the C-V2X cellular technology [13].

The AVs circulate in the form of Platoons [10, 14]. A Platoon is a set of vehicles that follow each other (form a train), which want to travel together voluntarily or not. A leader vehicle that manages the communication between vehicles of a Platoon allows to travel with a small distance between the vehicles, subsequently the traffic will be increased and reduces energy consumption [15, 16]. Intra-platoon communication [10] is a key task for autonomous driving, but this communication is not sufficient. In order to have a reliable autonomous driving, it is necessary to study also the inter Platoon communication.

The challenges of inter-platoon communication, in addition to the sharing of radio resources between Platoons, include the phenomenon of interference [17, 18], which consists of the reuse of the same radio resources at the same time. Therefore, the objective

L. Barolli (Ed.): AINA 2024, LNDECT 199, pp. 207–215, 2024.
https://doi.org/10.1007/978-3-031-57840-3_19

of communication management is to allocate a communication channel for each Platoons and not to reuse this frequency for the adjacent Platoons. On a large scale, the sharing of radio resources in terms of time and frequency becomes very complicated, and there is a risk of having the phenomenon of interference, which frequently degrades the stability of the strings.

In this paper, we propose a solution to reduce transmission interference in Platoons. To minimize the interference, it is necessary to minimize the number of vehicles that are allowed to send messages between Platoons while ensuring that messages are shared with all vehicles. Our strategy is to merge the adjacent Platoons, where several Platoons that merge have only one leader and one tail. So instead of sending the message $n$ times (each sending leader), a single sending leader, where the phenomenon of interference is reduced, the delay time of transmission from end to end is also reduced.

The problem becomes how to merge the Platoons and how to choose a leader? How to manage the trajectories of each AV, if you want to join another route different from that of leader? These questions will be answered in this article. We will model the probability that each AV will follow the leader or change course with each road deviation.

This paper is structured as follows. In Sect. 2 is described the system modeling and is present our solution to reduce the interference. In Sect. 3, we validate our proposed solution with simulation results. Finally, Sect. 4 concludes the paper.

## 2    System Modeling

In this section, we will describe our proposed solution to reduce interference. We will present a method of managing AVs routes based on game theory.

### 2.1    Solution Description

The platonnage system is the circulation in the form of Platoons, as shown in Fig. 1. If we have a 3-lane highway, in each lane there are one or more Platoons. Each Platoon is composed of a leader vehicle: for the management of Platoon is considered the sharing of radio resources between vehicles inter-Platoons. It also allows to manage the communication between the Platoons. A tail vehicle allows the sending of traffic data to the next Platoon. Each vehicle is equipped with two transmitters/receivers: one for inter-Platoon communication and the other for intra-Platoon communication. Communication is the task of sharing radio resources for transmission. Interference is the phenomenon of reuse of the same radio resource by two different equipments at the same time. If two vehicles are transmitting on the same frequency, there will be interference. Then the message will be distorted and cannot be used. And since time is a key factor for AVs, the interference phenomenon has a negative effect on the rate of dropped messages. And if we make a time sharing of the radio resources between the Platoon, we will have end-to-end delays.

To solve this problem, our solution is to merge the adjacent Platoons, and make them a single Platoon with a single leader vehicle and a single tail vehicle as shown in Fig. 2. So we will not have vehicles using the same radio channel simultaneously. Since the DSRC communication and data broadcasting in its range, therefore the transmission

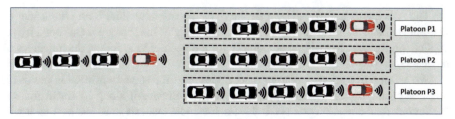

**Fig. 1.** Single way platoons.

**Fig. 2.** n-way platoons.

performance is improved. And the broadcasting of alert messages becomes faster since all vehicles are in the range of the leading vehicle.

Each Platoon vehicle can keep its position either to leave and join another Platoon or to form a new Platoon. To keep the stability of the Platoons it is necessary to keep the vehicles of the Platoon over a great distance.

The AVs are characterized by a strong dynamic of circulation. Being the leader of Platoon we have no advance idea about the trajectory of the Platoon members. In our application we want to assign a probability for the vehicle Vi to follow the trajectory of the leader.

## 2.2 Platoon Management by Game Theories

As the Platoons form a set of AVs, which desire to travel together, in a route, voluntarily or not. So we have to manage the route changes. To solve this problem, our solution is to merge the adjacent Platoons, and make them a single Platoon with a single leader vehicle and a single tail vehicle as shown in Fig. 2. So we will not have vehicles using the same radio channel simultaneously. Since the DSRC communication and data broadcasting in its range, therefore the transmission performance is improved. And the broadcasting of alert messages becomes faster since all vehicles are in the range of the leading vehicle.

Each Platoon vehicle can keep its position either to leave and join another Pla- toon or to form a new Platoon. To keep the stability of the Platoon, it is necessary to keep the vehicles of the Platoon over a large distance.

The AVs are characterized by a strong dynamic of circulation. Being the leader of Platoon we have no advance idea about the trajectory of the Platoon mem- bers. In our application we want to assign a probability for the vehicle Vi to follow the trajectory of the leader.

We assume that all vehicular network nodes have the same transmission radius rc, and that each node is characterized by a path to follow. All along its travel route it encounters road deviation points (RDP). Each vehicle requests in ad- vance that it will change direction. The leader will calculate the time (T) needed for the vehicle to change direction. If not, he plans the change of direction in the next deviation point. If two AVs ask at the same time to change direction, they will be penalized with a delay so that they can ask another time.

In game theory [19], a game consists of a set of players, a set of strategies for each player, and a set of corresponding utility functions. A strategy for a player is a complete plan of actions in all possible situations throughout the game.

We represent our game to manage the Platoon P set by the triplet (J, S, U). The different elements of the set are:

$-$ J $=AV_1, AV_2$. Each player has two possible actions (follow the leader's path or ask to change direction) at each road deviation point. If a player requests the change of direction late he will be penalized, and he cannot change direction. Let Si be the strategy adopted by AVi. Si is represented by the vector of actions of AVi at the level of RDP.

Without constraints, each player has N = 2RDP.

possible trategies.

Si $=$ (a1i, a2i, ..., aRDPi) with, abi $= 1$ means AVi decides to wait for the maxi- mum speed for the change point b; abi $= 0$ otherwise.

$-$ U denotes the set of utility functions that the players wish to maximize. For each player, the utility function Ui is a function of the strategy chosen by the player AVi and the strategies chosen by all other players in the game. The payoff is to change direction at the desired detour point. The reward is the time wasted at each detour point to make the desired change in direction.

$$U : S \rightarrow R$$
$$U = \{U_1, U_2, \cdots U_N\}$$
(1)

Each vehicle tends to maximize its utility function Ui (Si, S-i), where Avi chooses strategy Si and AV$-$i chooses strategy S$-$i.

$$U_i(\text{Si, S} - \text{i}) = \sum_{b=1}^{nbp} \frac{Instant}{T} \cdot D_i$$
(2)

Di: Delay time.

Instant: This is the instant of request for a change of direction.

T: estimated time needed to make the change of direction. T is given by the following equation:

$$T = \frac{v_{moy}}{d} + \sum_{i=0}^{nbr_{AV_{ob}}} t_{dec_{AV}} + t_t + t_{tr} \tag{3}$$

The $v_{moy}$: is the average speed to reach the required deflection point. The d is the distance needed to reach the RDP. While $nbr_{AV_{ob}}$ is the number of vehicles they must decline so that AVi can make the necessary change of direction.

$t_{dec_{AV}}$: Delay required for a AV to decline.

$t_t$: Change of direction request processing delay.

$t_{tr}$: Transmission delay of the change of direction request.

$$U_i = \begin{cases} \sum_{b=1}^{nbp} a_{b_i} \cdot \frac{Instant}{T} \cdot D_i \, if \, T \leq D_{max} \\ -\infty \, else \end{cases} \tag{4}$$

Dmax: estimated time remaining to complete the leader journey.

The flowchart in Fig. 3 summarizes the algorithm for changing the direction of a AV.

## 3  Simulation and Performance

For the validation of our strategy, we carried out a simulation. When a vehicle is started, the first one promises to be a leader and the others are followers. These $n$ vehicles represent a single Platoon P, as shown in Fig. 2. Then each of the vehicles that make up the Platoon must follow the leader's speed, which in turn runs with the dedicated maximum speed of the road. As the vehicles that make up the Platoon do not have the same route. So we have to manage the changes of direction.

Through the use of SUMO simulator for road traffic simulation, and the use of Omnet ++ simulator for communication network management. In Fig. 4, we obtain a comparison, between our solution of Platoon with n voice and classical Platoon, of the rate of messages transmitted successfully according to the number of Platoons in circulation for different number of vehicles in Platoon. Figure 4 shows that the rate of transmission deteriorates sharply, for the case of a single Platoon, when the number of Platoons increases. This degradation is due to interference. While by using our solution (Platoon with n-ways), degradation is very low. This shows that our proposed solution reduces interference phenomena.

As we have already mentioned in Sect. 2, so we must manage the changes of direction of vehicles that wish to join another Platoon. For this we have shown in Fig. 5, the time required to make the change of direction.

Figure 5, shows that the change time is low for 1-way Platoons, while time is important to make the change of direction for n-way Platoons. And this time depended on the position (1-st way, 2-nd way,… n-th way), and on the number of vehicles that make up

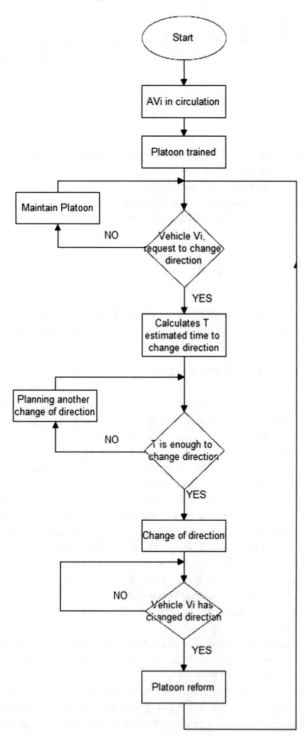

**Fig. 3.** Algorithm for managing the change of direction of a vehicle Vi.

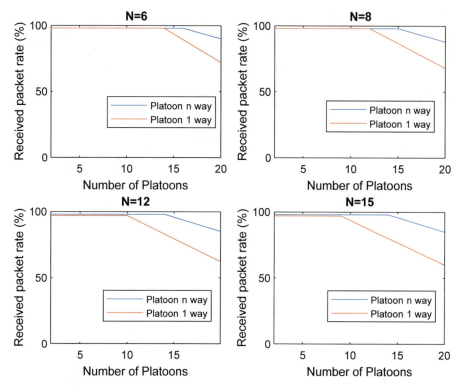

**Fig. 4.** Successful packet transmission rate based on number of Platoons.

the Platoon. To make the change of direction without any risk, it is necessary to ask at least 30 s, for a Platoon of 15 vehicles.

Our proposed solution minimizes interference, but to change direction it is necessary to inform the leader before at least 30 s. While in Platoon's one-way method the time of change of direction is weak, of envirent 6 s. But interference phenomena are common, which reduces the rate of packets transmitted.

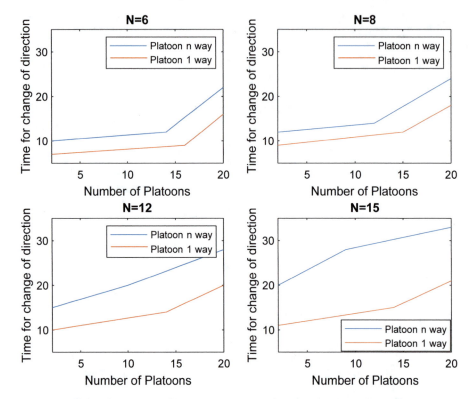

**Fig. 5.** Successful packet transmission rate based on number of Platoons.

## 4  Conclusions

In this paper we have presented a solution that helps to reduce interference during communication. Also we proposed a method based on game theory to manage changes in the direction of an AV in a Platoon. Our proposed solution is validated by simulation results, which shows the improved performance of our solution. This during our solution lacks the adjustment of Platoon when passing from n lanes to (n-1) lanes or (n+1) lanes. In the future work, we will study the change of number of paths using reinforcement learning.

## References

1. Gruyer, D., Magnier, V., Hamdi, K., Claussmann, L., Orfila, O., Rakotonirainy, A.: Perception, information processing and modeling: critical stages for autonomous driving applications. Annu. Rev. Control.. Rev. Control. **44**, 10 (2017)
2. Kwon, J.-W., Chwa, D.: Adaptive bidirectional platoon control using a coupled sliding mode control method. Intell. Transp. Syst. IEEE Trans. **15**, 2040–2048 (2014)
3. Behere, S., Torngren, M., Chen, D.-J.: A reference architecture for cooperative driving. J. Syst. Architect. **59**(10), 1095–1112 (2013)

4. Nunen, E., Koch, R., Elshof, L., Krosse, B.: Sensor safety for the European truck platooning challenge. 11 (2016)
5. Vanholme, B., Gruyer, D., Lusetti, B., Glaser, S., Mammar, S.: Highly automated driving on highways based on legal safety. IEEE Trans. Intell. Transp. Syst.Intell. Transp. Syst. **14**(1), 333–347 (2013)
6. Jiao, J.: Machine learning assisted high-definition map creation. In: 2018 IEEE 42nd Annual Computer Software and Applications Conference (COMPSAC), vol. 1, pp. 367–373 . IEEE (2018)
7. Massow, K., et al.: Deriving HD maps for highly automated driving from vehicular probe data. In: 2016 IEEE 19th International Conference on Intel-ligent Transportation Systems (ITSC), pp. 1745–1752. IEEE (2016)
8. Dabeer, W. et al.: An end-to-end system for crowdsourced 3D maps for autonomous vehicles: The mapping component. In: 2017 IEEE/RSJ International Conference on Intelligent Robots and Systems (IROS), pp. 634–641. IEEE (2017)
9. Naus, G., Vugts, R., Ploeg, J., van de Molengraft, R., Steinbuch, M.: Cooperative adaptive cruise control, design and experiments. In: Proceedings of the 2010 American Control Conference, pp. 6145–6150 (2010)
10. Boubakri, A., Mettali Gammar, S.: Intra-platoon communication in autonomous vehicle: a survey. In: 2020 9th IFIP International Conference on Performance Evaluation and Modeling in Wireless Networks (PEMWN), pp. 1–6 (2020)
11. IEEE 802.11p part11: Wireless LAN medium access control (mac) and physical layer (phy) specifications: Amendment 7: Wireless access in vehicular environment, july (2010)
12. Ucar, S., Ergen, S.C.: Ozkasap, O.: Security vulnerabilities of ieee 802.11p andvisible light communication based platoon. In: 2016 IEEE Vehicular Networking Conference (VNC), pp. 1–4 (2016)
13. Eckermann, F., Kahlert, M., Wietfeld, C.: Performance analysis of c-v2x mode 4 communication introducing an open-source c-v2x simulator. In: 2019 IEEE 90th Vehicular Technology Conference (VTC2019-Fall), pp. 1–5. IEEE (2019)
14. Liu, Y., Yao, D., Li, H., Lu, R.: Distributed cooperative compound tracking control for a platoon of vehicles with adaptive NN. IEEE Trans. Cybern. **52**, 7039–7048 (2021)
15. Boubakri, A., Gammar, S.M.: Speed control for autonomous vehicular in platoon. In: Barolli, L., Woungang, I., Enokido, T. (eds.) Advanced Information Networking and Applications: Proceedings of the 35th International Conference on Advanced Information Networking and Applications (AINA-2021), Volume 3, pp. 395–405. Springer International Publishing, Cham (2021). https://doi.org/10.1007/978-3-030-75078-7_40
16. Boubakri, A., Gammar, S.M.: Cooperative driving system based on artificial intelligence learning. In: 2021 International Wireless Communications and Mobile Computing (IWCMC), pp. 615–620. IEEE (2021)
17. Hong, C., et al.: A joint design of platoon communication and control based on lte-v2v. IEEE Trans. Veh. Technol., 1 (2020)
18. Leng, B., Gu, X., Zhang, L., Zhang, L.: Modeling and analysis of inter-platoon communication with stochastic geometry. In: 2019 22nd International Symposium on Wireless Personal Multimedia Communications (WPMC), pp. 1–5 (2019)
19. Lipovetsky, S., Conklin, M.: Analysis of regression in game theory approach. Appl. Stoch. Model. Bus. Ind.Stoch. Model. Bus. Ind. **17**(4), 319–330 (2001)

# Performance Improvement of DE Algorithm for Indoor Positioning in Wireless Sensor Networks

Shu-Hung Lee[1], Chia-Hsin Cheng[2(✉)], Kuan-Hsien Lu[2], Yeong-Long Shiue[2], and Yung-Fa Huang[3(✉)]

[1] School of Intelligent Manufacturing and Automotive Engineering, Guangdong Business and Technology University, Zhaoqing 526020, Guangdong, China
[2] Department of Electrical Engineering, National Formosa University, Yunlin 632301, Taiwan
{chcheng,syl}@nfu.edu.tw
[3] Department of Information and Communication Engineering, Chaoyang University of Technology, Taichung 413310, Taiwan
yfahuang@cyut.edu.tw

**Abstract.** Target positioning in wireless sensor networks (WSNs) are necessary for real applications. In this research paper, we employ the Differential Evolution (DE) algorithm to enhance the effectiveness of K Nearest Neighbor (KNN) algorithms integrated with receive signal strength indicator (RSSI) for indoor positioning. We examine the performance of random and fixed sensor deployment strategies in both simple and complex environments in relation to target positioning. Specifically, we investigate the impact of reference point numbers, where $K = 4$ for simple environments and $K = 5$ for complex environments. The simulation results demonstrate that setting the K value to 4 yields the highest average correct rates in simple and complex environments, reaching 99.54% and 99.7% respectively. Moreover, the performance improvement between using 5 and 4 reference points is less than 1% in all cases, except for a 2.02% increase observed in the complex environment with random deployment.

## 1 Introduction

Wireless sensor networks (WSNs) have attracted significant attention in various fields due to their ability to establish self-organizing networks using wireless communication. These networks consist of wireless data collectors and numerous sensors, and they find applications in areas such as disaster response, celestial investigation, woodland surveillance, safety setups, and industrial mechanization [1]. Diverging from conventional networks which heavily rely on fixed infrastructure, Wireless Sensor Networks (WSNs) thrive on direct node-to-node communication within their coverage range, forming an ad hoc network. Collaborative efforts among nodes play a critical role in promoting ad hoc communication, adapting to network modifications, and upholding a dynamic network topology.

The sensors in WSNs are equipped with computing and wireless transmission devices, allowing them to detect environmental factors like temperature and light. The

L. Barolli (Ed.): AINA 2024, LNDECT 199, pp. 216–226, 2024.
https://doi.org/10.1007/978-3-031-57840-3_20

collected data is then transmitted wirelessly to data collectors. This flexible framework enables sensors and data collectors to be placed randomly, reducing deployment costs and facilitating adaptation to various environments [2, 3].

By utilizing wireless communication and self-organization, WSNs provide a versatile solution for data collection and monitoring in diverse settings. Their ability to operate without the need for fixed infrastructure makes them particularly valuable in situations where traditional networks are not feasible or practical. As technology continues to advance, the applications and capabilities of WSNs are expected to expand further. In recent years, there has been a growing number of target positioning methods, such as Time of Arrival (TOA), Angle of Arrival (AOA), Time Difference of Arrival (TDOA), or Received Signal Strength Indicator (RSSI) [4–7]. However, TDOA, TOA, and AOA methods are often costlier and more computationally complex compared to RSSI. Among these options, RSSI stands out for its cost-effectiveness and straightforward implementation, as it eliminates the need for additional hardware expenses.

Hence, this study utilizes RSSI as a means to estimate the distance between the target point and the reference point. Nevertheless, RSSI is vulnerable to environmental fluctuations, which can introduce potential inaccuracies in the positioning process. Hence, the objective of this study is to address the issue of positioning errors caused by RSSI fluctuations. Indoor positioning systems require higher accuracy compared to outdoor systems. This creates greater challenges in achieving precise indoor positioning. The target location estimation for indoor localization has employed the K-nearest neighbor (KNN) algorithm [8]. Furthermore, sensor localization has been achieved by utilizing fingerprinting and KNN algorithms [9].

In this research, a pioneering wireless sensor network positioning technique is introduced. This method leverages the receive signal strength indicator (RSSI) channel model as the fundamental framework for localization. The method combines the K nearest neighbor (KNN) algorithm and the differential evolution (DE) algorithm to enhance accuracy and efficiency in network area positioning [10, 11]. This study focuses on exploring the deployment methods of sensor nodes, specifically examining the random deployment method and the fixed deployment method. The algorithm's performance is evaluated in terms of its positioning accuracy, shedding light on the effectiveness of these deployment strategies.

The remaining sections of this paper are structured as follows. Section 2 provides an in-depth description of the KNN and DE algorithms, setting the foundation for the proposed DE-KNN area positioning method introduced in Sect. 3. In Sect. 4, the simulation results evaluating the effectiveness of the area positioning method are presented and discussed. Finally, Sect. 5 concludes the paper by summarizing the key findings and offering insights for future research directions.

## 2  Background

### 2.1  K-nearest Neighbor Average Algorithm

In accordance with the concept of pattern matching, all training points are transformed into characteristic vectors and integrated into the positioning model, as $C_i = (c_1{}^i, c_2{}^i, .., c_j{}^i, \ldots, c_N{}^i)$. Concurrently, the target point's characteristic vector is defined as $S = (s_1, s_2, .., s_j, .., s_N)$. Subsequently, the computation of the Euclidean distance between the characteristic vector of the target and each training point is performed by

$$\|S, C_i\| = \sqrt{\Sigma_{i=1}^n \left(s_i - c_j^i\right)^2}. \tag{1}$$

After representing the positions of the K training points as coordinates, the x and y coordinates are averaged by

$$(x, y) = \left(\frac{\sum_{n=1}^K x_n}{k}, \frac{\sum_{n=1}^K y_n}{k}\right). \tag{2}$$

Drawing upon the average outcome, the estimated positioning suggests that the target point maybe located in region L5, as shown in Fig. 1.

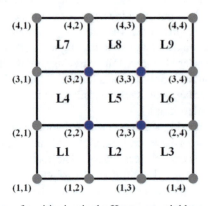

**Fig. 1.** Illustration of positioning in the K-nearest neighbor average algorithm.

## 2.2 Differential Evolution Algorithm

Differential Evolution (DE) is a prevalent population-based optimization algorithm employed for resolving optimization challenges. It is inspired by the process of natural evolution and works by iteratively improving a population of candidate solutions.

The DE algorithm is a multi-step process used for optimization problems. These steps are as follows:

Step 1 Initialization: Initially, a population of potential solutions is created in the search space. Each solution is represented as a vector.

Step 2 Mutation: Selecting three individuals randomly from the population, a mutant vector is formed by combining their information. This is achieved by scaling the difference between two individuals with a mutation factor and adding it to a third individual. A higher mutation factor leads to more exploration of the search space by causing larger changes to the candidate solutions. This can help the algorithm escape local optima and search for better solutions in the vicinity of the current solutions. However, too high a mutation factor may cause the algorithm to jump around the search space excessively, potentially hindering convergence.

Step 3 Crossover: During the evolution process, a crossover operation takes place between the mutant vector and the target vector, yielding a trial vector. This crossover operation typically transpires on an element-wise basis, implying that each element of the trial vector is selected from either the mutant vector or the target vector according to a pre-defined probability. The crossover rate, which generally varies between 0 and 1, determines the likelihood of crossover occurring between the mutant vector and the target vector.

Step 4 Selection: The trial vector undergoes a comparison with the target vector. If the trial vector enhances the fitness or objective function value of the target vector, it supersedes the target vector. In cases where the trial vector fails to improve the fitness, the target vector is retained for the subsequent generation.

Step 5 Steps 2–4 are iteratively repeated until a termination condition is satisfied. This condition can be defined as either reaching a predefined maximum number of iterations or attaining a satisfactory solution quality.

The DE algorithm presents several advantages, including simplicity, efficiency, and strong global search capabilities. Consequently, it has been widely utilized in solving optimization problems across various domains such as engineering, finance, and machine learning. By following these steps, the DE algorithm systematically explores and refines potential solutions, allowing it to converge towards an optimal or satisfactory solution. Its broad acceptance highlights its effectiveness and versatility in in tackling complex optimization challenges.

## 3    DE-KNN Aera Positioning Methods

This section outlines our proposed method for target localization, which integrates the K-NN approach with the DE algorithm. Firstly, we initialize the ethnic group consisting of 100 groups with 100 variable vectors. The dimension of the vectors depends on the number of reference points for positioning, and they can be initialized based on a fixed point or a random distribution. In the generation process, KNN is employed to identify N accurately positioned RSSI values within the population, which are then averaged to determine the fitness value. The fitness value is determined by assessing the geometric distance between the optimal target vector within the initial population and every target point, facilitating accurate evaluation. The size of the fitness value indicates the degree of similarity between the target point and the best target point. A smaller fitness value indicates a higher degree of similarity, while a larger fitness value suggests a lower degree of similarity.

During the crossover phase, we employ binomial crossover to guide the selection of vector elements, using a crossover rate (CR) set at 0.9 as the determining parameter. In both the target and mutation vectors, each element undergoes comparison with a random number between 0 and 1, utilizing the crossover rate (CR) to decide which elements are selected. This process generates a trial vector, and its fitness value is subsequently evaluated. If the fitness value of the trial vector is lower than that of the original target vector, it replaces the target vector in the subsequent iteration. Conversely, if the fitness value is higher, the original target vector remains unchanged. The algorithm terminates its evolutionary process after 100 iterations, yielding the final result. This approach ensures the stepwise improvement of solutions based on fitness evaluation, with the crossover rate (CR) playing a crucial role in determining vector selection during the generation of trial vectors. After obtaining the outcomes from the Differential Evolution (DE) algorithm, we apply the K-Nearest Neighbors (K-NN) method for local positioning in a single iteration. Subsequently, we analyze the simulated experimental findings to identify patterns and gain insights. Finally, the flowchart of the target positioning method combining K-NN and DE (DE-KNN) is shown in Fig. 2.

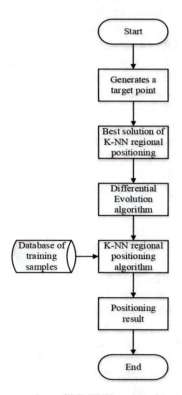

**Fig. 2.** The procedure of DE-KNN aera positioning method.

## 4 Simulation Results

This study aims to assess and compare the performance of the DE-KNN algorithm in indoor area positioning using a wireless sensor network. The network covers a 2D area measuring 3 m × 3 m, which is divided into 25 small regions. Each of these regions undergoes 100 simulated positioning instances, and the resulting data are statistically analyzed for evaluation purposes. Two simulation configurations are employed: one with 4 simulated positioning reference points and another with 5. n the former scenario, the four reference points are positioned at the central points of the small areas situated at the four corners of the 3 m × 3 m square. In the latter scenario, an extra reference point is included in the central small area, resulting in a total of five reference points. This setup allows for a comprehensive examination of the algorithm's performance under varying reference point densities. Statistical analysis of the simulated positioning instances provides insights into the algorithm's efficacy and suitability for indoor area positioning applications, contributing valuable knowledge to the field of wireless sensor network-based positioning systems.

First a total of 66 training points were collected to establish a sample database. We gather these training points by utilizing specific channel model parameters that emphasize the Received Signal Strength Indicator (RSSI) received from reference points. Additionally, we obtain RSSI readings from the reference points corresponding to the target point requiring localization. Following this, we compare these readings with the information in the sample database to determine the positions of the K training points that closely resemble the target point. By utilizing a majority decision-making process based on the identified training points, we estimate the location of the target point. Table 1 provides a comprehensive overview of the simulation parameters for the RSSI channel model. It is worth noting that a value of $\sigma = 2$ represents a simple environment, while a value of 10 represents a more complex environment. Additionally, Table 2 outlines the parameters for the Differential Evolution (DE) algorithm. The information presented in this table offers valuable insights into the setup and configuration of the simulation environment used to evaluate the target localization approach.

**Table 1.** Parameters of the RSSI channel model.

| Parameters | Value |
|---|---|
| Transmission Power | 2 mW |
| Carrier Frequency | 2.4 GHz |
| Path Loss Exponent | 4.5 |
| Reference Distance | 0.2 m |
| Antenna gains $G_t$, $G_r$ | 1 |
| Standard Deviation of Shadowing Fading | 2 dBm, 9 dBm |

This methodical approach ensures that data collection and analysis are carried out in a systematic manner, thus enabling a thorough assessment of the performance of the algorithm under varying environmental conditions and parameter settings. By conducting extensive simulations and statistical analyses utilizing the aforementioned parameters,

**Table 2.** DE algorithm parameters.

| Parameters | Value |
| --- | --- |
| Number of Population $NP$ | 100 |
| Mutation Weight Factor $F$ | 0.5 |
| Crossover Rate $CR$ | 0.9 |

the objective of this study is to assess and contrast the efficacy and precision of the DE-KNN algorithm proposed for indoor positioning. The insights gained from these simulations offer valuable contributions towards enhancing and advancing indoor positioning technologies. Ultimately, these findings aim to improve overall indoor positioning systems across diverse scenarios.

This study explores the DE-KNN area positioning method through simulations, examining both random and fixed deployment topologies. Various K values (3, 4, 5, and 6) are tested in the KNN algorithm to evaluate its performance under different scenarios.

Table 3 displays the average correct rate of area positioning in a simple environment with $\sigma = 2$ for 4 reference points (RPs), while Figs. 3 and 4 exhibit the convergence curve of fitness values in the DE algorithm for both fixed deployment and random deployment respectively. Similarly, in a complex environment with $\sigma = 10$, Table 4 presents the average positioning correct rate.

For 5 RPs, the average correct rate of area positioning in a simple environment with $\sigma = 2$ is provided in Table 5, and the results in a complex environment with $\sigma = 10$ are listed in Table 6.

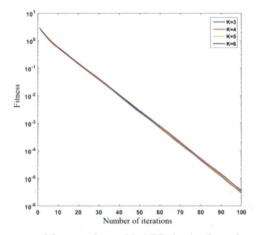

**Fig. 3.** Convergence curve of fitness values with 4 RPs in simple environment $\sigma = 2$ by fixed deployment.

By analysing the data from Table 3 to 6 and calculating the average correct rate for both random and fixed deployment, we can determine the average correct rate for different environments and reference points (RPs) for each K value. After further averaging the results for 4 and 5 RPs, we can identify the optimal K value that yields the highest average correct rate in both simple and complex environments, as presented in Table 7. This finding suggests that when implementing the DE-KNN positioning algorithm, setting the K value to 4 in the KNN algorithm can achieve satisfactory positioning performance without indiscriminately increasing the K value, which would only lead to unnecessary computational complexity.

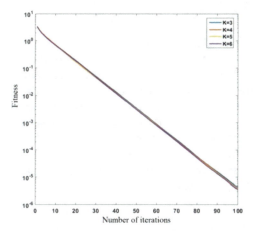

**Fig. 4.** Convergence curve of fitness values with 4 RPs in simple environment $\sigma = 2$ by random deployment.

**Table 3.** Average correct rate of random and fixed deployments with 4 RPs in a simple environment $\sigma = 2$.

| K value | Random deployment | Fixed deployment | Average |
|---------|-------------------|------------------|---------|
| 3 | 99.20% | 99.20% | 96.08% |
| 4 | 98.64% | 99.68% | 99.16% |
| 5 | 99.88% | 99.20% | 99.54% |
| 6 | 97.64% | 97.60% | 97.62% |
| Average | 97.28% | 98.92% | |

**Table 4.** Average correct rate of random and fixed deployments with 4 RPs in a complex environment $\sigma = 10$.

| K value | Random deployment | Fixed deployment | Average |
|---------|-------------------|------------------|---------|
| 3 | 93.44% | 97.24% | 95.34% |
| 4 | 99.12% | 99.72% | 99.42% |
| 5 | 97.20% | 99.96% | 98.58% |
| 6 | 97.12% | 98.16% | 97.64% |
| Average | 96.72% | 98.77% | |

Additionally, by averaging the results for each K value from Tables 3 to 6, the performance differences between using 4 RPs and 5 RPs in different environments can be obtained, as shown in Table 8. The results indicate that in the simple environment, using 5 RPs improves the performance by 0.32% for random deployment and 0.10% for fixed deployment compared to using 4 RPs. In the complex environment, the improvements are 2.02% and 0.49% respectively. It is evident that when using the DEKNN positioning method, adopting 5 RPs only yields a performance improvement greater than 1% for random deployment in complex environments, while the improvements in other cases are less than 1%. Therefore, considering the cost of system implementation, using 4 RPs can achieve satisfactory performance.

**Table 5.** Average correct rate of random and fixed deployments with 5 RPs in a simple environment $\sigma = 2$.

| K value | Random deployment | Fixed deployment | Average |
|---------|-------------------|------------------|---------|
| 3 | 92.08% | 96.68% | 94.38% |
| 4 | 99.88% | 99.96% | 99.92% |
| 5 | 98.76% | 100% | 99.38% |
| 6 | 99.68% | 99.44% | 99.56% |
| Average | 97.60% | 99.02% | |

**Table 6.** Average correct rate of random and fixed deployments with 5 RPs in a complex environment $\sigma = 10$.

| K value | Random deployment | Fixed deployment | Average |
|---------|-------------------|------------------|---------|
| 3 | 96.56% | 97.20% | 96.88% |
| 4 | 99.96% | 100% | 99.98% |
| 5 | 98.48% | 100% | 99.24% |
| 6 | 99.96% | 99.84% | 99.90% |
| Average | 98.74% | 99.26% | |

**Table 7.** Average Correct Rates of different K values in both simple and Complex Environments.

| K value | Simple environment $\sigma = 2$ | | | Complex environment $\sigma = 10$ | | |
|---------|-------|-------|---------|-------|-------|---------|
| | 4 RPs | 5 RPs | Average | 4 RPs | 5 RPs | Average |
| 3 | 96.56% | 97.20% | 95.23% | 95.34% | 96.88% | 96.11% |
| 4 | 99.96% | 100% | **99.54%** | 99.42% | 99.98% | **99.70%** |
| 5 | 98.48% | 100% | 99.46% | 98.58% | 99.24% | 98.91% |
| 6 | 99.96% | 99.84% | 98.59% | 97.64% | 99.90% | 98.77% |

**Table 8.** Difference of average correct rates between 5 and 4 RPs for random and fixed deployments in both Simple and Complex Environments.

| Deployment | Simple environment $\sigma = 2$ | | | Complex environment $\sigma = 10$ | | |
|------------|-------|-------|------------|-------|-------|------------|
| | 4 RPs | 5 RPs | Difference | 4 RPs | 5 RPs | Difference |
| Random | 97.28% | 97.60% | 0.32% | 96.72% | 98.74% | 2.02% |
| Fixed | 98.92% | 99.02% | 0.10% | 98.77% | 99.26% | 0.49% |

## 5 Conclusions

The research investigates the efficacy of integrating the DE algorithm with the K-NN algorithm for indoor area positioning within wireless sensor networks. The simulations indicate that the accuracy of target positioning correlates with the number of reference points (RPs) and the value of K employed in the K-NN algorithm. However, in both simple and complex environments, the highest average correct rates are 99.54% and 99.7% respectively when the K value is 4. This result suggests that there is no need to use higher K values to improve the performance of the KNN algorithm, as it would increase computational complexity. Additionally, in different environments with fixed or random sensor deployment, the performance improvement between using 5 and 4 reference points is less than 1% in all cases, except for a 2.02% increase observed in the complex environment with random deployment. This finding implies that using four reference points is sufficient to achieve satisfactory performance while considering system implementation costs.

**Acknowledgments.** This research was funded by Ministry of Science and Technology (MOST), R.O.C., with grant number NSTC 112–2221-E-324-010 and MOST-1112637-E-150-001.

# References

1. El Khediri, S.: Wireless sensor networks: a survey, categorization, main issues, and future orientations for clustering protocols. Computing **104**, 1775–1837 (2022)
2. Corke, P., Wark, T., Jurdak, R., Hu, W., Valencia, P., Moore, D.: Environmental wireless sensor networks. In: Proceedings of the IEEE, vol. 98, no. 11, pp. 1903–1917 (2010)
3. Kandris, D., Nakas, C., Vomvas, D., Koulouras, G.: Applications of wireless sensor networks: an up-to-date survey. Appl. Syst. Innovation **3**(1), 14 (2020)
4. Chugunov, A., Petukhov, N., Kulikov, R.: ToA positioning algorithm for TDoA system architecture. In: Proceedings of International Russian Automation Conference (RusAutoCon), pp. 871–876 (2020)
5. Ahmed, S., Abbasi, A., Liu, H.: A novel hybrid AoA and TDoA solution for transmitter positioning. In: Proceedings of International Conference on Indoor Positioning and Indoor Navigation (IPIN), pp. 1–7 (2021)
6. Tamer, Ö.: Relative localization of wireless sensor nodes using RSSI and ToA-based distance estimations. Dokuz Eylül Üniversitesi Mühendislik Fakültesi Fen ve Mühendislik Dergisi **25**(75), 647–658 (2023)
7. Yoshitome, E.H., da Cruz, J.V.R., Monteiro, M.E.P., Rebelatto, J.L.: LoRa-aided outdoor localization system: RSSI or TDoA? Internet Technol. Lett. **5**(2), e319 (2022)
8. Li, X., Dai, Z., He, L.: An indoor fingerprint location method based on coarse positioning circular domain and the highest similarity threshold in k-nearest neighbor algorithm. Measur. Sci. Technol. **34**(1), 015108 (2022)
9. Peng, X., Chen, R., Yu, K., Ye, F., Xue, W.: An improved weighted K-nearest neighbor algorithm for indoor localization. Electronics **9**, 2117 (2020)
10. Chakraborty, S., Saha, A.K., Ezugwu, A.E., Agushaka, J.O., Zitar, R.A., Abualigah, L.: Differential evolution and its applications in image processing problems: a comprehensive review. Arch. Comput. Methods Eng. **30**(2), 985–1040 (2023)
11. Rosić, M.B., Simić, M.I., Pejović, P.V.: An improved adaptive hybrid firefly differential evolution algorithm for passive target localization. Soft. Comput. **25**, 5559–5585 (2021)

# Terahertz Antenna Array for 6G CubeSat Inter-satellite Links

Chamseddine Oueslati$^{(\boxtimes)}$, Mondher Labidi, and Nabil Dakhli

Innov'Com Research Laboratory, Sup'Com, Ariana, Tunisia
chamseddine.oueslati@supcom.tn

**Abstract.** This study investigates the design and implementation of Terahertz (THz) antenna array designed for 6G inter-satellite communications in CubeSat constellations, emphasizing the vital role that these constellations play in contemporary satellite communication systems. This study describes a Terahertz TX/RX antenna system that is intended to execute beam steering patterns for the transmission and reception of cubesat constellations across intersatellite links (ISL). The system consists of a single antenna with a circulator. It is discovered that a $16 \times 16$ array is required for the antenna after link budget calculations. The transmitter/receiver antenna system operate at specific frequencies of approximately 0.628 THz, achieving a realized gain of 25.44 dB and a directivity of 24.12 dBi for the Tx link, and 1.316 THz with a realized gain of 30.18 dB and a directivity of 32.29 dBi for the Rx link. They have respective bandwidths of around 52 GHz and 158 GHz for the uplink and downlink.

## 1 Introduction

CubeSat constellations have become increasingly significant in transforming space-based applications because to their small size, affordability, and collaborative character. There is an increasing need for high-speed, low-latency communication solutions in space as we move closer to the era of 6G technology. To address these expectations, establishing effective intersatellite connectivity inside CubeSat constellations becomes essential. Future satellite communication systems could potentially benefit from using terahertz frequencies, which are recognized for their ability to provide high data rates and make extensive use of spectrum resources. In general, terahertz is described as a frequency that falls between infrared and millimeter waves, namely in the range of 0.1 to 10 THz. Its unique location dictates that its characteristics set it apart from other wave bands [1]. The following are the benefits of terahertz communication over microwave communication: The benefits of using Terahertz waves in intersatellite links include:

- Greater communication capacity and an abundance of bandwidth resources.
- Enhanced privacy and interference resistance occur as the atmosphere completely absorbs the waves, preventing their reach to the Earth's surface.
- The wave length in terahertz is comparatively short [2–4].

L. Barolli (Ed.): AINA 2024, LNDECT 199, pp. 227–238, 2024.
https://doi.org/10.1007/978-3-031-57840-3_21

Smaller antennas and simpler transceivers can do the same task while maintaining the same functionality, which helps to make the communication system weigh less and be easier for aerial transportation. Terahertz communications offers the following benefits over laser communication:

- More energy-efficient and appropriate for space platforms with limited energy.
- The relative's terahertz signal wide beam reduces the complexity of the capture and tracking system design. As a laser, it is also used in defense and security [5].

Distributed Antenna Systems (DAS) are frequently used in 5G and 6G networks to increase data capacity. The establishment of dependable and optimal connections between tiny cells and the macro network is greatly aided by these methods [6,7]. To guarantee seamless connectivity, these devices must be carefully designed and managed. When it comes to active antenna array design, precisely adjusting the phases and amplitudes of antenna elements is necessary to achieve a three-dimensional beam scan. For this, there are two different approaches that can be used:

- by employing digital/analog attenuators and phase shifters in conjunction with analog signals.
- By utilizing digital signal processing and converters that convert analog to digital or digital to analog.

The latter steps in this complex process, which is frequently carried out using FPGA/ARM microcontrollers [8], are the optimization, calibration, and implementation of coefficients related to the antenna elements. The design and simulation of the antenna elements as well as an array of $16 \times 16$ antenna are covered in detail in the next sections of this study, which also explore the subtleties of their evolution and performance. The plan involves conducting a technical analysis encompassing an examination of the Inter-Satellite Link Budget, parameters, and specifications. It further includes a detailed link budget analysis while considering the duplexer for both transmission at 0.628 THz and reception at 1.316 THz. The subsequent step involves drawing conclusions regarding the required antenna gain. Following this, the focus shifts to antenna design for terahertz frequencies, encompassing a THz unit cell antenna and an in-depth analysis of THz Antenna Arrays. The entire process culminates in a comprehensive conclusion summarizing the analysis and findings.

The structure of this paper is as follows: Sect. 2 describes the technical study, which includes the Inter-Satellite Link budget, parameters, and specifications, as well as a detailed link budget analysis that takes into account duplexer concerns. Section 3 focuses on the construction of ISL antennas for Terahertz frequencies, analyzing the THz unit cell antenna and providing a detailed study of the THz Antenna Array. Section 4 of the study provides a detailed conclusion that summarizes the technical assessments and design concerns discussed in the preceding sections.

## 2   Technical Analysis

The laboratory spectroscopic measurement of water vapor [9] consider that the terahertz-wave propagation model offer for Low Earth Orbit (LEO) satellites a better inter-satellite crosslink performance and less free-space path loss [10]. The two satellites are 100–1000 km apart and in the same orbit, both of which have an altitude of 550 to 880 km. The Ka band link, shown in Fig. 1, is the satellite-to-earth link that is selected to overcome the hung loss.

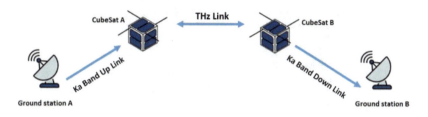

**Fig. 1.** THz Application in LEO-LEO link.

### 2.1   Analysis of Inter-satellite Link Budget

For designing terahertz band antenna array for a ISLs, it is necessary to measure certain requirements and some specifications has to be calculated as well. Because the link budget determines the total antenna gain, in the first step, it has to be specified. This study explores half-duplex transmission within the THz-band frequency range using a duplexer, aiming to ascertain the required antenna gains for reliable inter-satellite communication. However, it's important to note that the primary focus of this paper isn't centered on the detailed calculation of the link budget.

### 2.2   Parameters and Specifications

For optimal intersatellite communication, the Low Earth Orbit (LEO) satellite communication system's standards were meticulously set up and refined. Table 1 show the parameters and specifications for a system operated in the Terahertz (THz) band at an altitude of 600 km, with precise transmission and reception frequencies set at 0.628 THz and 1.316 THz, respectively. With a transponder producing 1 W (30 dBm) of power, the satellites may communicate on upgraded ISLs up to 200 km away. Interestingly, the satellites' angular separation (As) remained constant at 19°. In this satellite communication network, critical performance parameters like duplexer insertion loss, received power, path loss, and Signal-to-Noise Ratio (SNR) were meticulously tested and calibrated to guarantee the best possible signal strength and quality [11–13].

**Table 1.** Parameters and specifications.

| Parameters | Specifications |
|---|---|
| Satellite path | Low Earth Orbit (LEO) |
| Altitude from ground | 600 km |
| ISL Band | THz-band |
| Transmission Frequency | 0.628 THz |
| Reception Frequency | 1.316 THz |
| Output from Transponder | 1 W (30 dBm) |
| Updated ISL Distance | 200 km |
| Angular separation (As) | 19° |
| Transmitted power | 1 W (30 dBm) |
| Updated Path loss | 106 dB |
| Received power | −64 dBm |
| SNR | 20.5 dB |
| Duplexer Insertion Loss | 5 dB |

## 2.3  Link Budget Analysis with Duplexer Consideration

The study includes a half-duplex transmission scenario utilizing a circulator, influencing the required antenna gains for reliable communication. Here are the pertinent formulas used in the analysis:

### 2.3.1  Transmission at 0.628 THz

The analysis considered the following link budget equation for transmission:

$$P_{received}(\text{dBm}) = P_{transmitted}(\text{dBm}) + G_{transmitter}(\text{dB}) + G_{receiver}(\text{dB}) - \text{PL}(\text{dB}) - \text{IL}_{duplexer}(\text{dB}) \tag{1}$$

where:

- $P_{received}$ is the received power in decibels (dBm).
- $P_{transmitted}$ is the transmitted power in decibels (dBm).
- $G_{receiver}$ is the receiver antenna gain in decibels (dB).
- $G_{transmitter}$ is the transmitter antenna gain in decibels (dB).
- PL represents the path loss in decibels (dB).
- $\text{IL}_{duplexer}$ denotes the insertion loss of the duplexer in decibels (dB).

This led to the calculation of the antenna gain requirements for transmission as it is mentioned in Table 2:

**Table 2.** Link Budget Analysis with Duplexer consideration (Transmission).

| Parameters | Values |
|---|---|
| Transmitted Frequency | 0.628 THz |
| Updated Path Loss | 111 dB |
| Transmitted Power | 1 W (30 dBm) |
| Received Power | −64 dBm |
| Duplexer Insertion Loss | 5 dB |
| Calculated Antenna Gain | 19 dB |

### 2.3.2 Reception at 1.316 THz

Similarly, the analysis for reception frequency incorporated the following link budget equation:

$$P_R(\text{dBm}) = P_T(\text{dBm}) + G_{\text{transmitter}}(\text{dB}) + G_t(\text{dB}) - PL(\text{dB}) - IL_D(\text{dB}) \qquad (2)$$

where:

- $P_R$ represents the received power in decibels (dBm).
- $P_T$ denotes the transmitted power in decibels (dBm).
- $G_R$ signifies the receiver antenna gain in decibels (dB).
- $G_{\text{transmitter}}$ is the transmitter antenna gain in decibels (dB).
- $PL$ represents the path loss in decibels (dB).
- $IL_D$ denotes the insertion loss of the duplexer in decibels (dB).

Table 3 lists the calculation results of the antenna gain requirements for reception:

**Table 3.** Link Budget Analysis with Duplexer consideration (Reception).

| Parameters | Values |
|---|---|
| Received Frequency | 1.316 THz |
| Updated Path Loss | 111 dB |
| Transmitted Power | 1 W (30 dBm) |
| Received Power | −64 dBm |
| Duplexer Insertion Loss | 5 dB |
| Calculated Antenna Gain | 19 dB |

### 2.4 Conclusion on Antenna Gain Requirements

The estimated antenna gain needed, about 19 dB for both frequencies of transmission and reception, is obtained in the context of a half-duplex transmission with a duplexer.

## 3    ISL Antenna Design for THz Frequencies

Achieving a minimum gain of 19 dB within the authorized Terahertz frequency band is the main goal in constructing the THz antenna for Inter-Satellite Links (ISL). Optimizing radiation characteristics, minimizing losses, and ensuring compatibility with ISL's high-speed data transmission needs are the goals of this study. The THz antenna that is intended for use with Inter-Satellite Links (ISL) functions as a half-duplex system, using a single antenna construction for both reception (RX) and transmission (TX), incorporating a duplexer [14] and already defined components from the budget link computations facilitates this integration. Antenna can switch between TX and RX modes while effectively using the same radiating structure thanks to the duplexing component as shown in Fig. 2. Through the utilization of other system components' established characteristics, like received power, path loss, ISL distance, and transponder output, this integrated approach maximizes the performance of the antenna while maintaining compliance with the particular requirements mentioned in the previous budget link analysis. By optimizing the use of the THz antenna for bidirectional communication within the limitations of a half-duplex system, this approach simplifies the ISL setup.

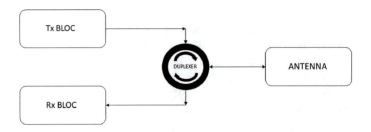

**Fig. 2.** Transceiver Block Diagram.

### 3.1    THz Unit Cell Antenna

The dimensions of the circular split ring (CSRR) resonator antenna are closely related to its operating frequency band, as well as to the substrate's thickness and composition. Under the ground layer, a 0.0035 mm thick silicone shape [15] is added in an attempt to improve the antenna's performance. In addition to providing mechanical support, this silicone additive reduces substrate losses and improves the antenna's impedance matching capabilities. The inclusion of this thin layer of silicone serves as an extra dielectric, modifying the effective dielectric constant and lowering substrate losses, both of which enhance radiation efficiency and antenna performance overall.

As seen in Fig. 3, the antenna itself consists of a split circular ring placed on a dielectric substrate on top of a conductive ground plane. An SMA feed-line is used to excite the patch, which transforms the electrical signal into an electromagnetic wave. This excitation causes a charge distribution in the patch, which generates attractive and repulsive interactions between the patch and the ground. Radiation is produced at the patch margins as a result of these interactions' fringing effect.

Figure 3 illustrates the geometric parameters of the THz circular SRR antenna. With a length Ls, width Ws and thickness of 190 μm × 190 μm × 22 μm. The substrate is made of FR4 material (tan = 0.025 and Er = 4.3). The SRR is 1.4 μm wide (a) and has an outside (OutR) and inner radious (InnerR) of 75 μm and 59 μm, respectively. The antenna makes use of the SMA feeding approach, and Electromagnetic simulations using full-wave EM software CST Microwave studio, are run to examine its performance attributes. The addition of a thin layer of silicone under the ground layer is a novel technique to improve the performance and efficiency of the antenna in the THz frequency range for ISL applications.

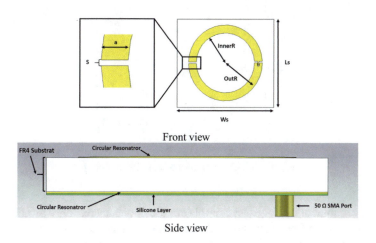

Front view

Side view

**Fig. 3.** Unit Cell Structure.

The S11 curve, shown in Fig. 4 of the proposed terahertz antenna, gives useful information about the antenna's performance at different frequencies. The S11 reflection magnitude reads −21.4 dB at 0.628 THz, showing a rather strong and efficient performance in terms of signal reception. As the frequency rises to 1.316 THz, the S11 reflection coefficient improves marginally but remains low at −16.35 dB. This pattern

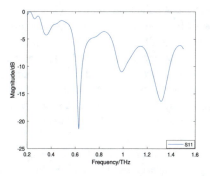

**Fig. 4.** Unit Cell Reflection Magnitude.

demonstrates the antenna's capacity to tolerate higher frequencies with a slightly lower reflection magnitude, implying flexibility and resonance at these specific terahertz frequencies.

As shown in Fig. 5, the split (S) inside (SRR) structure is crucial for adjusting the resonant frequency of the antenna and changing its adaption properties. By varying the gap width or the split position, the split's existence modifies the resonator's effective electrical length and permits frequency tuning.

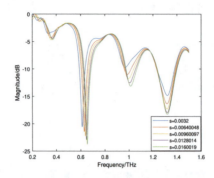

**Fig. 5.** Unit Cell Reflection Magnitude tuning.

These variables allow for exact control over the frequency response by adjusting the antenna's resonance frequency. Furthermore, changes made to the split size can have a big effect on the S11 parameters of the antenna, especially the reflection coefficient and impedance matching.

This capability to manipulate the split dimensions showcases the antenna's versatility and tunability, facilitating adaptability to varying frequency requirements and optimizing its performance for specific applications within the Terahertz frequency spectrum.

For ISL conditions, the realized gain of the antenna structure at both the transmission (TX) and reception (RX) frequencies, 0.628 THz and 1.316 THz, respectively, is critical. Figure 6 and Fig. 7 show that the single unit cell construction has a realized gain of 4.572 dB at 0.628 THz and 3.307 dB at 1.316 THz. However, these improvements are less than the required values calculated by the preceding budget connection study.

To overcome this, a shift to an array arrangement of these unit cell structures is required. The switch to an array design attempts to greatly increase the realized gain of the antenna while remaining consistent with the ISL budget link requirements. By combining separate parts to produce stronger directional gain and beam steering capabilities, arrays increase collective performance. The array architecture tries to attain the required gain levels expected by the ISL conditions by utilizing the aggregate contributions of numerous unit cells, guaranteeing reliable and efficient communication between satellites.

This strategic switch from a single unit cell to an array structure corresponds with the ISL budget link analysis's stated aim of exceeding the prescribed gain thresholds at

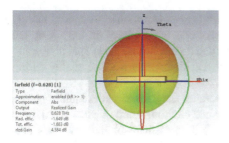
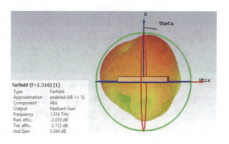
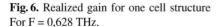

**Fig. 6.** Realized gain for one cell structure For F = 0,628 THz.

**Fig. 7.** Realized gain for one cell structure For F = 1,316 THz.

both TX and RX frequencies. The array's ability to dramatically increase realized gain bodes well for achieving the demanding performance standards required for smooth inter-satellite communication in the Terahertz frequency range.

## 3.2   THz Antenna Array Analysis

The THz antenna array section presents $8 \times 8$ structure (Fig. 8) with unit cell space-shifted by $0.83 \times \lambda$, resulting in significant performance improvements. As shown in Figs. 9 and 10, the array has significant improvements, with realized gains of 19.48 dB at 0.628 THz and 23.12 dB at 1.316 THz. These gains above the levels specified in the budget link criteria, suggesting that the array can meet the gain requirements for ISL applications.

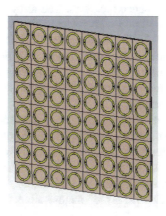

**Fig. 8.** $8 \times 8$ THz Antenna Array.

However, possible different types of attenuations or interferences that may occur in the spatial domain that were not accounted for in the first budget connection study must be considered. In real-world deployment circumstances, factors like as spatial fading, multipath propagation, and other environmental impacts may have an impact on the antenna's performance. Despite achieving estimated gain limits, these spatial-specific attenuation processes may have an influence on real signal intensity and quality during inter-satellite communication.

As a result, while the THz antenna array matches the defined gain requirements, a thorough study that handles spatial attenuations and environmental impacts is required. This extended study will provide a more detailed assessment of the antenna's performance under real-world ISL settings, including potential spatial-related attenuation effects that may impair communication reliability and efficiency between satellites.

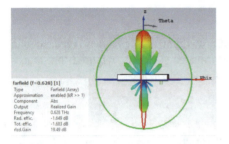 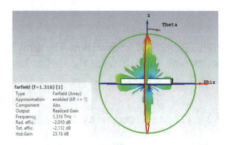

**Fig. 9.** 8 × 8 THz Antenna array gain For F = 0,628 THz.

**Fig. 10.** 8 × 8 THz Antenna array gain For F = 1,316 THz.

To avoid all this, The THz antenna array has been upgraded to a 16 × 16 structure, which gave significant performance increases. The array has a realized gain of 25.43 dB at 0.628 THz, as shown in Fig. 11 with power beam width of about 8°. Furthermore, at 1.316 THz, as seen in Fig. 12, the gain exceeds 30.14 dB with a beam width of 3.7°.

The shift to a 16 × 16 array arrangement show considerable enhancement in gain performance over earlier structures. These gains above the budget link conditions' criteria, showing the array's improved capacity to meet the gain requirements for ISL applications for both frequencies.

The increase in gain from an 8 × 8 array to a 16 × 16 structure demonstrates the array's scalability and potential to attain higher directional gain and enhanced performances throughout the Terahertz frequency band. This improvement puts the THz antenna array in a good position for reliable and efficient inter-satellite communication within the defined frequency ranges taking into account the strict constraints imposed by the standard dimensions of nanosatellite format, which are less than 10 cm × 10 cm × 10 cm for 1U cubesat, by not exceeding 0.304 cm, or 16 times the width of the cross section (Ws) of the antenna.

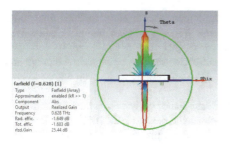

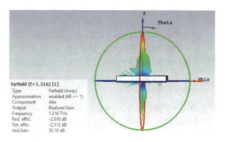

**Fig. 11.** 16 × 16 THz Antenna array gain For F = 0,628 THz.

**Fig. 12.** 16 × 16 THz Antenna array gain For F = 1,316 THz.

## 4   Conclusion

In summary, the terahertz antenna array that have been suggested for 6G inter-satellite communications in CubeSat constellations represent a major advancement in space communication. These antennas are essential for fast data transfer between satellites, improving connectivity that is needed for next space missions. We met strict performance objectives by carefully estimating the budget link needs, creating a unit tunable cell antenna, and integrating it into an array antenna MIMO arrangement, which resulted in amplified gain. By facilitating smooth data interchange and real-time monitoring, this technology promises to transform space exploration in addition to improving communication.

## References

1. Song, H.J., Nagatsuma, T.: Present and future of terahertz communications. IEEE Trans. Terahertz Sci. Technol. **1**, 256–263 (2011)
2. Akyildiz, I.F., Jornet, J., Han, C., et al.: Ultra-broadband communication networks in the terahertz band. IEEE Wirel. Commun., 130–135 (2014)
3. Ostmann, T.K., et al.: A review on terahertz communications research. J. Infrared Milli Terahz Waves **32**, 143–171 (2011)
4. Huang, K.C., Wang, Z.C., et al.: Terahertz terabit wireless communication. IEEE Microwave Mag., 108–116 (2011)
5. Liu, H., Zhong, H., Karpowicz, N., Chen, Y., Zhang, X.: Terahertz spectroscopy and imaging for defense and security applications. Proc. IEEE **95**(8), 1514–1527 (2007)
6. De Cola, T., Bisio, I.: QoS optimisation of eMBB services in converged 5G-satellite networks. IEEE Trans. Veh. Technol. **69**(10), 12098–12110 (2020)
7. Tariq, F., et al.: A speculative study on 6G. IEEE Wirel. Commun. **27**(4), 118–125 (2020)
8. Dastkhosh, A.R., Naseh, M., Lin, F.: K/Ka slotted stacked patch antenna and active array antenna design for a 5G/6G satellite mobile communication system. In: 2021 13th Global Symposium on Millimeter-Waves & Terahertz (GSMM) 23 May 2021, pp. 1–3. IEEE (2021)
9. Seta, T., Mendrok, J., Kasai, Y.: Laboratory spectroscopic measurement of water vapor for the terahertz-wave propagation model. In: Proceedings URSI Chicago General Assembly, pp. 1–4 (2008)

10. Walker, C.K., et al.: Terahertz astronomy from the coldest place on earth. In: 2005 Joint 30th International Conference on Infrared and Millimeter Waves & 13th International Conference on Terahertz Electronics, pp. 3–4 (2005)

11. Zhang, H., Shamim, A.: Gain enhancement of millimeter-wave on-chip antenna through an additively manufactured functional package. IEEE Trans. Antennas Propag. **68**(6), 4344–4353 (2020)

12. Alqaraghuli, A.J., Abdellatif, H., Jornet, J.M.: Performance analysis of a dual terahertz/Ka band communication system for satellite mega-constellations. In: 2021 IEEE 22nd International Symposium on a World of Wireless, Mobile and Multimedia Networks (WoWMoM). IEEE (2021)

13. Saisandeep, B., Morla, V.R., Ahmadsaidulu, S., Prasanna, C.S., Tirumalasetty, V.R.: High gain tapered slot antenna design and link budget analysis for inter-satellite link at 60 GHz. Int. J. Eng. Technol. **7**(1.1), 627-630 (2018)

14. Cui, D., Duar, W., Song, R.: The progress of terahertz communication for LEO satellite. In: 2021 IEEE 21st International Conference on Communication Technology (ICCT). IEEE (2021)

15. Leal-Sevillano, C.A., et al.: Compact duplexing for a 680-GHz radar using a waveguide orthomode transducer. IEEE Trans. Microwave Theory Tech. **62**(11), 2833–2842 (2014)

# Adaptive Admission Control for 6G Network Slicing Resource Allocation (A2C-NSRA)

Fadoua Debbabi[1,2(✉)], Rihab Jmal[2,3], Lamia Chaari Fourati[2,3], Raouia Taktak[2,3], and Rui Luis Aguiar[4,5]

[1] ISITCom, University of Sousse, 4011 Hammam Sousse, Tunisia
`debbabi.fadoua@gmail.com`
[2] SM@RTS Laboratory, Digital Research Center of Sfax (CRNS), Sakiet Ezzit, Tunisia
[3] Higher Institute of Computer Science and Multimedia of Sfax (ISIMS), University of Sfax, Sfax, Tunisia
[4] Instituto de Telecomunicações, 3810-193 Aveiro, Portugal
[5] Departamento de Electrónica, Telecomunicações e Informática, Universidade de Aveiro, Aveiro, Portugal

**Abstract.** The expansion of sophisticated and varied digital services is predicted to usher in the era of huge and extremely heterogeneous Sixth Generation (6G). The Network Slicing (NS) paradigm emerges as a crucial enabler, facilitating the development of diverse virtual network infrastructures required for dynamic behavior. In this context, addressing economic challenges becomes possible through NS, which can significantly impact the market model for various verticals. Particularly, inter-slice communication provides essential capabilities needed to execute slice operations while ensuring a high Quality of Experience. To address these considerations, we propose an Adaptive Admission Control (A2C) mechanism designed for revenue optimization within the 6G virtualized infrastructure. The primary goal of this study is to increase the acceptance rate of user requests by implementing a two-round admission control system. We will assess the effectiveness of the proposed model in a multi-service environment and compare its performance with classic admission control (AC) and random models.

**Keywords:** Sixth Generation (6G) · Network Slicing (NS) · Resources Allocation · Adaptive Admission Control · Revenue Optimization

## 1 Introduction

From the first generation to the Fifth Generation (5G) of mobile technology, each one was designed to meet the evolving demands of both network operators

This work is supported by the European Union/Next Generation EU, through Programa de Recuperação e Resiliência (PRR) [Project Nr. 29: Route 25 (02/C05-i01.01/2022.PC645463824-00000063)].

L. Barolli (Ed.): AINA 2024, LNDECT 199, pp. 239–250, 2024.
https://doi.org/10.1007/978-3-031-57840-3_22

and end-users. However, modern societies are undergoing increased computerization, reliance on data, and a shift toward being data-centric. The development of 5G entails navigating trade-offs related to factors such as latency, energy consumption, pricing, hardware complexity, throughput, and end-to-end reliability. Distinct 5G network architectures are tailored to serve the requirements of mobile broadband and ultra-reliable low-latency communications, reflecting the diverse needs of users. In contrast, the upcoming Sixth Generation (6G) [1] goes beyond merely improving Key Performance Indicators (KPIs). It is envisioned as a comprehensive response to the network requirements expected in the economic, social, technological, and environmental landscape of the 2030 era. 6G aims to deliver ultra-high reliability, increased capacity, enhanced efficiency, and low latency, positioning it as a transformative step forward in mobile technology. 6G networks are predicted to include unique features which are mainly: *1) Space networks* represented by air networking such as Unmanned Aerial Vehicles (UAVs), Low Earth Orbit (LEO) satellites, and ground networks like cellular Base Stations (BSs). All features are unified into a Space Air Ground Integrated Network (SAGIN) to afford on-demand services and global coverage. *2) Advanced network virtualization* that offers flexible network management through resource virtualization using Network Slicing (NS) techniques, as well as end-user virtualization using digital twin techniques. *3) Intelligence that covers all corners of networks* from end-users and network edges to the distant cloud in order to achieve ubiquitous intelligence. Today, delivering a wide range of services with diverse performance requirements is one of the main goals of 6G network. From the International Mobile Telecommunications (IMT-2030) [2], these services are categorized into six broad use cases and applications envisioned in Fig. 1.

**Fig. 1.** 6G slices

***Ubiquitous connectivity*** is closely related to super-enhanced mobile broadband (SeMBB). It represents an extension or advancement of the eMBB concept, indicating a more sophisticated and optimized version of eMBB services within 6G Network Slicing.

***Hyper reliable low-latency communications (HRLLC)*** is an extension of URLLC that implies a heightened level of reliability and low latency.

***Sensing and communication*** can be integrated to provide new applications and services that require Super Precision (SP) sensing capabilities.

***Immersive communication*** refers to Super Immersive (SI) experiences, often involving augmented reality (AR), virtual reality (VR), and Health applications.

***Massive communication*** is an extension of mMTC that involve the creation of dedicated slices for efficient management of a large number of IoT devices.

***AI and communication*** is the integration of Intelligence (IT) in 6G Network Slicing which entails utilizing AI and ML algorithms [3] to improve various facets of network management, optimization, security, and service delivery.

Accordingly, the communication between slices and the resource allocation remain still challenging issues considering the diverse demands of users. For this purpose, we consider the problem of multiple inter-slice resource allocation in the 6G scenario, which, to the best of our knowledge, has not been addressed before. For the computational results, we consider different characteristics of slices based on the standards. The main contributions are represented as follows:

- We present the system model for 6G inter-slice resource allocation, incorporating the A2C-NSRA module.
- We introduce an adaptive admission control (A2C) solution where the admission process occurs adaptively in two phases.
- We assess the A2C mechanism using various metrics and compared it with one-phase resource allocation called the classic admission control, as well as the random baseline method.

The manuscript is outlined as follows. Section 2 presents closely related works. The system model and problem statement are given in Sect. 3. Section 4 describes the A2C-NSRA mechanism that we propose. The experiment results and evaluation is discussed in Sect. 5. Finally, conclusions are drawn in Sect. 6.

## 2 Related Work

The inter-slice resource allocation problem arises from the need to optimally allocate shared resources, such as computing power, storage, and spectrum, among multiple slices to meet their service requirements. In this context, several studies in the literature have tackled this issue in Network Slicing for B5G/6G scenarios. Han et al. [4] addressed the slicing admission problem related to diverse tenant requests by proposing a statistical optimization model within a multi-queuing system based on AC utility. In another paper [5], an inter-slice allocation policy was introduced to minimize the disruption between slices experienced by the virtual network operator. The problem was modeled as a binary quadratic programming formulation, yielding various logical spectrum slices and generating interfering Base Stations. The authors in [6] proposed a pricing mechanism for multiple virtualized resources, including wireless network connectivity and cache resources. They addressed an optimization problem and proved that it is NP-hard by considering social welfare among users. They suggested a strategy for an auction-based solution that facilitates the coexistence of users in a market of virtualized heterogeneous resources. Furthermore, the authors in [7] introduced an optimization framework addressing pricing mechanisms, resource allocation, and admission control. This framework offers fixed resources restricted to only three services: eMBB, URLLC, and mMTC. To jointly solve the issues of NS and slice AC, paper [8] proposed a multi-agent Deep Reinforcement Learning (DRL) approach. This method tackles the problem by dynamically learning the slice-request traffic. Haque et al. [9] proposed two complementary methods to address the slice admission control problem. They used a queue-based optimization algorithm to dynamically assess and admit a group of slices that generate more income. Additionally, they automate the admission decision using Reinforcement Learning (RL). ADAPTIVE6G [10] is proposed as a novel approach for NS architecture that considers resource management and load prediction in data-driven B5G/6G wireless systems. This approach relies on machine and deep learning techniques, as well as transfer learning techniques, to enhance knowledge for load prediction across slices. Recently, in [11], the authors discussed the admission control (AC) of network slice requests to increase long-term revenue from admitted requests in an online scenario. They modeled the slice AC problem as an online multidimensional knapsack problem and present two reservation-based policies, along with their corresponding algorithms, as solutions to the problem. Although the recent work contributed to improving the resource allocation challenge, they didn't respond to this issue sufficiently mainly regarding the innovation and the evolving aspect of 6G networks.

## 3   System Model and Problem Statement

### 3.1   System Model

The system model for the A2C mechanism based on inter-slice resource allocation is illustrated in Fig. 2. This model is derived from the architecture presented in [12]. A Base Station (BS) serves a 6G infrastructure using wireless virtualization techniques to support a set of slices. The deployed services cater to specific use cases, each demanding a particular bandwidth.

The A2C module manages incoming requests adaptively to achieve an efficient allocation procedure that responds to end users' needs. It is important to note that we are interested in an A2C scheme where the allocation procedure occurs in two rounds: first, the more demanding services are accepted. Second, we address another subset of demands based on service type, bandwidth, and system availability. An adaptive slice can be created based on the more rejected requests. For example, the SeMBB slice accepts services from all sub-categories of SI, HRLLC, and IT, where SI and SP are sub-categories of HRLLC. This allocation ensures that user requirements are not violated. The remaining requests after the two rounds will then be rejected.

### 3.2   Problem Statement

Addressing the 6G inter-slice resource allocation problem is essential for realizing the full potential of diverse applications and services in the future wireless

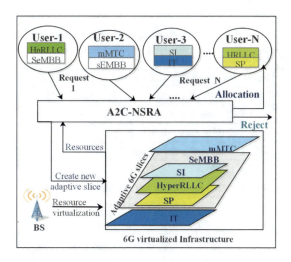

**Fig. 2.** System Model

communication landscape. Efficient resource allocation contributes to improved network performance, better utilization of resources, and enhanced user experience across a wide range of use cases. We model this problem using mathematical programming. For this, we first give in Table 1 some notations that will be used throughout the paper. We define two families of binary decision variables. For each demanded request $r \in Req$, and each slice $s \in S$, let $x_r^s = 1$ (resp. $y_r^s = 1$) if slice $s$ serves request $r$ retained in the first round (resp. second round), and 0 otherwise. Given these variables, the revenue per slice can be computed by $Rev^s = \sum_{r \in Req} P_r b_r (x_r^s + y_r^s)$, where $P_r$ is the price of each allocated bandwidth unit [13], and $b_r$ is the bandwidth demanded in each request $r$. Hence, the 6G infrastructure provider revenue can be calculated as the total revenue of distributed slices $Rev_{InP} = \sum_{s \in S} Rev^s$.

The problem is modeled through the following Integer Linear Programming (ILP) formulation.

$$\text{Maximize } Rev_{InP} \tag{1}$$

$$\sum_{r \in Req} b_r (x_r^s + y_r^s) \leqslant B^s \qquad \text{for all } s \in S, \tag{2}$$

$$\sum_{s \in S} (x_r^s + y_r^s) \leqslant 1 \qquad \text{for all } r \in Req, \tag{3}$$

$$\sum_{s \in S} \sum_{r \in Req} b_r (x_r^s + y_r^s) \leqslant Ar_{InP} \tag{4}$$

$$x_r^s \in \{0,1\} \qquad \text{for all } r \in Req, s \in S, \tag{5}$$

$$y_r^s \in \{0,1\} \qquad \text{for all } r \in Req, s \in S. \tag{6}$$

**Table 1.** Definitions and Notations

| Notation | Definition |
|---|---|
| $x_r^s$ | binary allocation variable for the first round |
| $y_r^s$ | binary allocation variable for the second round |
| $Req$ | set of all demand users requests |
| $r$ | demand user request, $r \in Req$ |
| $S$ | set of the slices used in the 6G infrastructure |
| $s$ | a slice of $S$ |
| $Rev_{InP}$ | revenue obtained from the whole 6G infrastructure |
| $Rev^s$ | revenue obtained for slice $s \in S$ |
| $P_r$ | bandwidth unit price assigned to the user |
| $Ar_{InP}$ | total bandwidth resources available in 6G environment |
| $B^s$ | maximum available bandwidth in a slice $s \in S$ |
| $b_r$ | bandwidth user requirements, $r \in Req$ |
| $B_{Req}$ | total bandwidth demanded |
| $b_{acc}^s$ | first round accepted bandwidth in a slice $s \in S$ |
| $R_{reject}$ | set of the most rejected requests from the first round |
| $T$ | set of types of all rejected requests from the first round |
| $t$ | type of a rejected request from the first round |
| $r_t$ | request of type $t$ rejected from the first round |
| $n_t$ | number of rejected requests of type $t$ from the first round |

Constraints (2) ensure that the assigned bandwidth resources in the two rounds to users are less than the available bandwidth in a specific slice. Constraints (3) express the fact that a user's request that is retained in one of the two rounds must be allocated to only one slice. Constraints (4) say that the sum of assigned resources of all slices through the two rounds does not exceed the available resources provided by the InP. Constraints (5) and (6) are the integrality constraints of the decision variables.

## 4 A2C-NSRA Mechanism

### 4.1 Admission Control for Bandwidth Slice

Various 6G applications exhibit distinct bandwidth requirements influenced by three factors: data rates, reliability, and latency. Determining whether a new user's request would be accepted within the allocated bandwidth for a specific network slice relies on the admission control mechanism for that slice's bandwidth. The efficiency of this control mechanism is pivotal in maintaining the quality of service (QoS) and ensuring utilization of the network's resources. The criteria for admitting bandwidth slices are formulated in inequality (7).

$$\sum_{r \in Req} b_r x_r^s \leqslant B^s, \qquad \text{for all } s \in S. \tag{7}$$

## 4.2 Admission Control for Slice Creation

The control for creating an adaptive slice helps prevent overloading the network or introducing services that could negatively impact the performance of the existing ones. This is particularly important in 6G networks where diverse applications with varying requirements coexist. Adaptive slices are generated based on the availability of resources, influenced by two primary factors: Factor 1: no additional slices are available to allocate resources in the second round when the slices in the first round have reached maximum usage (as defined by inequality (8)); Factor 2: the adaptive slice is created based on the more rejected requests $R_{reject}$ in the first round, as well as their respective types $T$ (see equation (9)).

$$b_{acc}^s \leq B^s, \qquad \text{for all } s \in S, \tag{8}$$

where $b_{acc}^s$ represents the bandwidth of accepted user requests in the first round for a slice $s \in S$.

$$R_{reject} = \bigcup_{\substack{r_t \in Req: \\ t=t_{max}}} r_t, \quad \text{where } t_{max} = \arg\max_{t \in T}(n_t). \tag{9}$$

Recall here that $T$ is the set of types of rejected requests from the first round (see Table 1 for further notations).

## 4.3 Admission Control Based on Resource Availability

The admission process depends on bandwidth resource monitoring (Sect. 4.1), evaluation of the incoming request (Sect. 4.2), and comparison between the demand and the available resources presented in the following inequality.

$$\sum_{s \in S} B^s \leq Ar_{Inp} \tag{10}$$

Accordingly, this process aims to prevent network congestion and avoid performance degradation from more resource usage. It helps to maintain a balance between the various requirements of applications. Algorithm 1 represents the details of the A2C-NSRA mechanism.

---

**Algorithm 1:** A2C-NSRA Algorithm

---

**Input** : $Ar_{Inp}$, $B^s$ for $s \in S$, $b_r$ and $P_r$ for $r \in Req$
**Output:** $Rev_{InP}$

1 **Begin** Preprocessing of synthetic 6G data for all requests $r \in Req$ based on their type
2 **repeat**
3    Consider $Req \in \{$SEmbb, SI, SP, HRLLC, mMTC, Intelligence$\}$
4                    ▷ slice is selected based on highest priority
5    **repeat**
6      | Assign $r \in Req$ to $s \in S$ using inequality (7)
7    **until** *inequality (8) is satisfied to maximum;*
8                     ▷ No more available slice
9    **while** $Ar_{Inp}$ *is not exceeded and still some requests not yet assigned* **do**
10      | Create adaptive slice using equation (9)
11    **end**
12    **repeat**
13      | Assign $r \in R_{reject}$ to $s \in S$
14    **until** *all the slides are explored to maximum;*
15 **until** *all Req are explored in the two rounds and $Ar_{Inp}$ is used to maximum;*
16 **End**

---

During the preprocessing phase (*line 2*), the 6G users' requests undergo classification based on their slice type and the corresponding service price, denoted as $P_v$. This price falls within a specific interval covering both Operational Expenditure (OPEX) and Capital Expenditure (CAPEX). The outcome of this process is the selection of requests that meet the criteria. In *line 4*, the system selects the most efficient slice based on the highest priority, considering the demands of the more resource-intensive requests for allocation. *Lines 6 to 8* describe the initial resource allocation round, where an admission control mechanism is applied, expressed by inequality (7). Requests rejected in the first round proceed to the second round of resource allocation. Upon achieving maximum resource usage for all slices, determined by inequality (8), new adaptive slices are dynamically created in *line 11*. This creation is based on the current resource availability within the system, as indicated by equation (9). The second round of resource allocation, detailed in *line 14*, enables the potential acceptance of requests rejected in the initial round. The A2C-NSRA algorithm concludes its execution when both existing slices and newly created ones reach maximum usage ($Ar_{InP} \approx 100\%$) and the available resources in the environment are insufficient, as stated in *lines 15 and 16*.

## 5   Experimental Results and Evaluation

We employed the PyCharm environment to implement various approaches, including the random, classic admission control, and the A2C mechanism. Since there is no publicly available 6G dataset in the literature, we conducted tests using synthetic data for 6G slices obtained from industry standards [14]. Our

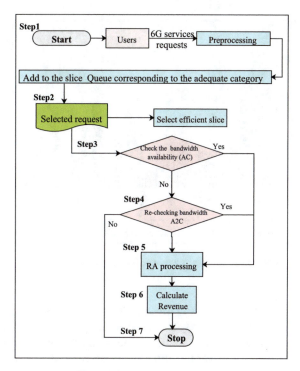

**Fig. 3.** A2C-NSRA Flow Chart

dataset focuses on bandwidth capacity and price per unit as well as the service type. In our scenario, we assumed the existence of a wireless virtualized infrastructure comprising a Base Station (BS) with a capacity of 10 Gbps. The BS supports a dynamic set of 6G slices ($|S| = 32$), generated based on the available bandwidth. Additionally, we set the minimum service price $Pv = 7$, and minimum data rate $= 20$ Mbps corresponding to SI slice. Figure 3 offers detailed clarification of the implementation. The incoming requests are generated from a CSV dataset that has been created. During the preprocessing phase, we classify the selected requests by associating the slice Queue with the corresponding 6G category using a dictionary. The efficient slice is then chosen based on the more demanding requests. Subsequently, we apply steps 2 and 3 as admission control, followed by a second round of admission control. Requests rejected during the initial admission control phase serve as input for the second round of admission control. The resource allocation process continues iteratively until the slices reach maximum utilization. The A2C-NSRA mechanism produces numerical results illustrated in the following curves. We evaluate the performance of the A2C-NSRA mechanism by comparing it with classic admission control and baseline random approaches implemented in the same environment. Figure 4 depicts the admitted requests from various methods based on the demanded user's requests. Our mechanism exhibits the highest acceptance rate (up to 0.80) and the lowest

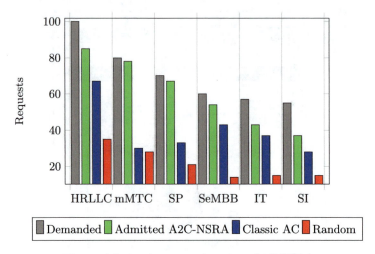

**Fig. 4.** Admitted requests based on A2C-NSRA

rejection rate (less than 0.20) across diverse categories. This favorable outcome is attributed to the effective strategy employed in our A2C mechanism, specifically the creation of adaptive slices to enhance the admission rate. Resource utilization is a crucial metric for evaluating the effectiveness of a resource allocation scheme. Therefore, mitigating inefficient resource usage is a primary factor in achieving this objective, which is the main goal of the A2C mechanism. Figure 5 shows that our approach attains maximum resource utilization, resulting in a heightened rate of at least $\approx 94\%$. This is attributed to the efficient use of resources in the A2C mechanism, achieved by accepting a maximum number of requests. As a result, the A2C mechanism achieves a maximum slice utilization.

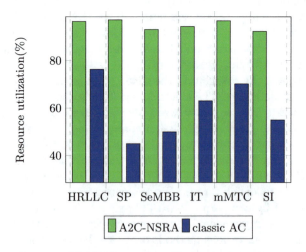

**Fig. 5.** Resource utilization rate

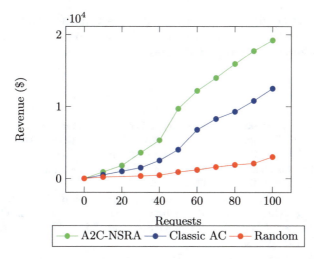

**Fig. 6.** Comparison in terms of revenue maximization

We also examine the effectiveness of the A2C-NSRA in terms of revenue maximization. Figure 6 demonstrates that our mechanism achieves the maximum revenue compared to the classic AC and random benchmark approaches. The high preference for gained revenue is attributed to the maximum number of admitted requests, relying on the two rounds of acceptance, the high level of resource utilization, and the slice priority based on the more demanding requests.

Additionally, we compare our A2C with other existing approaches from the literature such as the online admission control in terms of average revenue. The online AC gains an average revenue of less than 30% [11], while our approach achieves a minimum average revenue of over 40%.

Finally, we assess the scalability of the A2C-NSRA solution concerning convergence time by improving the resource availability and increasing the number of slices. To this end, we considered different simulation scenarios with various existing and new 6G slices, $|S| = 18$, $|S| = 25$, and $|S| = 32$. The convergence time was $4.79(s)$, $7.02(s)$, and $9.1(s)$, respectively.

## 6    Conclusion

This paper addresses the inter-slice resource allocation problem, proposing an A2C-NSRA mechanism with high performance in terms of acceptance rate, resource utilization, and revenue optimization. Consequently, we accept a maximum number of requests that yield the highest gain while efficiently utilizing all available slices. A functional and dependable communication network depends on effective admission management for bandwidth in a network slice. It guarantees that bandwidth is allocated and used with particular guidelines and regulations of each network slice, resulting in a smooth and efficient user experience. In the future, we aim to provide a strategy with stricter criteria for selection in the second round of admission control taking into account the latency constraint. Additionally, we plan to enhance the simulation by incorporating multiple industry scenarios for the 6G virtualized infrastructure.

# References

1. 6G: The Next Horizon, From Connected People and Things to Connected Intelligence, Huawei Technologies Co., Ltd. White Paper - Huawei (2021)
2. Recommendation, ITUR, framework and overall objectives of the future development of IMT for 2030 and beyond, International Telecommunication Union (ITU) Recommendation (ITU-R) (2023)
3. Debbabi, F., Rihab, J., Chaari, L., Aguiar, R.L., Gnichi, R., Taleb, S.: Overview of AI-based algorithms for network slicing resource management in B5G and 6G. In: 2022 International Wireless Communications and Mobile Computing (IWCMC), pp. 330–335. IEEE (2022)
4. Han, B., Sciancalepore, V., Feng, D., Costa-Perez, X., Schotten, H.D.: A utility-driven multi-queue admission control solution for network slicing. In: IEEE Conference on Computer Communications (INFOCOM), pp. 55–63 (2019)
5. Zambianco, M., Verticale, G.: Interference minimization in 5G physical-layer network slicing. IEEE Trans. Commun. **7**, 4554–4564 (2020)
6. Ndikumana, A., et al.: Pricing mechanism for virtualized heterogeneous resources in wireless network virtualization. In: International Conference on Information Networking (ICOIN), pp. 366–371. IEEE (2020)
7. Ben-Ameur, W., Cano, L., Chahed, T.: A framework for joint admission control, resource allocation and pricing for network slicing in 5G. In: IEEE Global Communications Conference (GLOBECOM), pp. 1–6 (2021)
8. Sulaiman, M., Moayyedi, A., Salahuddin, M.A., Boutaba, R., Saleh, A.: Multi-agent deep reinforcement learning for slicing and admission control in 5G C-RAN. In: IEEE/IFIP Network Operations and Management Symposium (NOMS), pp. 1–9 (2022)
9. Haque, Md.A., Kirova, V.: 5G network slice admission control using optimization and reinforcement learning. In: IEEE Wireless Communications and Networking Conference (WCNC), pp. 854–859 (2020)
10. Thantharate, A., Beard, C.: ADAPTIVE6G: adaptive resource management for network slicing architectures in current 5G and future 6G systems. J. Netw. Syst. Manag. **31**(1), 9 (2023)
11. Ajayi, J., Di Maio, A., Braun, T., Xenakis, D.: An online multi-dimensional knapsack approach for slice admission control. In: IEEE 20th Consumer Communications & Networking Conference (CCNC), pp. 152–157 (2023)
12. Debbabi, F., Jmal, R., Chaari, L., Aguiar, R.L.: An overview of inter-slice & intra-slice resource allocation in B5G telecommunication networks. IEEE Trans. Netw. Serv. Manage. **19**(4), 5120–5132 (2022)
13. Prince, M.: The relative cost of bandwidth around the world. https://blog.cloudflarecom/the-relative-cost-of-bandwidth-around-the-world/. Accessed 15 Dec 2018
14. Liu, R., Lin, H., Lee, H., Chaves, F., Lim, H., Skold, J.: Beginning of the journey toward 6G: vision and framework. IEEE Commun. Mag. **61**(10), 8–9 (2023)

# A Filtering Method for Machine Learning Utilization of ADS-B Data

Koichi Kakimoto[1], Takahiro Immaru[2], Makoto Ikeda[2(✉)] 📷, and Leonard Barolli[2] 📷

[1] Graduate School of Engineering, Fukuoka Institute of Technology, 3-30-1 Wajiro-Higashi, Higashi-Ku, Fukuoka 811-0295, Japan
mgm23104@bene.fit.ac.jp
[2] Department of Information and Communication Engineering, Fukuoka Institute of Technology, 3-30-1 Wajiro-Higashi, Higashi-Ku, Fukuoka 811-0295, Japan
makoto.ikd@acm.org, barolli@fit.ac.jp

**Abstract.** The development of low-cost aircraft surveillance systems based on Automatic Dependent Surveillance-Broadcast (ADS-B) technology has gained considerable interest, leading to many applications. Our interest is particularly in harnessing these data for developing flight prediction models and their applications considering the information in ADS-B signals. In this paper, we propose a filtering method and assess its effectiveness for processing ADS-B data to enhance the accuracy of predicted flight location coordinates. The evaluation results demonstrate that the proposed method can successfully separate single routes for Machine Learning purpose. Also, the predicted data are very close to learning data and the observed errors are quite small.

**Keywords:** ADS-B · Filtering method · Machine Learning · ML

## 1  Introduction

To realize carbon neutrality a multifaceted approach is essential, encompassing the enhancement of aircraft operations, the promotion of eco-friendly airports, and the advancement of biojet fuel technologies. The automated Air Traffic Management (ATM) systems play a critical role in elevating navigation, communication, surveillance, and passenger safety [18]. These systems are pivotal to the implementation of Automatic Dependent Surveillance-Broadcast (ADS-B) technology [13, 19, 23, 25, 26, 28].

In recent years, Machine Learning (ML) has been increasingly applied to forecast the arrival time of various transportation methods such as buses, trains and aircraft [1–3, 5, 12, 16, 17]. Data from regular operations, like daily routines, can be compiled and utilized as training data for ML and neural network models [8, 9, 15, 21]. This research concentrates on a filtering method for collected aircraft data to predict future aircraft positional coordinates.

In our previous work [10, 11], we proposed a ML-based flight prediction system that utilized location coordinates obtained from our ADS-B receiver data. However, this system did not incorporate any filtering method.

In this paper, we propose a filtering method and assess its effectiveness for processing ADS-B data to enhance the accuracy of predicted flight location coordinates. The

L. Barolli (Ed.): AINA 2024, LNDECT 199, pp. 251–260, 2024.
https://doi.org/10.1007/978-3-031-57840-3_23

evaluation results demonstrate that our proposed method can successfully separate single routes for ML purpose. Also, the predicted data are very close to learning data and the observed errors are quite small.

The overall structure of the paper is as follows: Sect. 2 offers an overview of related work. In Sect. 3 is introduced the design of the flight prediction system. Section 4 presents the evaluation results. The paper concludes with Sect. 5, which provides conclusions and outlines future work.

## 2    Related Work

The Airport Surveillance Radar (ASR) consists of two key components: Primary Surveillance Radar (PSR) [22] and Secondary Surveillance Radar (SSR) [24]. The installed radar systems depend on the size and function of the airport.

The PSR is an integrated radar system featuring a transmitter and a receiver. However, it lacks the capability to transmit altitude data and aircraft identification. While, the SSR system involves a transmitter sending a query signal to the aircraft transponder, which then returns a reply signal. This communication can be disrupted if the transponder is damaged, rendering monitoring ineffective.

The ADS-B offers a more advanced approach to aircraft surveillance. It continuously transmits data such as aircraft location, and other relevant information, which can be received by anyone using an ADS-B receiver [20].

Passive bistatic and multi-static primary surveillance radars, which utilize PSR, have been explored in various studies [4,6,7,14,27]. These systems are characterized by separate transmitter and receiver units, allowing signals to be received from multiple locations. This setup is anticipated to expand the monitoring coverage area and enhance the frequency of updates.

## 3    Filtering Function for Flight Prediction System

When collecting ADS-B data, the data from the same aircraft registration number often shows back-and-forth movements between two points. In such cases, when the aircraft code is used to simply learn the movement history, multiple flights may be interpreted as a single trajectory. This can lead to problems when using the data for movement prediction or other applications. Therefore, it is important to process the data considering the time delay differences.

### 3.1    Filtering Procedure

We have developed a system using Raspberry Pi to collect signals at $1,090$ MHz from an ADS-B receiver. The data received from ADS-B comprises eight different messages, with the third message specifically containing the location coordinates. Each message is composed of 22 fields. For ML training, we utilized the fields that include the longitude, latitude, and the time of message transmission.

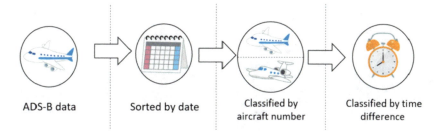

**Fig. 1.** Filtering procedure.

The filtering procedure is shown in Fig. 1. The original data are sorted by date and classified by aircraft number. Then, for each aircraft number the time difference is calculated based on the first message received.

The International Civil Aviation Organization (ICAO) requires the use of a Mode S code for the ICAO aircraft number. In our previous work [10, 11], learning was conducted with the data in its original form, which sometimes led to significant errors in the prediction results. Therefore, in this paper, we separate the data considering the substantial time difference.

### 3.2  Filtering Function

Figure 2(a) presents an example of ADS-B data sorted by date. While, Fig. 2(b) presents the results of extracting the data and coordinates, and classifying them by aircraft identification number. These data are captured from aircrafts that passed near Fukuoka Airport on October 24, 2023. The $Time_{diff}$ indicates the difference between the first received time and the received time $t_n$, where $n$ indicates the sequence number of the received messages.

We calculate the time difference between the previous received time $t_{n-1}$ and $t_n$ for each aircraft and the identification code is represented by $Time_{delta}$ (see Fig. 3).

$$Time_{delta} = t_{n-1} - t_n$$

In the traditional system, the data in Fig. 3(a) to Fig. 3(c) were treated as a single flight data set. While, by our proposed method, they have been filtered as three distinct paths. The extremely large initial values of $Time_{delta}$ in Fig. 3(b) and Fig. 3(c) indicate the presence of return flights. They are about 52 min and 180 min, respectively.

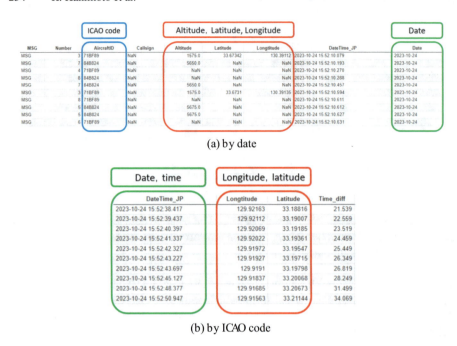

(a) by date

(b) by ICAO code

**Fig. 2.** Filtering procedures.

## 4   Evaluation Results

### 4.1   Mapping

There are three phases according to the proposed method shown in Fig. 3. In Fig. 4, the red and black lines indicate the flight from Okinawa to Fukuoka. While, the blue line indicates the flight from Fukuoka to Okinawa. From these mapping results, we successfully separated single routes for ML purpose.

By using the proposed filtering method, users can select two rectangular areas to filter only the flight routes that travel between these two points. This function is designed to facilitate learning for specific flight paths. Figure 5 illustrates the mapping of both the training and predicted routes, utilizing this newly added filter functionality. The data represented in this figure are from a flight traveling from Fukuoka to Korea.

### 4.2   Prediction Performance

In Fig. 6, we show the prediction results for two-dimensional flight coordinates. The vertical axis represents longitude in the upper graph and latitude in the lower graph, while the horizontal axis indicates the number of data. The orange line represents one set of the training data, while the blue line indicates predictions obtained through training. It can be seen that the predicted data are very close to learning data. There are slight discrepancies between two lines, but the observed errors are quite small.

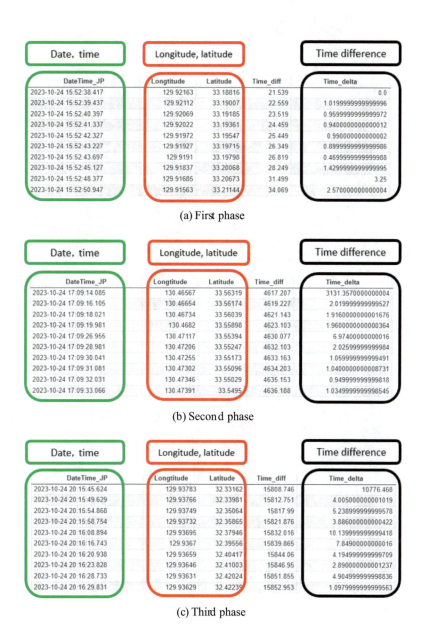

(a) First phase

(b) Second phase

(c) Third phase

**Fig. 3.** Sample of filtered ADS-B data.

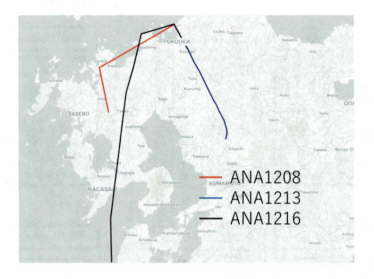

**Fig. 4.** Mapping diagram after filtering.

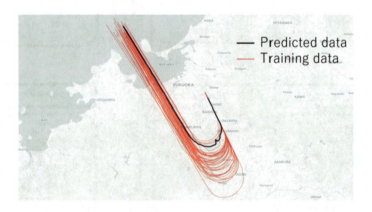

**Fig. 5.** Mapping diagram of training and predicted data.

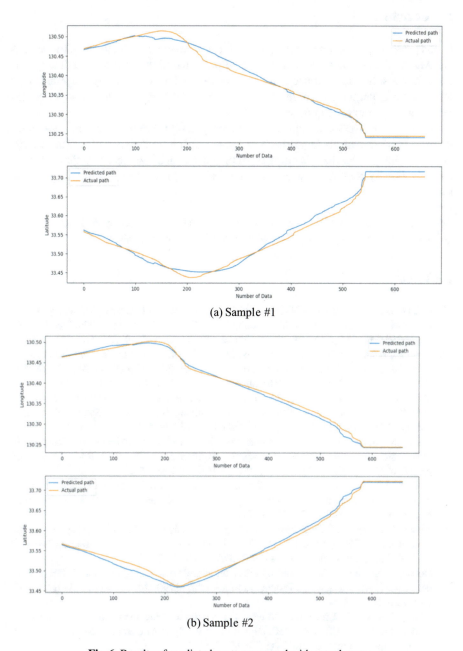

(a) Sample #1

(b) Sample #2

**Fig. 6.** Results of predicted route compared with actual route.

# 5  Conclusions

In this paper, we proposed a filtering method and assessed the effectiveness for processing ADS-B data to enhance the accuracy of predicted flight location coordinates. The evaluation results demonstrated that the proposed method successfully separated single routes for ML purpose. Also, the predicted data were very close to learning data and the observed errors were quite small.

In the future work, we will consider incorporating weather and wind conditions into the training data to improve the accuracy of the predicted routes. The filtering process developed for the utilization of ADS-B data can be applicable to the Automatic Identification System (AIS), which provides information on ship identifiers, types, and locations.

**Acknowledgment.** The ADS-B data used in this research are provided by the Electronic Navigation Research Institute (ENRI). The authors thank ENRI for the support.

# References

1. Cai, Q., Alam, S., Duong, V.N.: A spatial-temporal network perspective for the propagation dynamics of air traffic delays. Engineering **7**(4), 452–464 (2021). https://doi.org/10.1016/j.eng.2020.05.027. https://www.sciencedirect.com/science/article/pii/S2095809921000485
2. Choi, S., Kim, Y.J., Briceno, S., Mavris, D.: Prediction of weather-induced airline delays based on machine learning algorithms. In: 2016 IEEE/AIAA 35th Digital Avionics Systems Conference (DASC), pp. 1–6 (2016). https://doi.org/10.1109/DASC.2016.7777956
3. Duan, Y., Yisheng, L.V., Wang, F.Y.: Travel time prediction with LSTM neural network. In: 2016 IEEE 19th International Conference on Intelligent Transportation Systems (ITSC), pp. 1053–1058 (2016). https://doi.org/10.1109/ITSC.2016.7795686
4. Edrich, M., Schroeder, A.: Design, implementation and test of a multiband multistatic passive radar system for operational use in airspace surveillance. In: 2014 IEEE Radar Conference, pp. 12–16 (2014). https://doi.org/10.1109/RADAR.2014.6875546
5. Gui, G., Liu, F., Sun, J., Yang, J., Zhou, Z., Zhao, D.: Flight delay prediction based on aviation big data and machine learning. IEEE Trans. Veh. Technol. **69**(1), 140–150 (2020). https://doi.org/10.1109/TVT.2019.2954094
6. Honda, J., Otsuyama, T., Watanabe, M., Makita, Y.: Study on multistatic primary surveillance radar using DTTB signal delays. In: 2018 International Conference on Radar (RADAR), pp. 1–4 (2018). https://doi.org/10.1109/RADAR.2018.8557295
7. Honda, J., Otsuyama, T.: Feasibility study on aircraft positioning by using ISDB-T signal delay. IEEE Antennas Wirel. Propag. Lett. **15**, 1787–1790 (2016). https://doi.org/10.1109/LAWP.2016.2536725
8. Kim, Y.J., Choi, S., Briceno, S., Mavris, D.: A deep learning approach to flight delay prediction. In: 2016 IEEE/AIAA 35th Digital Avionics Systems Conference (DASC), pp. 1–6 (2016). https://doi.org/10.1109/DASC.2016.7778092
9. Martínez-Prieto, M.A., Bregon, A., García-Miranda, I., Álvarez Esteban, P.C., Díaz, F., Scarlatti, D.: Integrating flight-related information into a (big) data lake. In: 2017 IEEE/AIAA 36th Digital Avionics Systems Conference (DASC), pp. 1–10 (2017). https://doi.org/10.1109/DASC.2017.8102023

10. Matsuo, K., Ikeda, M., Barolli, L.: A machine learning approach for predicting 2D aircraft position coordinates. In: Proceedings of the 24th International Conference on Network-Based Information Systems (NBiS-2021), pp. 306–311, September 2021. https://doi.org/10. 1007/978-3-030-84913-9_30

11. Matsuo, K., Ikeda, M., Barolli, L.: A ML-based system for predicting flight coordinates considering ADS-B GPS data: problems and system improvement. In: Barolli, L., Kulla, E., Ikeda, M. (eds.) EIDWT 2022. LNDECT, vol. 118, pp. 183–189. Springer, Cham (2022). https://doi.org/10.1007/978-3-030-95903-6_20

12. Moreira, L., Dantas, C., Oliveira, L., Soares, J., Ogasawara, E.: On evaluating data prepro-cessing methods for machine learning models for flight delays. In: 2018 International Joint Conference on Neural Networks (IJCNN), pp. 1–8 (2018). https://doi.org/10.1109/IJCNN. 2018.8489294

13. Nijsure, Y.A., Kaddoum, G., Gagnon, G., Gagnon, F., Yuen, C., Mahapatra, R.: Adap-tive air-to-ground secure communication system based on ADS-B and wide-area multilat-eration. IEEE Trans. Veh. Technol. 65(5), 3150–3165 (2016). https://doi.org/10.1109/TVT. 2015.2438171

14. O'Hagan, D.W., Baker, C.J.: Passive bistatic radar (PBR) using FM radio illuminators of opportunity. In: 2008 New Trends for Environmental Monitoring Using Passive Systems, pp. 1–6 (2008). https://doi.org/10.1109/PASSIVE.2008.4787000

15. Olive, X., et al.: OpenSky report 2020: analysing in-flight emergencies using big data. In: 2020 AIAA/IEEE 39th Digital Avionics Systems Conference (DASC), pp. 1–10 (2020). https://doi.org/10.1109/DASC50938.2020.9256787

16. Pamplona, D.A., Weigang, L., de Barros, A.G., Shiguemori, E.H., Alves, C.J.P.: Supervised neural network with multilevel input layers for predicting of air traffic delays. In: 2018 Inter-national Joint Conference on Neural Networks (IJCNN), pp. 1–6 (2018). https://doi.org/10. 1109/IJCNN.2018.8489511

17. Peters, J., Emig, B., Jung, M., Schmidt, S.: Prediction of delays in public transportation using neural networks. In: International Conference on Computational Intelligence for Modelling, Control and Automation and International Conference on Intelligent Agents, Web Technolo-gies and Internet Commerce (CIMCA-IAWTIC 2006), vol. 2, pp. 92–97 (2005). https://doi. org/10.1109/CIMCA.2005.1631451

18. Post, J.: The next generation air transportation system of the United States: vision, accomplishments, and future directions. Engineering 7(4), 427–430 (2021). https:// doi.org/10.1016/j.eng.2020.05.026. https://www.sciencedirect.com/science/article/pii/ S209580992100045X

19. Schäfer, M., Strohmeier, M., Lenders, V., Martinovic, I., Wilhelm, M.: Bringing up Open-Sky: a large-scale ADS-B sensor network for research. In: IPSN-14 Proceedings of the 13th International Symposium on Information Processing in Sensor Networks, pp. 83–94 (2014). https://doi.org/10.1109/IPSN.2014.6846743

20. Sciancalepore, S., Alhazbi, S., Di Pietro, R.: Reliability of ADS-B communications: novel insights based on an experimental assessment. In: Proceedings of the 34th ACM/SIGAPP Symposium on Applied Computing, SAC 2019, pp. 2414–2421. Association for Computing Machinery, New York, NY, USA (2019). https://doi.org/10.1145/3297280.3297518

21. Shi, Z., Xu, M., Pan, Q., Yan, B., Zhang, H.: LSTM-based flight trajectory prediction. In: 2018 International Joint Conference on Neural Networks (IJCNN), pp. 1–8 (2018). https:// doi.org/10.1109/IJCNN.2018.8489734

22. Skolnik, M.I.: Introduction to Radar System, 3rd edn. Mcgraw-Hill College, New York, NY, USA (1962)

23. Smith, A., Cassell, R., Breen, T., Hulstrom, R., Evers, C.: Methods to provide system-wide ADS-B back-up, validation and security. In: 2006 IEEE/AIAA 25th Digital Avionics Sys-tems Conference, pp. 1–7 (2006). https://doi.org/10.1109/DASC.2006.313681

24. Stevens, M.C.: Secondary Surveillance Radar. Artech House, Norwood, MA, USA (1988)
25. Strohmeier, M., Lenders, V., Martinovic, I.: On the security of the automatic dependent surveillance-broadcast protocol. IEEE Commun. Surv. Tutorials **17**(2), 1066–1087 (2015). https://doi.org/10.1109/COMST.2014.2365951
26. Strohmeier, M., Schäfer, M., Lenders, V., Martinovic, I.: Realities and challenges of nextgen air traffic management: the case of ADS-B. IEEE Commun. Mag. **52**(5), 111–118 (2014). https://doi.org/10.1109/MCOM.2014.6815901
27. Willis, N.J.: Bistatic Radar, 2nd edn. Artech House (1995)
28. Yang, A., Tan, X., Baek, J., Wong, D.S.: A new ADS-B authentication framework based on efficient hierarchical identity-based signature with batch verification. IEEE Trans. Serv. Comput. **10**(2), 165–175 (2017). https://doi.org/10.1109/TSC.2015.2459709

# PvFL-RA: Private Federated Learning for D2D Resource Allocation in 6G Communication

Richa Kumari[✉], Dinesh Kumar Tyagi, and Ramesh Babu Battula

Department of Computer Science and Engineering, Malaviya National Institute of Technology,
Jaipur, India
richa.kumari7171@gmail.com, {dktyagi.cse,
rbbattula.cse}@mnit.ac.in

**Abstract.** Next-generation mobile networks (NGMN) have ushered in unprecedented demands for efficient data transfer and low-latency applications. Within the realm of 6G technology, device-to-device (D2D) communication emerges as a pivotal solution to address these challenges. Despite the promise of D2D in enabling massive data transfer with ultra-low latency, efficient radio resource allocation is required for D2D communication. This research introduces a novel decentralized private federated learning mechanism for D2D resource allocation (PvFL-RA) to tackle the privacy and resource management challenges in 6G. PvFL-RA integrates intelligent resource management methods with private federated learning, aiming to optimize the allocation of resources in a privacy-preserving manner. A novel D2D underlay communication is proposed, incorporating non-interference channel state information (CSI). PvFL-RA leverages CSI to extract channel gain concerning beam direction and distance, accurately determining non-interference zones for user equipment (UE) communication. The constructed CSI dataset trains a federated learning model incorporating local differential privacy (LDP) for predicting transmission power. Comparative analysis with the traditional centralized resource allocation model (CN-RA) demonstrates PvFL-RA's ability to accurately predict data rate and transmission power. Additionally, an abolition study contrasts PvFL-RA with federated learning-based resource allocation (FL-RA), revealing a privacy and accuracy tradeoff between the two models. Results underscore that the proposed decentralized PvFL-RA model significantly diminishes the overhead associated with centralized nodes while efficiently allocating near-optimal resources with enhanced privacy. This research contributes valuable insights into the evolving landscape of 6G D2D communication and resource allocation.

## 1 Introduction

In the era of 6G, the number of smart devices and data is tremendously increasing worldwide. To satisfy the aspirations of the NGMN, device-to-device communication is considered a vital component added in $3^{rd}$ Generation Partnership Project (3GPP) [6]. The bottleneck situation occurs in the traditional cellular mode of up-link and down-link data transmission due to the massive and rapid data transfer requirements for virtual game playing, metaverse communication, public safety, and immersive content sharing between nearby devices. Therefore, In-band D2D communication (IBD) [7] is one of the

© The Author(s), under exclusive license to Springer Nature Switzerland AG 2024
L. Barolli (Ed.): AINA 2024, LNDECT 199, pp. 261–273, 2024.
https://doi.org/10.1007/978-3-031-57840-3_24

6G technologies to improve resource re-usability. In IBD, the sharing of frequency spectrum between D2D user equipment (DUE) and cellular user equipment (CUE) improves the data rate by managing interference. The underlay and overlay D2D communication are part of IBD. The underlay D2D communication has become quite popular for the efficient spatial reuse of resources. In underlay D2D communication, proximity users transmit the data directly using a shared link without the involvement of the base station (BS). The underlay D2D communication enhances cellular network throughput by utilizing effectively licensed spectrum, capitalizing on rapid radio spectrum access with proximity gain, reuse gain, and paring gain [8]. However, interference between DUEs and CUEs is unavoidable due to sharing a single resource among massive DUEs. Consequently, the transmission power of DUEs needs to be adjusted appropriately. For D2D resource allocation, the beacon packet is needed to transfer between UEs and BS. This beacon packet consists of private information of UES like: the position, maximum transmission power, distance, etc. This private information is required to protect it from attackers. The differential privacy-enabled resource management mechanisms are designed and implemented with centralized and decentralized strategies that depend on the transmission power of the devices in the D2D 6G environment. In the CN-RA centralized model, a central macro base station (MBS) itself allocates the resources to UEs of a small base station (SBS) based on the CSI dataset. Using the FL-RA decentralized model, central MBS synchronously collects the gradients from all available local SBS, performs computation and provides an accurate model to the SBS node for resource allotment. In the PvFL-RA decentralized model, MBS aggregates local differential privacy (LDP)-based gradients and provides an efficient model for accurate resource allocation. PvFL-RA appropriately allocates the transmission power resources to the UEs with privacy constraints and substantially reduces the overhead of the central macro-base station-gNodeB (MBS-gNB) node. The primary contribution of this article is summarized as follows:

- Formulated a resource allocation mechanism for an underlay D2D communication network as a non-linear programming problem.
- Proposed a private federated learning model for D2D resource allocation (PvFLRA) provides a secure optimum resource allocation with reduced burden from the central node.
- The performance of the proposed model is validated through simulation. Furthermore, an ablation study compares PvFL-RA with the FL-RA model.

## 2  Related Work

Researchers focus on D2D communication resource allocation using mathematical models, game theory, graphs, and machine learning techniques. Current methods fall into two categories: centralized [9, 10] and distributed [11–13]. In the centralized approach [14, 15], the BS manages resource allocation for CUE and DUEs. Distributed methods [16, 17] aim to minimize signaling overhead. The use of deep learning (DL) in resource assignments has received attention recently. Without explicitly solving the complex problem, the optimal resource allocation problem can be achieved quickly using DL. Furthermore, Q-learning and Lagrange's dual decomposition-based joint resource and power allocation approach are proposed. The methodology in [18] combines the Equally Reduced Power (ERP) heuristic with a deep neural network to improve Quality of Service (QoS). It aims to ensure a minimum data rate for D2D users while minimizing interference to the base station (BS). In [19], the cellular mode switching problem is addressed with dynamic D2D mode using Markov Decision Process (MDP) to maximize long-term energy efficiency. In [20], the multi-agent deep reinforcement learning (MADRL) approach is introduced using a Stackelberg game (SG) framework. This approach is designed for smart power control and channel allocation, demonstrating faster convergence to optimal solutions than traditional MADRL methods. In [21], a deep reinforcement learning (DRL) based method is proposed for channel matching and power allocation to achieve maximum throughput and energy efficiency in the network. In [1], optimize the offloading decisions in federated learning, minimizing the overall utility cost of user equipment within a federated learning-based edge intelligence system. In [2], an energy-efficient federated learning framework addresses latency, bandwidth, and outage constraints under two scenarios: no CSI and partial CSI at the server. In the 6G industrial internet of things (IIoT), the author proposed [3] a D2D-aided DTEN model for efficient communication by IIoT devices in edge networks. Our decentralized resource allocation, combining federated learning and RL, maximizes D2D user capacity while ensuring QoS for all network users.

## 3 PvFL-RA: Private Federated Learning for D2D Resource Allocation in 6G Communication

In the 6G network era, smart devices and data are extensively increasing. D2D communication in 6G plays a key role in handling massive devices and huge amounts of data. Managing resources and ensuring device privacy are significant challenges in D2D communication. Thus, the private federated learning for D2D resource allocation (PvFL-RA) model is proposed to provide optimum resource allocation and privacy. The PvFL-RA algorithm goes through some phases that are described in Subsect. 3.1 and 3.2. Table 1 provides explanations for the notations used in further sections.

### 3.1 System Model

An underlay D2D communication is exemplified in Fig. 1, where CUE $u$ and DUE $v$ are under the coverage of SBS-gNB $s$. We assume only uplink cellular frequencies ($f_1$ and $f_2$) are shared with $u$ and $v$. The effect of interference on the receivers $v^{(tx)}$ and $s$ depends on $[\theta_{bd}^u, \theta_{bw}^u]$ and $[\theta_{bd}^{v^{(tx)}}, \theta_{bw}^{v^{(tx)}}]$, respectively. The direction of $\theta_{bd}^u$ and $\theta_{bd}^{v^{(tx)}}$ should not overlap $\theta_{bw}$ depends on the radiation power, which comprises the transmission power plus antenna gain. Accordingly, we need to select the adequate transmission power to decide the proper beam width for $u$ and $v^{(tx)}$ so they do not interfere with each other's signal. Consequently, we need to identify the channel gain and signal-to-noise plus interference ratio (SINR) to calculate the achievable data rate to maintain the quality-of-services (QoS) at the receiver side.

A. Channel Model

**Table 1.** Notation Table

| Symbols | Description |
|---|---|
| $M = \{1, 2, ...m\}$ | Number of CUE |
| $N = \{1, 2, ...n\}$ | Number of DUE pair |
| $L = \{1, 2, ...l\}$ | Number of available channel |
| $B = \{1, 2, ...b\}$ | Number of SBS-gNB clients |
| $u$, $v$, and $s$ | Represents CUE, DUE pair, and SBS-gNB, respectively |
| $v = [v^{(tx)}, v^{(rx)}]$ | $v$ comprises a pair of transmitter and receiver DUE |
| $\theta_{bd}$ and $\theta_{bw}$ | Beam direction and Beam width, respectively |
| $\theta_{bd}^u$ and $\theta_{bd}^{v^{(tx)}}$ | Beam direction of CUE $u$ and DUE $v^{(tx)}$ |
| $u$ $\theta_{bw}^u$ and $\theta_{bw}^{v^{(tx)}}$ | Beamwidth of CUE $u$ and DUE $v^{(tx)}$ |
| $\Upsilon_{s,u_m}^l$ and $\Upsilon_{v_n^{(rx)},v_n^{(tx)}}^l$ | Datarate of CUE and DUE on cellular link $l$, respectively |
| $\xi_{s,u}^l$ | SINR between SBS-gNB $s$ and CUE $u$ |
| $\xi_{v^{(rx)},v^{(tx)}}^l$ | SINR between $v^{(rx)}$ and $v^{(tx)}$ DUE pair |
| $\rho_u$ and $\rho_v$ $\rho_u^{max}$, $\rho_v^{max}$ | Current transmission power of CUE and DUE pair, respectively Maximum transmission power of CUEs and DUE pair |
| $K$ and $J$ | Number of global iterations and epochs, respectively |

The macrocell propagation model is used to characterize the path loss between $u$ and $s$. Path loss $PL_{s,u}(dB)$ between $u$ and $s$ is defined as [4]:

$$PL_{s,u}(dB) = 128.1 + 37.6 \times log_{10}(d) \tag{1}$$

Here, $d$ is the separation between $u$ and $s$ in kilometers (km). Similarly, the path loss $PL_{s,v(tx)}(dB)$ and $PL_{v(rx),u}(dB)$ is determined using Eq. (1). The path loss $PL_{v^{(rx)},v^{(tx)}}(dB)$ between DUE pair $v^{(rx)}$ and $v^{(tx)}$ is calculated by Eq. (2):

$$PL_{v(rx),V(tx)}(dB) = 148 + 40 \times log_{10}(d') \tag{2}$$

Here, $d'$ is the distance between the DUE receiver $v^{(rx)}$ and transmitter $v^{(tx)}$ in km. Now, the channel gain between $s$ and $u$ is calculated as Eq. (3):

$$H_{s,u} = 10(-PL_{s,u}/10) \tag{3}$$

Likewise, the channel gain $H_{v^{(rx)},v^{(tx)}}$ is determined using Eq. (3) by putting the respective pathloss. Moreover, $\xi_{s,u}^l$ and $\xi_{v^{(rx)},v^{(tx)}}^l$ SINR are determined to compute datarate. Thus, the total data rate $R_{tot}$ $T_{tot}$ of the system is formulated as Eq. (4):

$$R_{tot} = \Sigma_{n=1}^N \Upsilon_{v_n^{(rx)},v_n^{(tx)}}^l + \Sigma_{m=1}^M \Upsilon_{s,u_m}^l \tag{4}$$

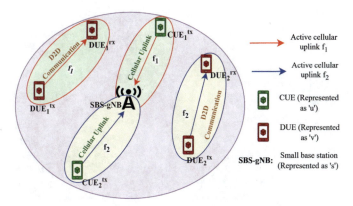

**Fig. 1.** Uplink-radio resource sharing between CUEs and DUEs

## 3.2  Problem Formulation

We aim to provide an efficient transmission power allocation strategy to maximize network throughput. Thus, we formulated the following optimization problem:

$$\psi = max_{\rho u,\rho v} \sum_{i=1}^{b} R_{tot}^{i}$$

$$s.t.\ C_1 : \xi_{s,u}^{l} \geq \xi_{s,u}^{l(req)}$$

$$C_2 : \xi_{s,u}^{l} \geq \xi_{v^{(rx)},v^{(tx)}}^{l(req)} \tag{5}$$

$$C_3 : 0 \leq p_u \leq p_u^{max}$$

$$C_4 : 0 \leq p_v \leq p_v^{max}$$

Here, $\Psi$ represents the total data rate of all $u$ and $v$ in the coverage of each $b$. Constraints $C_1$ and $C_2$ imply the transmission power of $u$ and $v$ should be higher than the channel gain to satisfy the required SINR. Constraints $C_3$ and $C_4$ indicate the transmission power of $u$ and $v$ should not be more than $p_u^{max\ x}$ and $p_v^{max}$, respectively. The formulated Eq. (5) is used to create a dataset that comprises the channel gain, positions of $u$ and $v$, distance, path loss, maximum transmission power, and available channels. This generated synthetic dataset is used for the proposed PvFL-RA model. Further, the proposed model aims to assure privacy with optimum accuracy as compared to the CN-RA model, as described in Eq. (6):

$$\Sigma_{it=1}^{k}\mathbb{L}^{PvFL}(\theta) \leq \Sigma_{j=1}^{J}\mathbb{L}^{CN}(\theta) \tag{6}$$

Here, $\theta$ represents the weight and bias of the model. $\mathbb{L}^{PvFL}$ and $\mathbb{L}^{CN}$ are the testing losses of the PvFL-RA and CN-RA models, respectively.

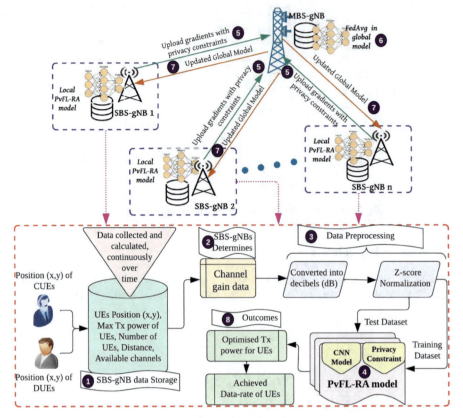

**Fig. 2.** Framework of PvFL-RA model

### 3.3 PvFL-RA Algorithm

The PvFL-RA model is a decentralized scheme of resource assignment. The process of the PvFL-RA approach is depicted in Fig. 2. SBS-gNB frequently collects information from CUEs and DUEs. Based on the received data, SBS-gNB calculates the channel gain matrix. In the data processing phase, convert the channel gain matrix unit into decibel milliwatts and perform zero mean-unit variance normalization. This data is first converted into a single-dimensional dataset. The proposed model comprises one convolution part with ReLu and one fully connected (FC) part. The convolutional part of a neural network is employed to extract essential features from the input data efficiently with minimal computational overhead. These extracted features are vital in determining the transmission power of CUEs and DUEs. Each SBS-gNB applies the PvFL-RA model to the local dataset and sends the gradients to the MBS-gNB. MBS-gNB performs the averaging of the gradients that are collected from all SBS-gNBs. The MBS-gNB reverts the updated gradients back to the local SBS-gNBs. Ultimately, this process goes on recursively until we receive enough accuracy from the model. The procedure of the PvFL-RA model is concluded in Algorithm 1.

---

**Algorithm 1 PvFL-RA using Local Differential Privacy (LDP)**

---

1: **Input:** MSB-gNB's global model $\mathbb{Q}$, Number of SBS-gNB clients $B$, Privacy Parameter $(\varepsilon, \delta)$, learning rate $\propto$, *clip*

2: **Output:** Updated MSB-gNB's global model $\mathbb{Q}$,

3: Initialize MSB-gNB's global model $\mathbb{Q}$,

4: **for** each global iteration **do**

5:         Select SBS-gNB clients $B$

6:         **for** each client $b$ in $B$ **do**

7:             Distribute $\mathbb{Q}$ to client $b$

8:             **for** each local epoch **do**

9:                 Train local model $\mathbb{Q}_b$ on the data of client $b$

10:                Compute predicted output: $y_{\text{pred}} = \mathbb{Q}_b(x)$                $\square$ $x$ channel gain input

11:                Compute MSE loss: $f = \frac{1}{2}(y_{\text{pred}} - y_{\text{true}})^2$

12:                Compute gradient of the MSE loss with respect to model parameters:

$$\nabla f_d = \frac{\partial f}{\partial \mathbb{Q}_b} = \frac{\partial}{\partial \mathbb{Q}_b}(\frac{1}{2}(y_{\text{pred}} - y_{\text{true}})^2)$$

13:                Determine $\sigma \Leftarrow= (\varepsilon, \delta)$

14:                Add Gaussian noise to the gradient: $\nabla f_b' = \nabla f_b + N(0, \sigma \times clip)$

15:                    Update local model parameters using the altered gradient $\mathbb{Q}_b' = \mathbb{Q}_b - \propto \cdot \nabla f_b'$

16:            **end for**

17:        **end for**

18:    Aggregateandupdateglobalmodel $\mathbb{Q} = \frac{1}{|B|}\sum_{b \in B} \mathbb{Q}_b'$

In Algorithm 1, the MSB-gNB's global model Q is initialized as a convolutional neural network (CNN) model. Then, the current global model Q is distributed to that SBS-gNB client $B$ in each global iteration. Subsequently, local model training occurs in epochs on the data specific to each SBS-gNB client. During local training, the SBS-gNB client computes the predicted output and the mean squared error (MSE) loss. The gradient of this loss with respect to the local model parameters is then calculated. Herein lies the crucial aspect of local differential privacy. A noise scale parameter $\sigma$ is determined based on the provided privacy parameters $(\varepsilon, \delta)$. Gaussian noise is introduced to the gradient during computation, deliberately altering individual updates to safeguard the privacy of local data. The local model parameters are then adjusted using the modified gradient. The global model is updated through aggregation after completing local training on all selected clients. This aggregated model becomes the updated global model Q. The entire process is repeated for a specified number of global iterations. At the end of the global iteration, we determine a single model that can work for all the SBS-gNB clients to determine the transmission power in a specific channel.

## 4   Performance Evaluation

The proposed PvFL-RA model's performance is compared with the CN-RA model in terms of learning operations and network performance. First, we compare the outcome of the PvFL-RA model with the MBS CN-RA. Then, perform the ablation study by comparing the PvFL-RA and FL-RA models. Table 2 summarizes the simulation parameters used to generate the dataset. Figures 3(a) and (b) depict the probability distribution of the normalized values within the dataset and the mean absolute error (MAE) of the MBS CN-RA model, respectively. Figure 4(a) and (b) indicate the MBS CN-RA model's training and testing losses and accuracy in each epoch, respectively. Figures 5(a) and (b) illustrate the loss and accuracy of the SBS-FL-RA and MBS-FL-RA models, respectively, after $1^{st}$ global iteration, wherein no differential privacy (DP) measures are applied in the FL process. Moreover, Figs. 6(a) and (b) showcase the loss and accuracy, respectively, of the SBS-PvFL-RA and MBS-PvFL-RA models following the $1^{st}$ global iteration incorporating DP in FL. In Fig. 6, the loss and accuracy of the PvFL-RA model are affected in comparison to the FL-RA model depicted in Fig. 5 for all SBS-gNB clients and MBS-gNB nodes. Hence, the performance of the PvFL-RA model is affected due to adding privacy constraints during the training process.

**Table 2.**  Simulation Parameter

| Parameters | Values |
|---|---|
| Area of simulation | 500 m radius |
| Carrier Frequency | 6GHz |
| $p_u^{max}$, $p_v^{max}$ | 23 dBm |
| Number of CUEs and DUE pairs in each SBS-gNB | 4 and 4, respectively |
| Number of SBS-gNB Clients $S$ | 4 |
| Maximum distance between two D2D users | 20 m |
| Learning rate ($\alpha$) and Batch size | 0.001 and 64 |
| *clip* | 0.1 |
| $\varepsilon$ and $\delta$ | 4 and $1 \times e^{-5}$, respectively |
| $K$ and $J$ | 20 and 100, respectively |

Figure 7(a) represents the loss and accuracy of SBS-FL-RA and MBS-FL-RA models after all global iterations where DP is absent. Furthermore, Fig. 7(b) demonstrates the loss and accuracy of SBS-PvFL-RA and MBS-PvFL-RA models after all global iterations in the presence of DP. Here, the total loss and accuracy of the PvFL-RA model are improved compared to the CN-RA model. Further, the performance of the PvFL-RA model is influenced by privacy restraints as compared to the FL-RA model. So, we can attain privacy while accepting a trade-off in accuracy. As illustrated in Fig. 8(a), employing the FL-RA model results in lower transmission power for DUEs and CUEs than the CN-RA model. This indicates that the FL-RA model excels in transmission

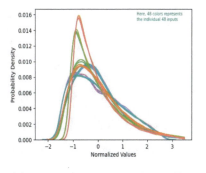
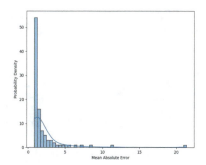

(a) Probability distribution of Normalized Values    (b) Probability distribution of mean absolute error

**Fig. 3.** Probability Density (PD) Graph

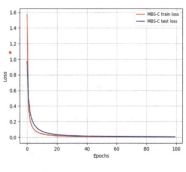
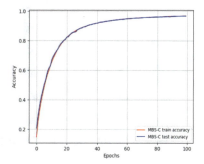

(a) Training and testing loss    (b) Training and testing accuracy

**Fig. 4.** Training and testing results of MBS-CN-RA model

power optimization. In Fig. 8(b), it is evident that the PvFL-RA model results in a lower transmission power for both CUEs and DUEs compared to the CN-RA model but higher than the FL-RA model. This highlights a nuanced trade-off, wherein the PvFL-RA model enhances privacy over the FL-RA model but at the expense of compromising little transmission power efficiency. As depicted in Fig. 9(a), the FL-RA model yields a higher data rate for DUEs and CUEs than the CN-RA model. This suggests the superior performance of the FL-RA model in data rate optimization. In Fig. 9(b), the PvFL-RA model achieves a data rate for both CUEs and DUEs that surpasses the data rate performance of the CN-RA model but falls short of the FL-RA model. This observation suggests that while the PvFL-RA model enhances privacy compared to the FL-RA model, there is a trade-off with network performance.

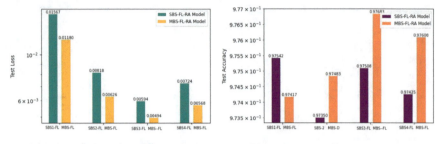

(a) SBS-FL-RA and MBS-FL-RA models test loss          (b) SBS-FL-RA and MBS-FL-RA model test accuracy

**Fig. 5.** SBS-FL-RA and MBS-FL-RA models after $1^{st}$ global iteration

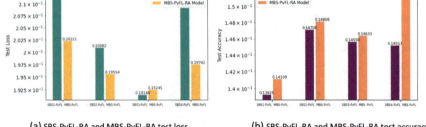

(a) SBS-PvFL-RA and MBS-PvFL-RA test loss          (b) SBS-PvFL-RA and MBS-PvFL-RA test accuracy

**Fig. 6.** SBS-PvFL-RA and MBS-PvFL-RA models after $1^{st}$ global iteration

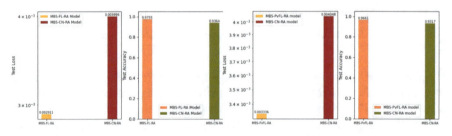

(a) MBS-CN-RA and MBS-FL-RA model          (b) MBS-CN-RA and MBS-PvFL-RA model

**Fig. 7.** Testing loss and accuracy of MBS's CN-RA Vs. FL-RA and PvFL-RA model

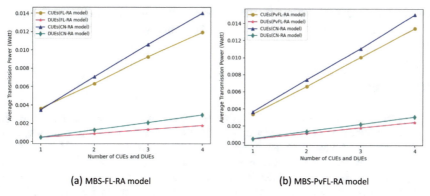

**(a)** MBS-FL-RA model          **(b)** MBS-PvFL-RA model

**Fig. 8.** Achieved Transmission Power using Federated Learning Approaches

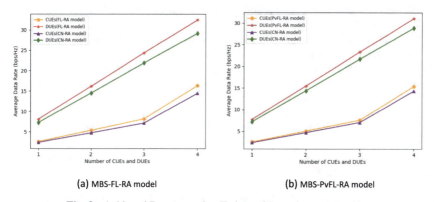

**(a)** MBS-FL-RA model          **(b)** MBS-PvFL-RA model

**Fig. 9.** Achieved Datarate using Federated Learning Approaches

## 5  Conclusion

D2D communication is a revolutionary technique to fulfill the requirements of high-speed data transfer with ultra-low latency for multimedia data exchange. Parallelly, user devices are increasing gradually. Thus, radio resource management and user privacy are significant challenges. The proposed PvFL-RA accurately predicts the non-interference zone with transmission power and provides the optimum data rate for devices as compared to the CN-RA model. In the context of 5G and beyond, the study considers the role of small base stations (SBS) possessing non-interference zone information for specific user equipment (UE) communications. Furthermore, the proposed mechanism introduces heterogeneity among SBS-gNB by adding local differential privacy (LDP) in federated learning to protect individual training data points. The simulation results show that PvFL-RA protects the identification of SBS-gNB's users at the cost of little accuracy as compared to the FL-RA model. Although we have demonstrated the effectiveness of our proposal, there are still more things to explore. For forthcoming research, we will consider more practical and massive scenarios of D2D communication, such as inter-cell interference, LoS and non-line-of-sight (NLoS) urban path loss models, and the mobility of users.

# References

1. Hu, B., Isaac, M., Akinola, O.M., Hafizh, H., Zhang, W.: Federated learning empowered resource allocation in UAV-assisted edge intelligent systems. In: IEEE 3rd International Conference on Computer Communication and Artificial Intelligence, pp. 336–341 (2023)
2. Mahmoud, M.H., Albaseer, A., Abdallah, M., Al-Dhahir, N.: Federated learning resource optimization and client selection for total energy minimization under outage latency, and bandwidth constraints with partial or no CSI. IEEE Open J. Commun. Soc. **4**(April), 936–953 (2023)
3. Guo, Q., Tang, F., Kato, N.: Federated reinforcement learning-based resource allocation for D2D-aided digital twin edge networks in 6G industrial IoT. IEEE Trans. Ind. Inf. **19**(5), 7228–7236 (2023)
4. Wang, D., Qiu, A., Zhou, Q., Partani, S., Schotten, H.D.: The Effect of Variable Factors on the Handover Performance for Ultra Dense Network (2023)
5. Yang, Y., Hui, B., Yuan, H., Gong, N., Cao, Y.: PrivateFL: accurate, differentially private federated learning via personalized data transformation. In: 32nd USENIX Security Symposium, USENIX Security, vol. 2023, no. 3, pp. 1595–1611 (2023)
6. Hoyhtya, M., Apilo, O., Lasanen, M.: Review of latest advances in 3GPP stand.: D2D communication in 5G systems and its energy consumption models. Fut. Internet **10**(1), 3 (2018)
7. Murtadha, M.K.: Adaptive D2D communication with integrated in-band and out-band spectrum by employing channel quality indicator. J. Eng. Sci. Technol. **17**(1), 0491–0507 (2022)
8. Kar, U.N., Sanyal, D.K.: A critical review of 3GPP standardization of device-to- device communication in cellular networks. SN Comput. Sci. **1**(1), 1–18 (2020)
9. Rashed, S.K., Asvadi, R., Rajabi, S., Ghorashi, S.A., Martini, M.G.: Power allocation for D2D communications using max-min message-passing algorithm. IEEE Trans. Veh. Technol. **69**(8), 8443–8458 (2020)
10. Budhiraja, I., Kumar, N., Tyagi, S.: ISHU: interference reduction scheme for D2D mobile groups using uplink NOMA. IEEE Trans. Mob. Comput. **21**(9), 3208–3224 (2021)
11. Li, X., Shankaran, R., Orgun, M.A., Fang, G., Xu, Y.: Resource allocation for underlay D2D communication with proportional fairness. IEEE Trans. Veh. Technol. **67**(7), 6244–6258 (2018)
12. Librino, F., Quer, G.: Distributed mode and power selection for non-orthogonal D2D communications: a stochastic approach. IEEE Trans. Cogn. Commun. Netw. **4**(2), 232–243 (2018)
13. Librino, F., Quer, G.: Channel, mode and power optimization for non-orthogonal D2D communications: a hybrid approach. IEEE Trans. Cogn. Commun. Netw. **6**(2), 657–668 (2019)
14. Sawyer, N., Smith, D.B.: Flexible resource allocation in device-to-device communications using stackelberg game theory. IEEE Trans. Commun. **67**(1), 653–667 (2018)
15. Yuan, Y., Yang, T., Hu, Y., Feng, H., Hu, B.: Two-timescale resource allocation for cooperative D2D communication: a matching game approach. IEEE Trans. Veh. Technol. **70**(1), 543–557 (2020)

16. Abrardo, A., Moretti, M.: Distributed power allocation for D2D communications underlaying/overlaying OFDMA cellular networks. IEEE Trans. Wireless Commun. **16**(3), 1466–1479 (2016)

17. Chen, Y., Ai, B., Niu, Y., Guan, K., Han, Z.: Resource allocation for device-to-device communications underlaying heterogeneous cellular networks using coalitional games. IEEE Trans. Wireless Commun. **17**(6), 4163–4176 (2018)

18. Lee, W., Lee, K.: Resource allocation scheme for guarantee of QoS in D2D communications using deep neural network. IEEE Commun. Lett. **25**(3), 887–891 (2020)

19. Zhang, T., Zhu, K., Wang, J.: Energy-efficient mode selection and resource allocation for D2Denabled heterogeneous networks: a deep reinforcement learning approach. IEEE Trans. Wireless Commun. **20**(2), 1175–1187 (2020)

20. Shi, D., Li, L., Ohtsuki, T., Pan, M., Han, Z., Poor, H.V.: Make smart decisions faster: deciding D2D resource allocation via stackelberg game guided multi-agent deep reinforcement learning. In: IEEE Trans. Mob. Comput. **21**(12), 4426–4438 (2022)

21. Yuan, Y., Li, Z., Yang, Y., Guan, X.: Double deep Q-network based distributed resource matching algorithm for D2D communication. IEEE Trans. Veh. Technol. **71**, 984–993 (2022). https://doi.org/10.1109/TVT.2021.3130159

# Adaptive SVC-DASH Video Streaming Using the Segment-Set-Based Backward Quality's Increment Control

Chung-Ming Huang$^{(\boxtimes)}$ and Han-I. Wang

Department of Computer Science and Information Engineering, National Cheng Kung University, Tainan, Taiwan
{Hungcm,wanghi}@locust.csie.ncku.edu.tw

**Abstract.** This paper proposed a Scalable Video Coding (SVC) based Dynamic Adaptive Streaming over HTTP (SVC-DASH) method using the Multi-access Edge Computing (MEC) technique over the vehicular network. Instead of using one video segment as the downing unit, the proposed method devises the concept of *Video Segment Set (VSS)*, for which a layer of those video segments delimited in a *VSS* is considered as a downloading unit. Then, the buffer-aware *VSS*-based bitrate adaptation through the backward quality's increment streaming control, which was executed in a MEC server, was proposed for SVC-DASH video streaming. Based on the results of performance evaluation, the proposed method can (i) enhance video quality, (ii) avoid video stalling and (iii) supplement video quality to the top level depending on the networking situation for vehicular network's SVC-DASH video streaming.

## 1 Introduction

In recent years, online video streaming service has become very prevalent. According to Ericsson Mobility Report, video traffic will reach 79% of global mobile data traffic by 2027 [1]. Scalable Video Coding (H.264/SVC), which is extended from H.264/AVC [2], separates a single-stream video into a multi-stream flow consisting of one base layer and some enhancement layers. The base layer has the lowest bit rate; the enhancement layers can increase the bit rate to improve the overall quality gradually [3]. Thus, a client selects and decodes different numbers of layers based on the condition of both the network and the device itself.

Dynamic Adaptive Streaming over HTTP (DASH) [4] adopts the HTTP protocol to achieve the reliable and bitrate-adaptive video streaming service. This work proposed the vehicular network -based SVC-DASH video streaming control method using the MEC technique [5], in which the main concern is that vehicle's mobility makes video streaming more complicated [6]. Instead of using one video segment as the downing unit, i.e., the downloading video segment can download k layers, the proposed method devises the concept of **Video Segment Set** *(VSS)*, for which the $k^{th}$ layer of those video segments delimited in a *VSS* is considered as a downloading unit. Three factors that control the adaptive video streaming of the proposed SVC-DASH method ve are (i)

L. Barolli (Ed.): AINA 2024, LNDECT 199, pp. 274–285, 2024.
https://doi.org/10.1007/978-3-031-57840-3_25

buffer occupancy, (ii) estimated bandwidth and (iii) quality variation. For the estimated bandwidth issue, it is derived based on the historical data that the client reports to the MEC server. For the buffer occupancy issue, to (i) avoid video stalling, (ii) enhance video quality and (iii) to supplement the video quality to the top level, it needs to set some thresholds for downloading's reference to decide which $VSS$'s video segments and which $VSS$'s $k^{th}$ layer(s) can be downloaded. For the quality variation issue, a layer of several video segments delimited in a $VSS$ becomes a downloading unit and a forwarding scheme for the $VSS$ is devised based on the estimated bandwidth and the buffer's situation. As a result, the proposed SVC-DASH method can improve the overall quality of service/experience (QoS/QoE).

The organization of the paper's remaining part is as follows. Related work is presented in Sect. 2. Details of the proposed SVC DASH method are described in Sect. 3. The results of performance evaluation are shown in Sect. 4. Conclusion remarks are given in Sect. 5.

## 2 Related Work

In [6], the authors proposed a DASH-based bitrate adaptation method called EDRA. EDRA's aim is to prevent video playback interruption and reduce the number of bitrate switches. Thus, EDRA adjusts each video segment's video bitrate dynamically by considering factors of (i) buffer level, (ii) current and previous bandwidth measurements and (iii) bitrate variation compared to currently downloaded segments. EDRA achieves a significantly higher accumulated played utility, ranging from 6% to 22%, and significantly reduces the number of bitrate switches by 30% to 77%.

In [7], an adaptation algorithm based on the DASH framework was proposed. The goal of the proposed method is to maximize the QoE of end-users by balancing three QoE matrices simultaneously: (i) avoid video stalling during the playback, (ii) minimize the quality variance in the downloaded video by controlling and smoothing the downloaded segments if the buffer fullness is sufficient, and (iii) maximize the aggregated video quality. To reduce the frequent video quality switching, the algorithm selects corresponding video segments' enhancement layers based on (i) the current network condition and (ii) two buffer thresholds $V_{up}$ and $V_{down}$ to reduce the quality difference and keep the high QOE for the downloaded segments that are stored in the buffer. Consequently, the devised adaptation algorithm can both (i) restrict re-buffering events and (ii) achieve quality's stability and high video quality at the same time.

In [8], the authors proposed an MEC-based framework and took the retention rate and popularity of video streams into consideration to achieve the maximum video bitrate. The defined optimization problem is maximizing the average video bitrate based on the limited computing capacity and the limited storage size. The proposed Online Iterative Greedy-base Adaptation (OIGA) algorithm adopts a greedy solution with strict limitations on the computing capacity and the storage size of the cache server. The experimental results shown that OIGA adapts with the change of the retention rate and video popularity to achieve the maximal video bitrate.

**Table 1.** The mapping between CQI and the corresponding data rate.

| CQI | Bitrate [Mbps] |
|-----|----------------|
| 1   | 2.768          |
| 2   | 2.768          |
| 3   | 4.432          |
| 4   | 7.548          |
| 5   | 10.32          |

**Table 2.** The mapping between SVC layers and corresponding data rate.

| Layer | Resolution | Avg. Bitrate [kb/s] | Total Bitrate [kb/s] |
|-------|------------|---------------------|----------------------|
| *BL 0* | 640*360   | 650                 | 650                  |
| *EL 1* | 1280*720  | 1550                | 2200                 |
| *EL 2* | 1920*1080 | 2800                | 5000                 |
| *EL 3* | 1920*1080 | 8500                | 13500                |

The works proposed in [6, 7] can be classified as the hybrid method considering both (i) the buffer's situation and (ii) the networking bandwidth. The work's proposed SVC-DASH method belongs to the hybrid method too. Nevertheless, our proposed method is also based on the MEC paradigm, i.e., the MEC-centric video streaming adopted in [8]. The proposed method's client reports its context to the MEC server periodically such that the MEC server can have related processing for the adaptive video streaming control. Then, the MEC server forwards the corresponding cached video segments' layers, to which the remote video server delivered, to the corresponding client. Consequently, client side's QoS/QoE of video streaming service over vehicular network can be improved.

## 3   The Proposed Method

Three aspects of the proposed method's goal are (i) avoiding the frequent quality switching and the extreme downgrading of the quality level, (ii) avoiding video stalling and (iii) maximizing the QoS/QoE of video streaming. The proposed SVC-DASH method selects (i) the suitable segments as a *VSS* and then (ii) the selected *VSS*'s corresponding video layers as the next downloading video unit based on the currently buffered situation and the currently available bandwidth. Thus, the proposed method consists of three control mechanisms: (1) the commencement mechanism, (2) the downloading control mechanism and (3) the playback control mechanism.

It needs to have enough buffered video data before commencing video presentation to avoid video stalling resulted from network jitter. Nevertheless, initially, since it does not have any historic downloading record, this work derives the initial data rate that a moving vehicle can have by referring 4G/5G cellular network's Channel Quality Indicator (CQI).

Table 1 depicts the relationship between CQI and the corresponding data rate [4]; Table 2 depicts the mapping between SVC layers and the corresponding data rate based on the given resolution. The proposed method's commencement control mechanism utilizes the CQI index currently sensed by the vehicle intending to have the service of SVC-DASH video streaming to get the associated network bit rate X and then the initially downloaded video segments' quality level Y is derived based on X.

Let the commencement threshold be $Buf_{commencement}$ and the buffered presentation time be $Buf_{cur}$. Let CQI be equal to 3, i.e., the corresponding network bandwidth is 4.432 Mbps referring to Table 1; referring to Table 2, the corresponding vehicle can download video layers of $BL + EL\ 1$ that are associated with the 1280*720 quality, which needs the bandwidth of 2200 kbps that is smaller than the estimated bitrate of 4.432 Mbps. Thus, the streaming can be commenced from the quality of $BL + EL\ 1$ when the corresponding vehicle's $Buf_{cur}$ reaches $Buf_{commencement}$.

The playout control mechanism functions are as follows. The proposed method defines three buffering buffer thresholds, i.e., $Buf_{low}$, $Buf_{mid}$ and $Buf_{high}$, to indicate the buffering situation. It can (1) trigger the video's playback and (2) maintain downloading video segments to playback in the future to increase (i) the buffered presentation time or (ii) the level of video quality when the buffered presentation time equals $Buf_{commencement}$, i.e., $Buf_{cur}$ equals $Buf_{commencement}$, for which $Buf_{commencement}$ is defined as $\lceil \frac{Buf_{low}+Buf_{mid}}{2} \rceil$.

Let (i) the currently playback segment be denoted as $S_p$, (ii) the currently downloading segment be denoted as $S_d$ and (iii) the buffer's last segment be denoted as $S_{last}$. During the video playback time, if $S_{p+1}$ equals $S_{last}$ and $S_d$ equals $S_{last}$, which means that the buffer's last segment is the next segment to playback and is in the status of downloading when the currently playback segment is ended, then it invokes the pause-resume control, i.e., suspend video playback and keep video downloading to quickly increase the buffered video's presentation time. When the currently buffered presentation time $Buf_{cur}$ equals threshold of $Buf_{resumed}$ that is set as equaling $Buf_{commencement}$, it can continue the video playback.

The devised downloading control mechanism functions are as follows:

(1) The Base Stage: If the buffer's situation is $Buf_{cur} \le Buf_{low}$, which represents that the buffer is toward empty and is in the "dangerous" zone. Consequently, this stage downloads only the base layer to replenish the buffer immediately to avoid stalling.
(2) The Enhancement Stage: If the buffer's situation is $Buf_{mid} < Buf_{cur} < Buf_{high}$, which represents the moderate situation and is in the "safe" zone to be able to avoid draining the buffer. Consequently, according to the estimated bandwidth, this stage can request base and/or several enhancement layers of the downloaded segments that are stored in the buffer to playback to increase these downloaded segments' quality.
(3) The Supplement Stage: If the buffer's situation is $Buf_{cur} \ge Buf_{high}$, which represents that it is in the "smooth" zone because enough buffered presentation time with enhanced quality is stored in the buffer. Consequently, those partially downloaded segments that are stored in the buffer can be supplemented to the full quality.

### 3.1 The Base Stage

An illustrated buffer's configuration because of the poor network's situation is depicted in Fig. 1. Referring to Fig. 1, when segment S97 that is $S_d$ and $S_{last}$ has been downloaded completely, segment S96 is to be the playback index $S_p$, i.e., the buffer only has two video segments and $Buf_{cur}$ is smaller than $Buf_{low}$. In other words, the buffer is to be empty and some follow-up segments' base layers need to be downloaded to quickly increase the buffered presentation time.

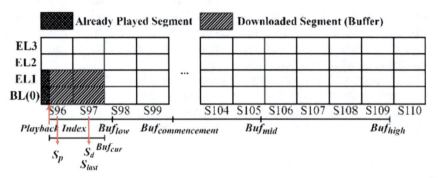

**Fig. 1.** An illustrated buffer's configuration for the Base stage.

When the next segment to playback is (i) the last segment and (ii) the currently downloading segment, i.e., $S_d$ equals $S_{last}$ and $S_{p+1}$ equals $S_d$, when the playback of $S_p$ is finished, which occurs in the extremely poor network's situation, then it triggers the pause-resume control process, i.e., the video stalling is launched, to download buffered video data quickly. When the buffered presentation time $Buf_{cur}$ reaches $Buf_{resumed}$, the video playback process is continued.

It can switch to the Enhancement stage from the Base stage when the situation of $Buf_{cur} \geq Buf_{mid}$ is reached, i.e., the buffered presentation time is equal to or greater than $Buf_{mid}$. The situation represents that a lot of segments' base layers have been downloaded because the network's situation is becoming better during the time of video playback.

### 3.2 The Enhancement Stage

The video streaming control can keep in the Enhancement stage unless one of the following two conditions happens: (a) $Buf_{cur} \leq Buf_{low}$, which needs to return to the Base stage; (b) $Buf_{cur} \geq Buf_{high}$, which can change to the Supplement stage.

The Enhancement stage's main concern is to estimate the available bandwidth to identify to download how many higher enhancement layers. This work devises Eq. (1), which considers the past networking situation to currently for estimating the available bandwidth $BW_e^t$ on time $t$ adopting the Exponentially Weighted Moving Average (EWMA) method:

$$BW_e^t = \alpha * BW_e^{t-1} + (1 - \alpha) * BW_m^t \qquad (1)$$

in which $BW_m^t$ denotes the measured bandwidth for a segment's video layer on time $t$, and $\alpha$, $0 < \alpha < 1$, is the weighting factor. The starting value $BW_e^0$ is derived based on the CQI value in the commencement control processing.

To reduce the frequence of quality switching in the Enhancement stage, a specific layer for every $N$ segments is selected as the *Video Segment Set (VSS)*. The *VSS*'s first/last segment is named as *VSS Header (VSSH)/VSS Tail (VSST)*. The Enhancement stage repeatedly executes the following six steps: (1) configuring *VSS*, (2) calculating estimated bandwidth $BW_e$, (3) selecting the follow-up video layer to download, (4) checking whether some more *VSS*'s video layers can be downloaded or not, (5) starting to download selected video layers of the segments delimited in *VSS*, and (6) shifting the range of *VSS*.

Step 1 adjusts *VSSH* to the segment next to segment $S_{low}$, which is derived using the threshold of the current $Buf_{low}$, i.e., *VSSH* equals $S_{low+1}$. Step 2 adopts Eq. (1) for estimating the available bandwidth $BW_e$.

Let segment $k$'s $BL$ to $EL$ $i$, $i = 1, 2, 3,.., n$, be in the buffer and $EL$ $i$ be denoted as segment $k$'s representation layer $RL(S_k)$. Thus, $RL(S_k)$ contains the index of segment $k$'s highest layer downloaded in the buffer. $RL(S_k)$ equals $-1$ when there is no buffered video layer of segment $k$. Step 3 finds $RL(S_k)$ of *VSS*'s every segment, for which the smallest one is denoted as $L_s$. Consequently, $L_s$ is the next video layer for downloading and $L_s + 1$ is the next higher video layer that is selected as the candidate one for downloading.

Step 4 compares $BW_e$ with the average bitrate of *VSS*'s layer $L_s + 1$ to decide to download which layer of which *VSS*'s segment. When (i) $BW_e$ is equal to or greater than the average bitrate of $L_s + 1$ and (ii) $L_s + 1$ is an $EL$, then execute Step 5 for downloading the remaining $(L_s + 1)^{th}$ layer of *VSS*'s segments, e.g., *VSS*'s segments from $S_{low+1}$ to $S_{low+N}$; otherwise, execute Step 6 for shifting the range of *VSS*.

Step 5 downloads the $(L_s + 1)^{th}$ layer from *VSS*'s corresponding segment to *VSST*. Let the downloading of *VSS*'s associated segments' $(L_s + 1)^{th}$ layer be called a round. During layer $L_s + 1$'s downloading round, the buffered presentation time is checked when a segment's downloading layer $L_s + 1$ is done: (i) it switches to the Base stage if $Buf_{cur} \leq Buf_{low}$, (ii) it switches to the Supplement stage if $Buf_{cur} \geq Buf_{high}$, (iii) otherwise, it keeps in the Enhancement stage. $BW_e$ needs to be estimated in the background after one segment's one layer is downloaded when it is kept in the Enhancement stage. During the downloading round of a *VSS*'s layer, if $S_{p+1}$ equals $S_d$, i.e., the next segment to playback is the currently downloading segment, when the playback of the $S_p$ segment is ended, then the interrupted situation may occur:

(1) $S_{p+1}$ equaling $S_d$ and $S_d$ equaling $S_{last}$: The situation represents that the last segment in the buffer is the next segment to playback, i.e., the networking situation becomes extreme poor, and it needs to trigger the pause-resume processing to continuously downloading $BLs$. When $Buf_{cur} = Buf_{resumed}$, it can continue the playback.

(2) $S_{p+1}$ equals $S_d$ and $S_d$ does not equal $S_{last}$: The situation represents that an $EL$ of the next segment, which is not buffer's last segment, to playback is still in the downloading status. It suspends the downloading of segment $S_d$ and checks the current buffered presentation time: (i) if it is the condition of $Buf_{cur} > Buf_{low}$, then *VSS* is re-configured, i.e., execute Enhancement stage's Step 1; (ii) if it is the condition of $Buf_{cur} \leq Buf_{low}$, then end the Enhancement stage and switch to the Base stage.

Step 6 adjusts (i) the original *VSST*'s next segment as the new *VSSH* and (ii) the segment pointed by the new *VSSH* + *N-1* as the new *VSST*. It executes Step 3 to select which segment's which layer to download after the range of *VSS* is shifted.

### 3.3 The Supplement Stage

The streaming control can switch to the supplement stage when one of the following conditions occurs: (i) *Buf $_{cur}$* ≥ *Buf $_{high}$*, which represents the currently buffered presentation time is equal to or greater than *Buf $_{high}$*, (ii) the highest enhancement level of some segments is not reached. The supplement stage contains three Steps: (1) determining the *Nominee VSS (NVSS)*, (2) supplementing the *nominee segments* in *NVSS* sequentially, and (3) checking the buffer situation.

Step 1 forms *NVSS* by selecting a subset of segments between $S_{low+1}$ and $S_{last}$. When the highest layer of a segment is not in the current buffered enhancement layers, the segment becomes a nominee. Let *NVSS* contain $k$ segments. Then, supplement each *nominee segment* from the *1$^{st}$ nominee segment* to the *$k^{th}$ nominee segment* in sequence with its un-downloaded video layers completely.

The currently buffered presentation time needs to be checked when the supplement of one layer of a *NVSS*'s nominee segment is done: (i) in the condition of $Buf_{cur} \geq Buf_{high}$: the supplement process can be continued; (ii) in the condition of $Buf_{cur} < Buf_{high}$: it switches to either the Enhancement stage or the Base stage. If the networking situation is so good such that (i) the highest layers of all segments between $S_{low} + 1$ and $S_{last}$ have been downloaded and (ii) the buffered presentation time equals $Buf_{high}$, i.e., $Buf_{cur}$ equals $Buf_{high}$, then it pauses the video downloading to avoid buffer overflow.

During the supplement process, if $S_p + 1$ equals $S_d$, i.e., the next segment to playback is the currently supplementing segment, when segment $S_p$'s playback is ended, it may result in the interrupted situation. The processing of the interrupted situation is the same as that in the Enhancement stage.

## 4 Performance Evaluation

The proposed method's performance evaluation results are presented in this Section.

The emulation environment for evaluating performance contains (1) one Universal Software Radio Peripheral (USRP), which is a low-cost eNB of software radio and widely used for academic universities' research, and (2) two computers. USRP adopts SrsRAN, which is an open-source LTE software suite. The Linux tc command is used to control the bandwidth between eNB and each moving vehicle through a computer. The proposed method and compared methods are coded as edge applications executed in the MEC server that is coded in Python 3.9. A mobility model using SUMO is defined for simulating vehicles' mobility. The adopted video contains 1 base layer and 4 enhancement layers, for which Table 3 shows each layer's video bitrate. Each segment's playback time is 2 s.

Three methods used to compare with the proposed SVC-DASH are as follows:

(1) The conventional SVC-DASH (CSVC) method, in which a client requests a video segment based only on the estimated network bandwidth calculated through Eq. (1).

**Table 3.** Different quality video bitrates

|        | Avg. Bitrate [kbps] | Cumulative Bitrate [kbps] |
|--------|---------------------|---------------------------|
| BL 0   | 600                 | 600                       |
| EL 1   | 2250                | 2850                      |
| EL 2   | 4820                | 7670                      |
| EL 3   | 6750                | 14420                     |
| EL 4   | 11250               | 25670                     |

(2) The Rate Adaptation SVC-DASH (RASD) method proposed in [7], which also adopts the backward quality increment way to increase the quality of the already downloaded segments when (i) the estimated network bandwidth is enough or (ii) the buffer fullness is in the safe status when it does not have the enough estimated network bandwidth. RASD defines two buffer thresholds $\alpha$ and $\beta$, for which it denotes the buffer is (i) in the safe status when the buffered presentation time is bigger than $\alpha$ and (ii) not in the safe status when the buffered presentation time is smaller than $\beta$, which may result in video stalling.

(3) The Video Quality Adaptation Framework (VQAF) proposed in [8], in which two contained phases are (i) the download phase and (ii) The smooth-out phase. The former one can request new video segments with the corresponding quality when the buffered presentation time does not reach $Th_{safe}$. The later one increases the quality level one by one for a previously downloaded segments' subset that is still stored in the buffer to playback repeatedly to reduce the possibility of layer's switching when the buffered presentation time is still bigger than $Th_{safe}$.

Three performance metrics used for comparison are as follows:

(1) **Average Segment Quality Level** ($Avg_{QL}$), which is derived by summing all of the downloaded segments' quality levels and then being divided by downloaded segments' amount. For convenient calculation, the quality level (i) of *BL 0, EL 1, EL 2, EL 3* and *EL 4* is denoted as *0,1,2,3,4* respectively, and (ii) is denoted as *-1* when it is in the video stalling situation.

(2) **Average Representation Bitrate** ($Avg_{RB}$), which is derived by summing all downloaded segments' cumulative bitrates and then being divided by downloaded segments' amount. The cumulative bitrate of a segment denotes the sum bit rate of all of the segment's downloaded video layers, for which possible values of the used video are depicted in Table 3.

(3) **Average of the Quality Level Switching** ($Avg_{Diff}$), which is derived by summing all of the absolute value of two adjacent segments' quality levels' difference and then being divided by the number of having quality level switching in adjacent segments. The quality level is set as -1 if video stalling occurs.

Let the min/max/average bitrate be 5780/13900/10040 respectively. Figure 2 depicts (i) the measured bandwidth's configuration and the estimated bandwidth's configuration

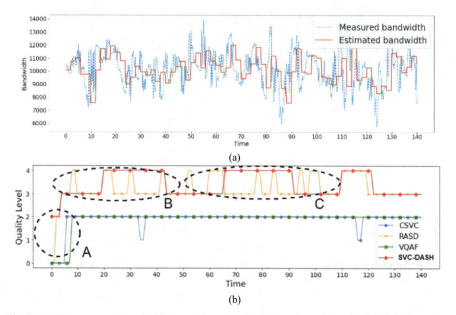

**Fig. 2.** (a) The measured bandwidth's configuration and the estimated bandwidth's configuration and (b) different methods' configurations of streaming quality during the simulation period.

using Eq. (1) and (ii) different methods' configurations of streaming quality during the simulation period.

Referring to Fig. 2-(b), it can be observed that CSVC has the lower quality level but more stable quality level switching than the proposed SVC-DASH method and RASD. Referring to Fig. 2-(a), the available bandwidth is between 5780 and 13900, Based on Table 3's depicted cumulative bitrate, since CSVC selects the next downloaded segment's quality level that CSVC selects is based only on the estimated bandwidth, which is in the range from 7579 kbps to 11998 kbps, CSVC can select to download (i) BL, (ii) BL + EL 1, and (iii) BL + EL 1 + EL 2. Since the highest estimated bandwidth is between the cumulative bitrate of BL + EL 1 + EL 2 and the cumulative bitrate of BL + EL 1 + EL 2 + EL 3, i.e., BL + EL 1 + EL 2 is the highest quality level that CSVC can download, CSVC has the more stable video quality. Additionally, CSVC keeps download the quality level of BL + EL 1 + EL 2 because most of the estimated bandwidth is enough to download BL + EL 1 + EL 2.

The quality level variation of the proposed SVC-DASH method is more stepwise than that of CSVC, RASD and VQAF, which can be observed in the region marked with text "A" of Fig. 2-(b). The reason is that the proposed SVC-DASH method adopts the CQI value to decide the initial quality level, but other methods just download BLs to minimize the commencement delay.

Referring to the region marked with "B" and "C" in Fig. 2-(b), the proposed SVC-DASH has the higher stability of video quality than RASD, which has many "hills", i.e., the quality level is increased and then decreased in a short period. It results from RASD's backwardly increasing quality's characteristic. RASD estimates the available bandwidth after downloading one BL. Thereafter, when the buffered presentation time is in the "safe" condition or there is enough estimated bandwidth, it downloads the higher quality layer of previously downloaded segments that are still stored in the buffer to playback in the horizontal way, i.e., increasing all of these segments' quality level. Since the proposed SVC-DASH method's downloading is based on one $VSS$'s video layer, i.e., same video layer of several segments, in contrast to one segment's all higher video layers, in the Enhancement stage, the proposed SVC-DASH method has the better stability of quality level.

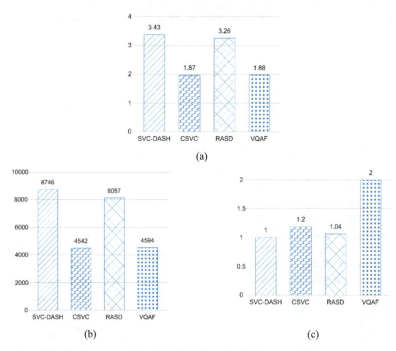

**Fig. 3.** Comparison results of (a) $Avg_{QL}$, (b) $Avg_{RB}$ and (c) $Avg_{Diff}$.

Comparing with SVC-DASH and RASD, VQAF shows the lower quality level but more stable quality level switching in Fig. 2-(b). The main reason is that VQAF keeps BLs' downloading to increase the buffer before the buffered presentation time being bigger than the lower bound threshold in the download phase. VQAF requests appropriate quality levels' new video segments based on the estimated bandwidth when the buffer presentation time equals the lower bound threshold.

Figure 3 depicts $Avg_{QL}$, $Avg_{RB}$ and $Avg_{Diff}$ of these methods. Considering the QoS aspect, the proposed SVC-DASH method has the higher $Avg_{QL}$ and $Avg_{RB}$ than the other methods, which can be observed in Fig. 3-(a) and (b). It is because SVC-DASH is able to (i) backwardly increase the quality level in the Enhancement stage and (ii) maximize the video quality of the downloaded segments that are still stored in the buffer to playback in the Supplement stage.

In addition to the aforementioned aspects, the proposed SVC-DASH method's $Avg_{QL}$ and $Avg_{RB}$ are much higher than that of VQAF. It is because even if it has enough estimated bandwidth to download the higher quality level's segments in the smooth-out phase, VQAF does not increase previously downloaded segments' video quality when adjacent segments have no quality level difference. That is, VQAF only increases previously downloaded segments' quality level when the adjacent segments' difference of the quality level is bigger than 0 in the smooth-out phase.

Figure 3-(c) shows that the proposed SVC-DASH method's $Avg_{Diff}$ is the lowest among these four methods. It is because, instead of in the unit of one segment's video layers, SVC-DASH's video downloading is in the unit of one video layer of those segments contained in a VSS in the Enhancement stage, which consequently results in the better stability of video quality.

## 5   Conclusions

The *VSS*-based SVC-DASH video streaming method that adopts backward quality's increment and the bitrate adaptation principle based on the MEC paradigm to provide vehicular network's SVC-DASH streaming service has been proposed in this work. That the proposed SVC-DASH method's adopted three factors to achieve stable quality and suitable bitrate adaptation are (i) buffer occupancy, (ii) estimated bandwidth and (iii) quality variation. The proposed SVC-DASH method, which is executed as an edge application in the MEC server, has (i) used three predefined thresholds as the downloading reference and (ii) considered to download video layers of a set of segments, instead of an individual segment's some video layers, at the same time to decide to download which video segments and which layers. In this way, it is able to (i) avoid buffer underflow, i.e., video stalling, (ii) enhance video quality and/or to supplement the quality to the maximum using the backward quality's increment process to achieve adaptive video streaming having the more stable video quality in the vehicular environment's various networking situation. The performance evaluation's results have proven that the proposed SVC-DASH method is able to achieve both (i) the higher average quality level and received bitrate, i.e., having the better QoS, and (ii) the higher quality stability, i.e., having the better QoE, than the other compared methods that also use the backward quality's increment processing.

**Acknowledgments.** This work presented in the work has the final support from the Ministry Of Science and Technology (MOST), Taiwan, for which the grant number MOST 111–2221-E-006 -117 -MY3.

## References

1. Wiegand, T., Sullivan, J.-G., Bjøntegaard, G., Luthra, A.: Overview of the H.264/AVC video coding standard. IEEE Trans. Circ. Syst. Video Technol. **13**(7), 560–576 (2003)
2. Unanue, I., et al.: A tutorial on H.264/SVC scalable video coding and its tradeoff between quality, coding efficiency and performance. Recent Advances on Video Coding (open access), Intechopen (2011)

3. Kua, J., Armitage, G., Branch, P.: A survey of rate adaptation techniques for dynamic adaptive streaming over HTTP. IEEE Commun. Surv. Tutorials **19**, 1842–1866 (2017)

4. Lee, G., Saad, W., Bennis, M., Kim, C., Jung, M.: An online framework for ephemeral edge computing in the Internet of Things. IEEE Trans. Wireless Commun. **22**, 1992–2007 (2023)

5. Vineeth, N., Guruprasad, H.: A survey on the techniques enhancing video streaming in VANETs. Int. J. Comput. Netw. Wirel. Mob. Commun. **3**, 37–46 (2013)

6. Togou, M.A., Muntean, G.M.: An elastic DASH-based bitrate adaptation scheme for smooth on-demand video streaming. In: Proceedings of 2022 IEEE International Symposium on Broadband Multimedia Systems and Broadcasting (BMSB), Bilbao, Spain (2022)

7. Lekharu, A., Kumar, S., Sur, A., Sarkar, A.: A QoE aware SVC based client-side video adaptation algorithm for cellular networks. In: Proceedings of the 19th International Conference on Distributed Computing and Networking (2018)

8. Tran, A.T., Dao, N.N., Cho, S.: Bitrate adaptation for video streaming services in edge caching systems. IEEE Access **8**, 135844–135852 (2020)

# How to Run GUI Applications on a Java Development Environment Based on JupyterLab

Yibao Liang[✉] and Minoru Uehara[✉]

Graduate School of Information Science and Arts, Toyo University, Saitama, Japan
{s3B102210018,uehara}@toyo.jp

**Abstract.** Programming education is important. Although Java is an important language, it is rarely used in introductory programming education. One reason for this is that there is no programming education environment suitable for Java. In previous research, we showed that running Java on JupyterLab is a suitable method for programming education. Furthermore, we used JupyterLab to build a browser-based Java development environment and verified whether the actual class code could be executed. We found that Java source code that does not use a graphical user interface (GUI), including code that performs communication, is almost 100% executable. However, code that uses a GUI was difficult to execute. In this paper, we propose a method to run Java code that uses a GUI using Jupyter Lab and CheerpJ. We found that GUI code that does not perform communication is almost 100% executable.

## 1 Introduction

Programming education is becoming increasingly important. In many countries, programming education is provided in elementary and junior high schools to cultivate programming thinking and problem-solving skills. However, the definition and evaluation criteria for programming proficiency are still vague. Although graphic languages such as Scratch and Blockly are often used in programming education in elementary education, they are not appropriate for higher education and vocational training [1, 2].

There are problems with introductory programming education in existing higher education practice. Although there is a large body of literature in the field of introductory programming education, there is little systematic evidence to support any particular approach, and hence no teaching method has yet been shown to be scientifically sound [3]. The pass rate for introductory programming education at higher educational institutions, such as universities, was previously found to be about 68% [4]. This pass rate was lower than that of other introductory science, technology, engineering, and math courses [5]. In addition, even after many years, the pass rate for introductory programming education in higher educational institutions is still only about 70% [6, 7]. The success rate is even lower for online classes such as massive open online courses [8], and even if learners complete the course content, they score poorly on programming foundation tests [2].

To improve the pass rate of introductory programming courses, the following problem-solving abilities and mathematical knowledge were found to be the most important: prior knowledge of programming learning, student motivation and involvement, and appropriate methods and tools [9]. In addition, student frustration is strongly related to prior knowledge. Although many educational methods and tools have been developed to handle student frustration, their spread is limited. Promoting the dissemination of these educational methods and tools and investigating their long-term effects is necessary [10].

The approach to introductory programming education focusing on data science has three characteristics [11, 12]: 1) it concentrates on data analysis using Python and does not emphasize traditional programming techniques such as object-oriented programming (OOP), 2) students are motivated to learn to solve specific problems in their own work through data analysis, and 3) it uses Jupyter, an easy-to-use tool suitable for data science. This educational approach is thought to enable a wide variety of students to quickly acquire programming techniques optimized for data science, but it lacks versatility. With reference to these data science-based approaches, we propose that computer science in general can also use Jupyter for introductory programming education. It is superior to traditional integrated development environments (IDEs) in providing an interactive programming environment, information dissemination, and ease of use.

Jupyter is a good way to realize a teaching approach. Education using Python is often done in a Jupyter (or Colab) environment. The pass rate of CS1 (computer science introductory) is not strongly related to language [4, 6], and language should depend on educational objectives [3]. Hence, depending on the educational policy of the university, languages other than Python may be taught. The Toyo University Faculty of Information Sciences and Arts also teaches Java. Unlike interpreted languages, Java requires compilation, and complex PC-based IDEs are used in regular classes. A browser-based Java development environment such as Jupyter reduces the cost of learning an IDE for beginners [3]. Programming lessons can be conducted in ordinary classrooms, any device that can open a browser can be used, and the bottleneck caused by the requirement for PC classrooms can be alleviated [1]. Embedding IJava in Colab allows anyone to run Java programs without a local server. However, Colab IJava cannot handle some topics, such as network computing and graphical user interfaces (GUIs).

In previous research, we proposed a browser-based Java learning environment using tools from the Jupyter series and evaluated it using code that is used in actual classes. Our conclusion is that it is possible to cope with a wide range of education, from non-GUI-based basic processing to processing using a network. Moreover, in this paper, we propose a method to execute Java code using GUI by using Jupyter Lab and CheerpJ. We found that a Java GUI application that does not perform communication is almost 100% executable.

This paper is organized as follows. Section 2 describes related work. Section 3 describes the design and implementation of the system. Section 4 evaluates the proposed approach, and Sect. 5 presents our conclusions.

## 2  Related Work

### 2.1  Inspiration from Data Science

Data science, which has become popular in recent years, also requires programming. Table 1 presents an example of the technology stack that modern data scientists must learn so that they can write code to obtain insights from data in a robust and reproducible manner [11]. Data scientists often learn these skills from fellow data scientists rather than formal computing instructors [11]. Dedicated programming education for data analysis can address the challenges posed by the high heterogeneity of the data science student population. It also lowers the barriers caused by the need for prior knowledge from elsewhere. Because data science course content is often compiled by data scientists from industry, the skills students learn during their education can cover a wider range of the skills they will need in future jobs. Tools such as Jupyter and Colab are already popular with data scientists. There are many cases in which these tools are used for Python education. Because much of the data science educational content and tools are from actual work and research sites, we found that data science teaching methods can deal well with the issues mentioned in Chapter 1. Therefore, we focus on Jupyter. Features such as ease of use, ease of sharing, and high computational reproducibility [14, 15] mean that Jupyter can be used not only for work and research, but also for learning.

However, programming specialized for data analysis has little versatility, and it can be assumed that it is difficult for data science students to use the programming techniques they have learned to develop programs for purposes other than data analysis.

**Table 1.** Example technology stack used in data science.

| Content |
| --- |
| Data modeling & visualization libraries (e.g., pandas, R tidyverse) |
| Base programming language (e.g., Python, R) |
| Computational narrative & workflow (e.g., Jupyter, RMarkdown) |
| Development support (e.g., Git, Python/R package managers) |
| Unix command line for cross-app scripting and sysadmin |
| Reproducibility infrastructure (e.g., Docker, virtual machines) |

In previous research, we proposed three effective educational characteristics for computer science. Moreover, these educational methods or tools should be realized in a form that is easy to disseminate. With respect to prior knowledge, programming specialized for data analysis assumes the "computation-as-computation" metaphor [16]. By contrast, Stein insists on replacing "computation-as-computation" with "computation-as-interaction" in CS1 [17], generating the syllabus presented in Table 2. From this point of view, programming is about composing a community of interacting processes, much like the structure of software today. There have been reports that this teaching method is effective in places other than the Massachusetts Institute of Technology [18].

The "computation-as-interaction" approach is very different from the approach of current mainstream CS1 teaching and is closer to the content after CS2. Some students (or "code warriors") resist this teaching method [18]. In any case, programming courses should incorporate different elements from different learning styles so as to be suitable for all programming novices [13]. In addition, to fill the gaps between the content of courses and supplemental detailed content that is not sufficient for one course, online micro-course-type public lectures are an effective way to supplement the prior knowledge of students.

**Table 2.** Course syllabus of interactive programming. This reconceptualization allows a radical shift that runs throughout the curriculum without requiring a significant restructuring of our course sequence [17].

| Content | Laboratories |
| --- | --- |
| - Introduction to Interactive Programming | Spirograph (Expressions and Statements) |
| - Expressions and Statements | - Nodes and Channels (Interactions) |
| - Objects and Classes | - Design Project |
| - Interfaces and Exceptions | - Balance (Classes) |
| - Self-Animating Objects | - Calculator (Procedures) |
| - Inheritance | - (Documentation Project) |
| - Object-Oriented Programming | - Scribble (Events) |
| - Dispatch Mechanisms | - Cat and Mouse (Systems of Systems) |
| - Procedural Abstraction | - Final Project (Networked Interactions) |
| - Events Driven Programming and java.awt | |
| - Event Delegation (and more java.awt) | |
| - Safety, Liveness, and Synchronization | |
| - Interfaces and Protocols: Composing Systems | |
| - Explicit Communication: java.io, java.net | |
| - Servers | |
| - Arbitration or RMI | |
| - Design Architectures | |
| - Interactive Programming as Program Design | |

With respect to student motivation, it is difficult to obtain or use teaching materials from typical data science workplaces. The use of problem-based learning(PBL)should increase the motivation of students by asking them to envision some specific problems and set the goal of solving them as they proceed with their studies. JeLL [19] is a good implementation of PBL. JeLL is a system that applies and supports collaborative learning by creating problems similar to programming problems. Computer science learning and performance can be improved through collaboration among students [20].

As for educational tools, we suggest using Jupyter for several reasons. First, Jupyter is programming language-agnostic and can support the incorporation of diverse learning styles and elements into the classroom [13]. The use of Jupyter does not preclude the use of traditional programming environments, but instead augments them [21]. Interactive programming in Jupyter seems to be well suited for the implementation of the "computation-as-interaction" approach [17]. In addition, Jupyter's accessibility, simplicity, functionality, and other features meet the requirements for improving student participation [21]. Beginners can easily recover from errors such as misspelled words or line syntax errors when editing code cells. Students enjoy adding different twists to the sample code, and all code is easy to edit and evaluate [21]. Finally, Jupyter is open software, supported over the long term, customizable, and retains learning records.

## 2.2 Jupyter also Suitable for Java

Depending on the educational policy of the university, languages other than Python may be taught, and traditional IDEs are often used in Java education. Traditional IDEs were developed to streamline the programming process. However, in the case of introductory programming education, there are many cases where the functions that make programming more efficient are unnecessary, and learning traditional IDEs then becomes a burden for beginners [3]. If a Jupyter-like method is used to construct the environment for learning, it is only necessary to build the environment once, and it can be used simply by accessing the built web page anytime (the concept of "build once, use anywhere"). Multi-device or low-performance facilities (e.g., a PC for library catalog search) can also be used for programming learning and practice. The physical barrier to learning is lowered, and the need for a PC classroom can be overcome [13]. Jupyter's interactive programming approach also matches the evolution of the Java language itself. Java's read-eval-print-loop shell interface, called JShell, was defined for teaching and debugging [22]. Java modifiers can be concisely described in the "Main" method to reduce the initial burden on beginners [23]. These description methods can be used in Jupyter. However, there is little research on Jupyter-like environments in computer science classes [21].

Java programs can be executed by embedding IJava [24] in Colab. It would be easy if the IJava kernel could be installed on Colab and used in class, but it would be difficult to implement this solution smoothly in a wide range of Java classes. In many cases, the scope of practical use is limited to basic content and does not meet the needs of the class. Michael et al. [21] reported their experience of using Jupyter Notebook (JN below) in CS1, but after changing the teaching language in CS1 from Java to Python, they found that JN is not compatible with CS2. Hence, modifying Colab or Jupyter so that they are more suitable for Java classes is a worthwhile task.

CheerpJ [25] is a WebAssembly-based Java virtual machine for the browser. It is 100% compatible with Java 8 and provides a full runtime environment for running Java applications, applets, libraries, and Java Web Start/JNLP applications in the browser without plugins. CheerpJ, which is easy to use and browser-based, seems to be compatible with Jupyter, so it can be considered to be used to run Java GUI Applications in JupyterLab.

## 3  System Design

### 3.1  Build System Using Docker

In this section, we will explain the principle of the Jupyter-based system proposed in our previous research and what we added for this study. The Jupyter-based system consists of JupyterHub, and Jupyter Notebook or JupyterLab. JupyterLab is a new version of Jupyter Notebook. Both these tools use a.ipynb file to run code or translocate calculated results. This feature is the most popular with data scientists to communicate their calculated procedures and results. It is even used in some paper presentations.

However, JupyterLab and Notebook are single-user systems. JupyterHub allows multiple users to use Jupyter. To execute Java code on Jupyter, the Java Jupyter kernel must be installed on Jupyter. Jupyter has the iPython kernel installed by default for executing Python code. IJava is a Jupyter kernel for Java. It mainly provides information transmission between JShell and the Jupyter server.

This system can build rapidly in Docker. Use JupyterHub as the base image. Then, install JDK9 (or a newer version), JupyterLab, and IJava. Some configurations can then yield a multi-user Jupyter-based Java development environment. Next, download CheerpJ2.3 from the official website and unzip it to a /tmp directory or another convenient place (if you use CheerpJ3, you can skip this step). Now, the system construction has been completed.

### 3.2   Use the System in the Browser

Execution methods other than the GUI can be carried out almost the same as in Python, so they will be omitted in this chapter. It is not possible to terminate input operation using Ctrl + Z, and JUnit evaluation was performed using CMD. Strictly, the above two points still remain, but as long as the instructor fully explains them to the students before class, this will not be a problem. The system configuration is shown in Fig. 1. If using CheerpJ 3, steps (3) and (4) are not necessary.

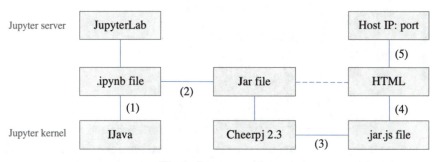

**Fig. 1.** System overview.

The steps to run a Java GUI application in a browser using a Jupyter-based system and CheerpJ are as follows:

(1) Write the Java GUI class in the Jupyter Lab cell, execute it, and call the main method. There is no actual argument check, so the check is done when calling the method.
(2) Extract the GUI class body and related sources from the .ipynb file in (1) and compile it into an executable .jar file.
(3) Compile the .jar file in (2) into a .jar.js file using the CheerpJ.py file from the path where CheerpJ was unzipped. For example: python3 /tmp/CheerpJ_2.3/CheerpJfy.py "$jar_file"
(4) Place the .jar and .jar.js files obtained in (2) and (3) in the same directory, and also create the HTML file for displaying the GUI application in that directory. Enter the source, such as the .jar path, in the HTML file.
(5) Expose the port in the directory in (5) (using methods such as python3 -m http.server 8002), and access http://host IP:8002/.

Steps (2), (3), and (4) are difficult tasks for beginners. In particular, step 2 is easy to fail. Therefore, to use it smoothly, steps other than (1) should be encapsulated in an automatic script. The ideal usage pattern for a beginner is to simply check the source code in the cell and successfully execute the script provided by the instructor. Although this method is adequate, it is insufficient in terms of ease of use, and creating an easy-to-use interface will be a challenge in the future.

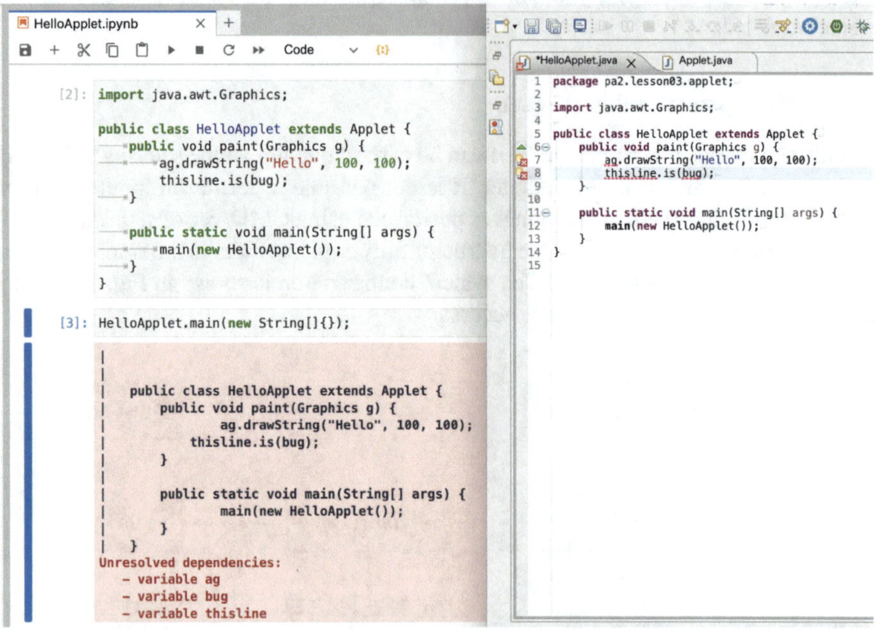

**Fig. 2.** There is no parameter verification function before compilation, and the actual parameters can only be confirmed by calling methods. The right side is an example of an IDE.

Note that the proposed system does not have a pre-compilation argument check function like an IDE, so the check is performed by executing the main function. What is more subtle is that if the user gets a GUI-related error saying "No X11 DISPLAY variable was set," it is probably not a compilation error. In Fig. 2, syntax errors, such as the actual argument of the HelloApplet class were correctly read into the cell and not detected until the main method was called. JShell also has these characteristics, so care must be taken when using them.

## 4   Evaluations

In our previous research, we evaluated Jupyter from three aspects: reproducibility (percentage of code used in the IDE that can be reproduced with Jupyter without modification), performance (usability), and usage patterns (e.g., processing using networks). We found that Jupyter can run applications executed on the console almost 100% of the time. In this paper, we mainly evaluate the reproducibility and performance of the proposed Jupyter-based system, focusing on Java GUI applications.

### 4.1 Reproducibility

Here, we evaluate reproducibility. The evaluation method is to minimally edit the code that would be executed in the IDE in the actual class and run it on the proposed system. We compare the results with the IDE's execution results. Using the Java GUI application code described in the previous section, we run it on the proposed system using the method described in Sect. 3.2. In addition, we review code that cannot yet be verified in the part of network computing (NC) The results are shown in Table 3. The numbers in parentheses are the recall rates of previous studies.

**Table 3.** Coverage of the sample code.

| Part | Classes | Coverage | Content |
|------|---------|----------|---------|
| Base | 50 | 100% | Data Types, Methods, Array, I/O, Exception, Java.io |
| OPP | 31 | 100% | Abstract class, Interface, Inheritance, Polymorphism |
| NC | 35 | 100% (97%) | Concurrency, Thread, Network, Process, Pipe, Prime, Rmi, Udp, Mqtt |
| GUI w/o NC | 75 | 100% (0%) | Swing, Components, Layout, Graphics |
| GUI w/ NC | 12 | 0% | Web application code template |

The source code for the GUI without the NC part was 100% reproducible with only a few source modifications. The slogan on the CheerpJ2.3 official website is that 100% of the code written in Java 8 can be compiled into JavaScript. However, since JShell has only been available since Java 9, it can be expected that CheerpJ2.3 will not be able to fully support it.

As a result of the evaluation, everything from simple parts such as drawing lines, buttons, and text boxes to more complex interactive GUI applications can be reproduced as long as it is not linked to network computing. Although there are some differences in visual effects compared with native Java applications, these do not hinder their use. An execution sample is shown in Fig. 3. The panel on the left is what was reproduced in the browser using Cheepj2.3, and the panel on the right is a Java native application. The source code used in the figure is a typical RPGPane, which uses various Java interfaces. For example, it uses an external image and JPanel to create a GUI and a Timer function and the keyboard to control a character. All GUI and character movements were able to be reproduced.

Some problems occurred during the evaluation. For example, when verifying MazePane, if the maze configuration used recursion and.this (self-class instance) programming techniques, a null pointer exception occurred, and the application failed. Hence, we used stack or dequeue instead of recursion and added error catching. The app was now able to run, but the null pointer error still occurred. When CheerpJ3 was

**Fig. 3.** A running RPGPane Java GUI application on the browser (left) and Java native application (right).

released, we tested the program using it, and it worked correctly without the null point exception mentioned above.

In the part of GUI with NC, the GUI itself could almost be displayed, but if a network-related function was called from event processing, such as when a button was pressed, some error would occur. Depending on the type of processing, errors such as Java-related connection lost (32109), Client is not connected (32104), and JavaScript-related errors related to cross-origin resource sharing were displayed on the browser console. Even using CheerpJ3, the network linkage function could not be reproduced. Note that it is difficult to debug problems that occur during execution in such browsers. CheerpJ is not open-source software and cannot be investigated from the compilation flow or JavaScript files. Therefore, the results regarding reproducibility showed that Java GUI applications that do not use network processing can be reproduced 100% in a browser.

In our previous research, applications created using ProcessBulider() were unable to run. In this evaluation, the problem was solved by properly preparing the class file in the process path. Note that, because of failure to introduce an inherited class to a cell a SPIResolutionException may appear instead of specific error information. Similar to Fig. 2, we also found that the proposed system was inferior to the IDE in terms of error display functionality.

### 4.2   Performance

Performance evaluation was performed on both the front and back ends. The hosts used were as follows. OS: macOS Monterey 12.5; Memory: 16 GB 3733 MHz LPDDR4X; Processor: 2 GHz 4-core Intel Core i5; Chrome version: 119.0.6045.159 (Official Build) (x86_64).

Java GUI applications compiled with CheerpJ are executed in a browser, so it is necessary to verify front-end performance. The method was to continue using the RPGPane example from Sect. 4.1 and observe the data in the Network panel of Chrome Developer Tools when the page loads. The evaluation results are shown in Table 4.

**Table 4.** Application page performance.

| Version | Req | Transf./MB | Res./MB | Finish/S | DCL/ms | Load/ms |
|---|---|---|---|---|---|---|
| CheerpJ2.3 | 188 | 16.86 | 67.2 | 10.21 | 111.6 | 520.88 |
| CheerpJ3 | 192 | 15.86 | 21.6 | 9.616 | 71.4 | 126.6 |

The size of the transferred file and resource size are larger than a normal page, and this is affected by the network, but the response time is within an acceptable range. In addition, it was found that the performance of CheerpJ 3 was superior to CheerpJ 2.3.

Table 5 summarizes the evaluation results on the back end. Since CheerpJ2.3 is compiled on the host, we were able to observe that memo usage increased by 100 MB in Docker stats during the compilation period. CheerpJ3 does not require host compilation, so it is omitted here.

**Table 5.** Performance and estimates.

| users | Cpu% | memo/MB | memo/per user | estimate/MB |
|---|---|---|---|---|
| 0 | 0.02 | 87.53 | 0 | 247.53 |
| 1 | 101.25 | 832.17 | 832.17 | 992.17 |
| 2 | 197.11 | 1370.646 | 685.323 | 845.323 |
| 3 | 300.97 | 1892.886 | 630.962 | 790.962 |
| 4 | 396.82 | 2349.59 | 587.397 | 747.397 |
| 5 | 491.73 | 2828.822 | 565.764 | 725.764 |
| 6 | 585.93 | 3306.006 | 551.001 | 711.001 |
| 7 | 685.14 | 3780.118 | 540.017 | 700.017 |
| 8 | 778.27 | 4360.726 | 545.09 | 705.09 |
| 9 | 891.11 | 4705.814 | 522.87 | 682.87 |
| 10 | 998.58 | 5159.446 | 515.94 | 675.94 |
| 11 | 1077.32 | 5632.534 | 512.05 | 672.05 |
| 12 | 1150 | 6007.318 | 500.61 | 660.61 |

Based on the evaluation results of previous research, we found that, when compiling using CheepJ2.3 and running two kernels at the same time, 1.2 GB of memory per student is sufficient for Java classes.

## 5  Conclusions

In previous research, we proposed, implemented, and evaluated a browser-based Java learning environment using Jupyter. It can be used for a wide range of education, from basic processing that does not use a GUI to processing that uses a network. It turns out that you can migrate from a PC-based IDE to Jupyter. Based on this, this paper found that GUI code that does not perform communication is almost 100% executable.

A future challenge is to develop a browser-based Java programming environment that can run GUI applications that can communicate. In addition, a GUI for Jupyter is required to teach GUI apps on Jupyter.

# References

1. Uehara, M.: Experiences with a single-page application for learning programming. In: Barolli, L., Takizawa, M., Enokido, T., Chen, H.-C., Matsuo, K. (eds.) BWCCA 2020. LNNS, vol. 159, pp. 55–66. Springer, Cham (2021). https://doi.org/10.1007/978-3-030-61108-8_6
2. Lee, M.J., Ko, A.J.: Comparing the effectiveness of online learning approaches on CS1 learning outcomes. In: Proceedings of the Eleventh Annual International Conference on International Computing Education Research. New York, NY, pp. 237–246. ACM (2015)
3. Pears, A., et al.: A survey of literature on the teaching of introductory programming. SIGCSE Bull. **39**(4), 204–223 (2007)
4. Bennedsen, J., Caspersen, M.E.: Failure rates in introductory programming. SIGCSE Bull. **39**(2), 32–36 (2007)
5. Simon, et al.: Pass rates in introductory programming and in other STEM disciplines. In: Proceedings of the Working Group Reports on Innovation and Technology in Computer Science Education (ITiCSE-WGR 2019). New York, NY, pp. 53–71. ACM (2019)
6. Watson, C., Li, F.W.B.: Failure rates in introductory programming revisited. In: Proceedings of the 2014 Conference on Innovation & Technology in Computer Science Education (ITiCSE 2014). New York, NY, pp. 39–44. ACM (2014)
7. Bennedsen, J., Caspersen, M.E.: Failure rates in introductory programming: 12 years later. ACM Inroads **10**(2), 30–36 (2019)
8. Jordan, K.: Initial trends in enrolment and completion of massive open online courses. Int. Rev. Res. Open Dis. **15**(1), 133–160 (2014)
9. Medeiros, R.P., Ramalho, G.L., Falcão, T.P.: A systematic literature review on teaching and learning introductory programming in higher education. IEEE Trans. Educ. **62**(2), 77–90 (2019)
10. Qian, Y., Lehman, J.: Students' misconceptions and other difficulties in introductory programming: a literature review. ACM Trans. Comput. Educ. **18**(1), 27 (2018)
11. Kross, S., Guo, P.J.: Practitioners teaching data science in industry and academia: expectations, workflows, and challenges. In: Proceedings of the 2019 CHI Conference on Human Factors in Computing Systems (CHI 2019). New York, NY, pp. 1–14. ACM (2019)
12. Brunner, R.J., Kim, E.J.: Teaching data science. Procedia Comput. Sci. **80**, 1947–1956 (2016)
13. Konecki, M.: Problems in programming education and means of their improvement. DAAAM Int. Sci. Book **2014**, 459–470 (2014)
14. Kluyver, T., et al.: Jupyter Notebooks—A publishing format for reproducible computational workflows. Elpub, pp. 87–90 (2016)
15. Perkel, J.M.: Why Jupyter is data scientists' computational notebook of choice. Nature **563**(7732), 145–147 (2018)
16. Stein, L.: Challenging the computational metaphor: implications for how we think. Cybern. Syst. Syst. **30**(6), 473–507 (1999)
17. Stein, L.: What we've swept under the rug: radically rethinking CS1. Comput. Sci. Educ. Sci. Educ. **8**(2), 118–129 (1998)
18. Weber-Wulff, D.: Combating the code warrior: a different sort of programming instruction. SIGCSE Bull. **32**(3), 85–88 (2000)

19. Uehara, M.: Programming learning by creating problems. In: 2020 Eighth International Symposium on Computing and Networking Workshops (CANDARW). Naha, Japan, pp. 272–276 (2020)
20. Porter, L., Guzdial, M., McDowell, C., Simon, B.: Success in introductory programming: what works? Commun. ACM. ACM **56**(8), 34–36 (2013)
21. Al-Gahmi, A., Zhang, Y., Valle, H.: Jupyter in the classroom: an experience report. In: Proceedings of the 53rd ACM Technical Symposium on Computer Science Education – Volume 1 (SIGCSE 2022), vol. 1, New York, NY, pp. 425–431. ACM (2022)
22. Zastre, M.: Jupyter notebook in CS1: an experience report. In: Proceedings of the western Canadian conference on computing education (WCCCE 2019). New York, NY, pp. 1–6. ACM (2019)
23. JEP 222: jshell: The Java Shell (Read–Eval–Print Loop). https://openjdk.org/jeps/222
24. JEP 445: Unnamed Classes and Instance Main Methods (Preview). https://openjdk.org/jeps/445
25. IJava: https://github.com/SpencerPark/IJava. Accessed 21 Nov 2023
26. Cheerp, J.: https://labs.leaningtech.com/cheerpj3. Accessed 21 Nov 2023

# Health Parameters Monitoring Through an Integrated Multilayer Digital Twin Architecture

Constantin Lucian Aldea, Razvan Bocu[✉], and Delia Monica Duca Iliescu

Department of Mathematics and Computer Science, Transilvania University of Brasov, B-dul Eroilor nr. 29, Brasov, Romania
{costel.aldea,razvan.bocu,delia.duca}@unitbv.ro

**Abstract.** This paper evaluates the possibility to use a digital twin architecture to support the management and improvement of the vital health parameters. The proposed approach considers an architecture that integrates machine learning and deep learning algorithms. The digital twins are automated programs that learn from training data and new data, and then assist the end user in various situations. The digital twin model that is proposed in this paper, which is called MyMLTwin, assists the user in monitoring and keeping the health parameters within the correct threshold values. Threshold values can be set as a target by users or suggested and modified by the digital twin's artificial intelligence core. Data are securely collected from the end users and properly anonymized. Furthermore, the data can be used to digitally assist the user in maintaining their health parameters by suggesting the proper nutrition plan, and physical activity patterns relative to the proposed digital twin model. The model includes elements of web semantics and appropriate ontologies, which are used to create a proper equivalent of the required nutrition plan. The proposed model integrates state of the art conceptual models and technologies that relate to machine learning, deep learning, and distributed systems. The paper describes a functional prototype that explores different types of features related to different algorithms and methods used to train and support the decisions of the digital twin.

## 1 Introduction

The increasingly complex structure of the industrial production processes determines a continued search for enhanced models, which integrate the physical and the digital dimensions. In this context, the concept of digital twin was proposed in 2003 [1]. Initially, digital twins were mainly considered in the military and aerospace sectors. Currently, the approach is considered in any context that requires a proper modeling of real-world entities and processes.

L. Barolli (Ed.): AINA 2024, LNDECT 199, pp. 298–309, 2024.
https://doi.org/10.1007/978-3-031-57840-3_27

The perspective on the digital twins has progressed from a conceptual app-roach to a fully functional implementation, which complements a wide range of real-world processes. Considering the physical parameters of the physical entity, the model of the digital twin can be defined by four perspectives: rule, physics, geometry, and behaviour [2]. Thus, the geometric model presents the geometric shape and structural or architectural features of the physical entity. The physical model depicts the physical parameters, features and constraints of the physical entity. Furthermore, the behavioral model relates to the dynamic behavior of the physical entity in response to the internal and external factors. The rule model represents the historical data, and it can process relevant knowledge. It is imme-diate to note that this structural and logical characteristic suggests a natural blend of the general digital twins approach with the tools and algorithms in the general scope of artificial intelligence and machine learning. The consideration of this multidisciplinary approach implies that the digital twin model may be considered in order to conduct cognitively complex tasks, such as estimation of the evolution, optimization of the informational flows, and general control of the implemented functional features.

Nevertheless, modelling a digital twin concerns several perspectives [3]. Thus, the design of the digital twin model relates to an integration of the knowledge in relevant related fields of study in order to design the fundamental digital twin model's structure, such as the unit level digital twin architectural model in a specific application field. Furthermore, digital twin models with complex architectures may be designed considering an assembly of elementary digital twin models relative to a spatial dimension. In particular cases, the spatially assembled model is not capable to completely describe the internal mechanisms of the modeled real-world entity. Consequently, further refinement of the model regarding its fusion in multiple disciplines is required following the model assem-bly phase. Furthermore, it is necessary to validate the constructed digital twin model. Nevertheless, if the digital twin model does not fully abstract the physical production requirements or any other relevant aspects of the real-world physical entity or complex of entities, supplementary changes of the implied digital twin model are mandatory to ensure the required level of structural and functional accuracy. Another significant aspect is represented by the management work-flows. These provide important auxiliary services, including the management of already created digital twin models, and also the processing and effective usage of data and knowledge generated during the digital twin modeling process. There-fore, it is immediate to note the complexity of the digital twin modeling and management. As a consequence, the next section analytically surveys the most relevant recent scientific contributions, which are identified in the existing liter-ature.

## 2    Relevant Existing Papers

The problematic of digital twins modeling is approached by a fair amount of papers. Nevertheless, only a subset addresses the problematic in a relevant, efficient, easy to follow and reproduce manner. Therefore, the following paragraphs survey the properly identified scientific literature. Additionally, the scientific survey methodology is presented.

The authors of article [4] intend to evaluate the conceptual specification and features of a digital twin, its functional links with relevant digital technologies considered in the construction of real estate buildings. Thus, the research process considers an extensive survey and semi-structured interviews with ten industry experts. The scientific survey assesses the merits and the significance of digital twins relative to existing digital technologies. It also evaluates potential design and implementation challenges. The data obtained from the semi-structured interviews are assigned to five thematic categories: implementation challenges, applications and benefits, definitions and enablers of digital twins, existing practical applications, and future development avenues. The research outcomes describe the general features of relevant approaches, which may be considered as the basis for future research developments.

Metaverse, as a concept that effectively defines a digital world, was described in 1992. During the past few years, digital twins have also been proposed as complementary tools in the realm of Metaverse. Thus, article [5] surveys definitions, applications, and general challenges of digital twins at the confluence with Metaverse. Furthermore, the paper defines a three-layer architecture, which is claimed to link the physical environment to the Metaverse through a user interface. Additionally, the paper describes an investigation concerning the security and privacy challenges that pertain to the usage of digital twins in the Metaverse.

There are articles, which report the application of digital twins to specific domains. As an example, paper [6] evaluated the outcomes of a survey on applications of digital twins in greenhouse-based horticulture. The survey presents eight articles that specifically address the mentioned subject, and another 115 papers that implicitly connect the digital twins concept to smart IoT-based systems. It is determined that the concept of digital twins is intensely studied in the context of greenhouse horticulture. Nevertheless, there are existing real-world use case scenarios, which are not yet completely assimilated to the scope of digital twins. Furthermore, there is a pregnant focus concerning the farming processes at the level of a greenhouse, such as concerning climate control, energy management and lighting. Approximately 9% of the analyzed articles create digital twins for the plants themselves, which indicates that the level of granularity is fundamentally limited. Interestingly, approximately 7% of the surveyed papers go beyond the study of individual greenhouses and biorhythms of regular plants. It is important to note that none of the surveyed papers address the potential perspective of the company. Practice suggests that applications should consistently approach monitoring and control of the state and behaviour of real-life objects. Furthermore, certain applications should include predictive and prescriptive algorithmic and functional features along their complete life cycle. Nevertheless, the study suggests that such use case scenarios are still insufficiently studied, although predictive digital twins continuously gain prominence in related research projects. It is interesting to note that article [7] suggested the consideration of digital twins in order to increase the efficiency of buildings construction in the context of Industry 5.0, while article [8] analyzes the application of digital twins to the general scope of smart manufacturing. Additionally, paper [9] extended the discussion concerning the utilization of digital twins to the field of healthcare. Thus, the paper proposes the concomitant consideration of artificial intelligence methods and digital twins, in order to efficiently generate virtual entities that reflect reality as accurately as possible. Nevertheless, it is determined that the inaccurate design and insufficient calibration of the involved artificial intelligence methods implies the generation of digital twins that do not reflect reality considering a sufficient level of accuracy. This problematic determines a particular field of research, which is partly approached in this paper. Moreover, article [10] analyzed the large amounts of data that are generated by the Internet of Things (IoT) devices, which are deployed in a smart city. The methodology uses the apparatus that is specific to big data analysis (BDA), which is enhanced by deep learning (DL), and convolutional neural networks (CNN). The performance evaluation suggests that the accuracy of the constructed system reaches 97.80%, which is claimed to be at least 2.24% greater than the DL-based algorithmic solutions considered by other researchers.

Healthcare determines a significant domain, which seems to benefit from the conceptual and practical advantages of digital twins. Thus, article [11] suggested that successes had already been reported concerning cardiovascular diagnostics and insulin pump control, among other real-world use cases. More advanced medical digital twins will be essential to turn precision medicine into a reality. This perspective article discusses about the importance of the immune system, which influences the dynamics of numerous health problems. Nevertheless, the relevant developments imply significant challenges, which are determined by the intrinsic complexity of the immune system. Furthermore, it is difficult to conduct an in vivo analysis of the immune system's state, considering all relevant biological parameters. Consequently, the perspective article described a plan for a comprehensive research effort that would address this problematic through the creation of an immune system digital twin. The proposed research process considers four stages, which commence with the specification of a concrete use case, goes through the construction of the digital twin model, which is consequently personalized, and continuously improved. This is important, as the model potentially catches and models the dynamics of the developments in the relevant medical sciences. It is interesting to note that additional contributions regarding the application of digital twins in healthcare are reported in articles [12–15].

The surveyed literature includes articles that couple the idea of digital twins to the application of proper cybersecurity mechanisms in various fields, including healthcare. As an example, article [16] presents an approach regarding the detection of potentially vulnerable functional features of healthcare digital twins.

The model involves the development of a deep neural network, which is suitable to detect bi-directional contextual relationships established between problematic code keywords. The experiments were conducted on a data set of collected data (ground truth), which includes tens of thousands of safe and problematic functions. The reported experimental results may suggest that the proposed model would efficiently detect potentially unsafe aspects of the designed digital twins. Interesting related contributions are reported in articles [17,18].

There are papers that study the non-technical aspects of this problematic. Thus, article [19] addresses relevant ethical problems. This study regards the ethical risks that are implied by the usage of digital twins relative to customized health care services. Initially, the article draws a functional definition concerning the application of digital twins to the customization of health care services. This may facilitate the upcoming developments regarding the ethical issues that relate to digital healthcare services. Consequently, the authors propose a process-oriented ethical map, which defines the principal ethical risks that manifest during the several data processing stages. Additionally, interesting related discussions are contained in articles [20–22].

Recently, digital twins have been considered in the medical field to fundamentally changes the accuracy and efficiency of particular processes. Thus, article [23] describes a medical diagnosis algorithm, which considers product-based neural network (FMCPNN) that is enhanced through the utilization of product-based neural networks (PNN). More precisely, the PNN component represents an end-to-end factorization machine (FM) algorithm, which addresses the issue of data sparsity. Nevertheless, the considered PNN model does not properly approach the low-order feature interaction, which determines an insufficient generalization ability. Additionally, the FMCPNN component considers the addition of a second-order interaction, which is part of FM, considering the fundamental support that is offered by PNN. This algorithmic approach enhances the general functional behaviour of PNN. Although interesting, the proposed model is not properly evaluated relative to its potential real-world performance. It is interesting to note that article [24] presents an integrative model, which combines Internet of Things (IoT) and digital twins, in order to propose an enhanced approach for the digitalization of relevant processes in healthcare.

The problematic of data security is also approached in the surveyed literature. Thus, article [25] discusses about possible methods that may be considered to detect spoofed medical records. Also, article [26] presents a blockchain-related secure digital twin framework, which is suitable for the specific environment of a smart city that aims to ensure proactive healthcare mechanisms to the inhabitants. Consequently, they analyze a real-world case study relative to the COVID-19 pandemic. They assert that the usage of digital twins ensures a proactive prevention of new cases, and it also allows for a personalization of the medical treatment to be implemented. The usage of the blockchain-based mechanism mediates the implementation of proper personal medical data privacy mechanisms.

The surveyed literature suggests that existing contributions may not fully fulfill the functional and logical requirements, which are specific to certain real-world use cases, such as the efficient and safe collection and processing of physiological data. Therefore, this paper presents an architectural and functional optimization strategy for a personal health data collection system, which considers the utilization of digital twins.

Following, the paper considers the following sections. The next section presents the considered use case scenario, and related architectural model. The necessary algorithmic and implementation details and provided. Additionally, relevant considerations concerning the used ontologies and considered semantics are made. Following, the proposed approach is validated through an experimental process, which is described. The last section discusses about the open research avenues, and it also concludes the paper.

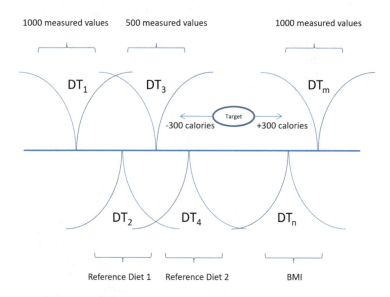

**Fig. 1.** The MyMLTwin model - reference data and nutritional programs

## 3 Remarks Concerning the MyMLTwin Concept

The MyMLTwin concept proposes the creation of a virtual identity for each user, which mediates the collection of personal nutritional data, such as physical activity, the quality of sleep, measurements of cardiac rhythm, measurements of blood sugar, and other physiological parameters. These data are acquired in a safe and privacy-preserving manner. This allows a proper model to be trained, which assists personal decisions concerning nutrition, daily physical activity, and also adequate scheduling of the sleep schedule, in order to ensure that the body is trained through a properly balanced routine. Initially, the health status assessment is evaluated considering two factors: body mass index, and the waist circumference.

The Fig. 1 illustrates the idea of data classification. Once the data is collected and anonymised, classification algorithms can incrementally classify the data and then decide for new data sets whether they belong to one category or another. This gives the user advice based on similarity to other real users who have the same habits and similar measurements. After users have contributed their anonymised data, new learning epochs can be run and classification accuracy can be improved. This improves the likelihood of being classified in the correct category, and the digital twin continues to learn. For the learning algorithms, the real data samples improve precision and accuracy and aren't directly associated with the person who provided them.

The paper presents and analyzes the intended use case scenarios. Furthermore, a software architecture is proposed, and the related software system is implemented. Consequently, its ability to be used as a personal assistant is evaluated, considering various informational contexts (Fig. 2).

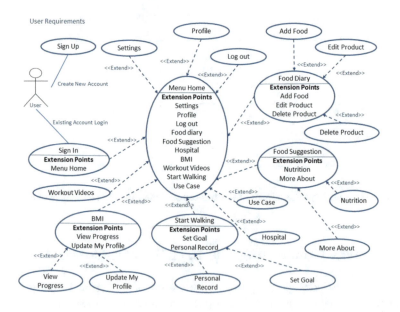

**Fig. 2.** Use cases related to the MyMLTwin system.

The user interacts with the ecosystem mainly through a mobile or web client. The user can create and manage their profile and collect measurements manually and automatically from the wearable devices. The user's data can be visualised.

The user's measurements can be compared to user defined thresholds. Depending on the option enabled, the data can be sent to the core services where the measurement is decoupled from the user and learning and feature extraction is performed. Nutrion data equivalences are represented by means of ontologies.

## 3.1 Data Considered by the Conceptual Model

The data that are collected from the enrolled end users are classified in the following categories.

- Age
- Height
- Body Mass
- The circumference of waist and hips
- Logged data concerning consumed food
- Logged data concerning conducted physical activity
- Logged data regarding sleep hours
- Food types that provoke allergies
- Medical conditions and illnesses
- Dietary and nutritional preferences
- Body mass, which is considered as a parameter that determines the evaluation of the health state

## 3.2 Relevant Use Cases

The research process that is reported identified and analyzed certain aspects that may be monitored and processed using a digital twin, through an application that continuously monitors the personal health status (Fig. 3).

1. Monitoring the cardiac rhythm - it involves the measurement of the heart-beats and pulse data. Thus, to determine the cardiac rhythm and pulse data, the application uses the mobile device's camera. Thus, the user is directed to place the index finger on the mobile device's photographic camera. Consequently, the application displays a diagram that illustrates the real-time cardiac rhythm and pulse data.
2. Monitoring the level of blood sugar - it is a very important parameter, particularly in the case of persons that suffer from diabetes.
3. The daily distance traveled - monitoring this variable is important for any person, as medical experts generally recommend, at least, thirty minutes of daily physical activity. Thus, an application that incorporates the ability to provide statistics regarding the distance traveled, may stimulate the end users to progressively adopt a healthier life style. The determination of the traveled distance involves that the application counts the number of bodily inclinations, which are generated by a normal pedestrian walk pattern. Considering that each step is characterized by approximately the same length, it is sufficient to count the number of steps, and the number of body inclinations. Consequently, the obtained distance is the result of a simple multiplication operation.

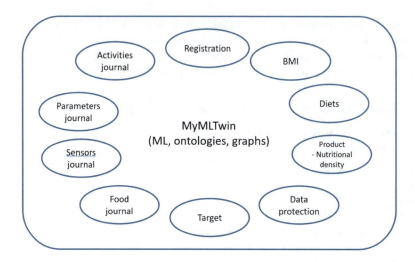

**Fig. 3.** Conceptual components of MyMLTwin

4. The total number of steps - it involves the determination of the total steps count considering a certain period of time (day, week, month). This is useful in order to stimulate enrolled users to adopt an as active as possible daily routine, inclusively by organizing competitions with family members or friends.

5. Gather relevant data and consequently compute the burned calories - this is particularly useful for the users that follow a dietary regimen, which is based on intake calories restrictions. The relevant calculations essentially consider the body mass, and the nature of the conducted physical activities.

6. Monitoring the daily intake of carbohydrates, proteins, and fats - this considers the ingredients of the respective food intake. This particular functional feature is realized considering a dietary journal, which records the particulars of the food intake over a certain period of time. Each day may be represented in the form of a chart, which illustrates the daily intake of proteins, carbohydrates and fats. Thus, deficiencies regarding the daily dietary style are determined, and proper recommendations are issued.

7. Monitoring the consumption of water - the importance of water for a healthy and balanced life style is well known. Thus, the system is capable to keep track of the amount of ingested water, and it also assesses the consumption of other liquids, such as coffee and tea.

8. Length of the physical training program - this is a significant performance parameter, particularly in the case of persons that adopt a certain daily training schedule. It determines the desire for self-improvement, through greater traveled distances, or through more intense training sessions.

9. Sleep monitoring - several studies demonstrate that a peaceful sleep is important for the overall state of health and wellbeing. The reported system considers several interesting concepts. It calculates the optimal time to wake up,

particularly during the "awakening" period. Thus, as an example, considering an awakening period of twenty minutes, and an alarm that is scheduled at 7 AM, if the application detects slight movements in the sleep ten minutes before 7 AM, then it will wake up the user at that moment in time. This is because ten minutes later, it is possible that the user mau have fallen back into a deep sleep. Apart from thus standard alarm function, the application helps to user to wake up at optimal moments in time, which also contributes to a sensation of "well rested".

10. Monitoring body weight - Considering the most recent studies, people that keep track of their weight often, are able to lose weight easier than people who do not weigh at all, or rarely weigh themselves. Thus, this justifies the importance of constant monitoring of body weight, as it may further calibrate the personal behaviour towards a healthier and more active life style.

11. Monitoring the sugar intake - it is a particularly important parameter for persons that suffer from diabetes, but it is also important for the general population. If a glucometer is used, the acquired data can be sent over to the presented application, which ensures a continuous monitoring process. This facilitates the identification of food samples that contain high amounts of sugar, which further supports the efforts to balance the consumption of dietary sugar.

12. Monitoring the consumption of caffeine - this is a useful feature, considering that the ingestion of high amounts of caffeine over extended periods of time is potentially detrimental for the human body.

13. The improvement or training of the mental state, through inputs like positive assertions. It is generally accepted that psychological reliability is a feature of successful people, which is sometime more useful than raw physical ability. Medical experts assert that the mental state can be trained the same way as physical condition is improved through practice. The relevant proposed exercises challenge the brain, which strengthen memory, the ability to concentrate, and other relevant cognitive functions. Psychologists assert that there are several mechanisms, which helps to improve mental resilience, the same way physical exercises contribute to the improvement of body muscles. Thus, relevant exercises that train memory and speech are proposed, which include puzzles, crosswords, and other types of exercises. Additionally, exercises that train logical thinking, and reasoning abilities, are also proposed.

14. Monitoring the number of cigarettes, which have not been smoked yet, in the case of persons that wish to give up smoking. It has been empirically demonstrated that such a support tool effectively helps to extend the period between two smoked cigarettes, which significantly helps the enrolled users to reach the final goal.

The proposed system properly implements all the relevant aspects and functional perspectives.

## 4    Conclusion and Future Developments

The paper presents an integrated multilayer digital twin architecture, which monitors the relevant health parameters in an efficient and precise manner. The initial prototype has been successfully assessed. Thus, it was determined that it accurately collects the data that pertain to all the relevant health parameters in a computationally efficient manner.

The future versions of the integrated system will consider the improvement of the implied machine learning and deep learning algorithmic models. Additionally, the performance of the system will be evaluated using a substantially expanded experimental infrastructure. This will further enhance the functional accuracy and computational efficiency of the integrated system.

## References

1. Grieves, M., Vickers, J.: Digital twin: mitigating unpredictable, undesirable emergent behavior in complex systems. In: Kahlen, F.-J., Flumerfelt, S., Alves, A. (eds.) Transdisciplinary Perspectives on Complex Systems, pp. 85–113. Springer, Cham (2017). https://doi.org/10.1007/978-3-319-38756-7_4
2. Tao, F., Liu, W., Zhang, M., Hu, T., Qi, Q., Zhang, H.: Five-dimension digital twin model and its ten applications. Comput. Integr. Manuf. Syst. **25**(01), 1–18 (2019)
3. Tao, F., Zhang, H., Qi, Q., Xu, J., Sun, Z., Hu, T.: Theory of digital twin modeling and its application. Comput. Integr. Manuf. Syst. **27**(1), 1–16 (2021)
4. Shahzad, M., Shafiq, M., Douglas, D., Kassem, M.: Digital twins in built environments: an investigation of the characteristics, applications, and challenges. Buildings **12**(2), 120 (2022)
5. Far, S., Rad, A.: Applying digital twins in metaverse: user interface, security and privacy challenges. J. Metaverse **2**(1), 8–15 (2022)
6. Ariesen-Verschuur, N., Verdouw, C., Tekinerdogan, B.: Digital Twins in greenhouse horticulture: a review. Comput. Electron. Agric. **199**, 107183 (2022)
7. Wang, W., Guo, H., Li, X., et al.: BIM information integration based VR modeling in digital twins in industry 50. J. Indust. Inform. Integrat. **28**, 100351 (2022)
8. Li, L., Lei, B., Mao, C.: Digital twin in smart manufacturing. J. Ind. Inf. Integr. **26**, 100289 (2022)
9. Kaul, R., Ossai, C., Forkan, A., et al.: The role of AI for developing digital twins in healthcare: the case of cancer care. Wiley Interdisc. Rev. Data Mining Knowl. Dis. **13**(1), e1480 (2023)
10. Li, X., Liu, H., Wang, W., Zheng, Y., Lv, H., Lv, Z.: Big data analysis of the internet of things in the digital twins of smart city based on deep learning. Futur. Gener. Comput. Syst. **128**, 167–77 (2022)
11. Laubenbacher, R., Niarakis, A., Helikar, T., et al. Building digital twins of the human immune system: toward a roadmap. npj Digital Medicine (2022)
12. Croatti, A., Gabellini, M., Montagna, S., Ricci, A.: On the integration of agents and digital twins in healthcare. J. Med. Syst. **44**, 1–8 (2020)
13. Volkov, I., Radchenko, G., Tchernykh, A.: Digital twins, internet of things and mobile medicine: a review of current platforms to support smart healthcare. Program. Comput. Softw. **47**, 578–90 (2021)

14. Elayan, H., Aloqaily, M., Guizani, M.: Digital twin for intelligent contextaware IoT healthcare systems. IEEE Internet Things J. **8**(23), 16749–57 (2021)

15. Hassani, H., Huang, X., MacFeely, S.: Impactful digital twin in the healthcare revolution. Big Data Cognitive Comput. **6**(3), 83 (2022)

16. Zhang, J., Li, L., Lin, G., Fang, D., Tai, Y., Huang, J.: Cyber resilience in healthcare digital twin on lung cancer. IEEE Access **8**, 201900–13 (2020)

17. Khan, S., Arslan, T., Ratnarajah, T.: Digital twin perspective of fourth industrial and healthcare revolution. IEEE Access **10**, 25732–54 (2022)

18. Sahal, R., Alsamhi, S.H., Brown, K.N.: Personal digital twin: a close look into the present and a step towards the future of personalised healthcare industry. Sensors **22**(15), 5918 (2022)

19. Huang, P.H., Kim, K.H., Schermer, M.: Ethical issues of digital twins for personalized health care service: preliminary mapping study. J. Med. Internet Res. **24**(1), e33081 (2022)

20. Popa, E.O., van Hilten, M., Oosterkamp, E., Bogaardt, M.J.: The use of digital twins in healthcare: socio-ethical benefits and socio-ethical risks. Life Sci. Soc. Policy **17**(1), 1–25 (2021)

21. Mittelstadt, B.: Near-term ethical challenges of digital twins. J. Med. Ethics **47**, 405–6 (2021)

22. de Kerckhove, D.: The personal digital twin, ethical considerations. Phil. Trans. R. Soc. A **379**(2207), 20200367 (2021)

23. Yu, Z., Wang, K., Wan, Z., Xie, S., Lv, Z.: FMCPNN in digital twins smart healthcare. IEEE Consumer Electr. Mag. **12**, 66–73 (2022)

24. Patrone, C., Lattuada, M., Galli, G., Revetria, R.: The role of internet of things and digital twin in healthcare digitalization process. Trans. Eng. Technol. World Congress Eng. Comput. Sci. **26**, 30–7 (2020)

25. Garg, H., Sharma, B., Shekhar, S., Agarwal, R.: Spoofing detection system for e-health digital twin using EfficientNet convolution neural network. Multimedia Tools Appli. **81**(19), 26873–88 (2022)

26. Azzaoui, A.E., Kim, T.W., Loia, V., Park, J.H.: Blockchain-based secure digital twin framework for smart healthy city. Adv. Multimedia Ubiquit. Eng. **716**, 107 (2021)

# Advancing Multi-writer Snapshots Algorithm

Sudhanshu Sharma$^{(\boxtimes)}$ and Dharmendra Prasad Mahato

National Institute of Technology, Hamirpur, India
22mcs022@nith.ac.in

**Abstract.** Maintaining consistency among multiple threads that access and modify shared data structures [1] is paramount in concurrent programming. Multi-writer snapshot algorithms capture consistent snapshots of shared data and enable reliable program operation. The Amram-Atiya-Touitou (AAT) algorithm [2] is a widely accepted technique for capturing consistent snapshots. However, relying on global snapshot locks limits [2] performance, especially in high contention scenarios. This study addresses the performance limitations of the original His AAT algorithm by introducing improvements that significantly increase its performance and efficiency. Proposed changes include fine-grained locking [3], lazy snapshot creation, and early lock release. Through these optimizations, the reformed AAT algorithm effectively addresses the performance bottlenecks of the original algorithm, leading to significant improvements in throughput and scalability. Benchmark results show that the reformed AAT algorithm outperforms the original AAT algorithm significantly, especially in highly competitive scenarios.

**Keywords:** Shared Data Structure · Multiwriter Snapshots

## 1 Introduction

In the dynamic realm of concurrent programming, ensuring consistency across multiple threads that access and modify shared data structures is fundamental to program correctness and integrity. Multiwriter snapshot algorithms [5] prove to be an important tool for achieving this consistency by taking consistent snapshots of shared data and enabling reliable and predictable program behavior. It has been proven. Among these algorithms, the AAT algorithm is widely recognized for its effectiveness in obtaining consistent snapshots [6].

Despite its popularity, his original AAT algorithm has performance limitations due to its inherent dependency on global snapshot locks. This global lock becomes a contention point that can impede concurrency and impact performance, especially in high-contention scenarios. To address these limitations, this study presents an innovative approach to reform the AAT algorithm and introduces improvements that significantly improve its performance and efficiency.

© The Author(s), under exclusive license to Springer Nature Switzerland AG 2024
L. Barolli (Ed.): AINA 2024, LNDECT 199, pp. 310–318, 2024.
https://doi.org/10.1007/978-3-031-57840-3_28

## 1.1   Our Contribution

The proposed changes aim to address performance bottlenecks related to global snapshot locks and improve the overall throughput and scalability of the AAT algorithm. These changes include:

1. Fine-grained locking: Replaces global snapshot locks with fine-grained locks associated with individual data items. This allows multiple threads to access and modify different parts of shared data simultaneously, reducing lock contention and increasing overall throughput.
2. Deferred snapshot creation: Snapshot creation is deferred until the global snapshot lock is no longer held, minimizing overhead when other threads do not take the snapshot.
3. Early lock release: Release fine-grained locks before acquiring global snapshot locks. This allows other threads to continue accessing and modifying the freed data while the snapshot is being taken.

The reformed AAT algorithm represents a breakthrough evolution in multi-writer snapshot technology, improving performance, efficiency, and scalability for concurrent programming applications [4]. The ability to minimize lock contention and maintain strong linearizability makes it an attractive choice for building scalable and efficient concurrent data structures. This study provides a comprehensive analysis of the proposed changes and a rigorous evaluation of the performance and accuracy of the reformed AAT algorithm. The results demonstrate the effectiveness of the proposed approach in improving the performance and efficiency of multi-writer snapshots and paving the way for more scalable and powerful concurrent programming applications.

To evaluate the performance and scalability of the proposed algorithm, we conducted extensive experiments with different workloads and system configurations. Our results show that EMSA outperforms existing algorithms in terms of latency and throughput, especially in high-contention [11] scenarios. Similarly, SMSA has better scalability compared to existing algorithms, ensuring low latency and high throughput even as the number of processes and shared memory size increases.

In addition to evaluating the performance of the algorithm, we also conducted a comprehensive analysis of the spatial complexity and communication overhead. Our results suggest that both EMSA and SMSA have spatial complexity comparable to existing algorithms and minimal communication overhead, making them suitable for practical applications.

Overall, our study presents two new multi-writer snapshot algorithms with significant improvements in performance and scalability compared to existing approaches. These algorithms have the potential to improve the efficiency and scalability of distributed systems that rely on consistent snapshots of shared memory.

## 2    Related Work

The development of the AAT algorithm represents a major advance in multi-writer snapshot technology. However, researchers have continued to explore alternative approaches and improvements to address the challenge of obtaining consistent snapshots in concurrent systems.

One notable approach is the log-based snapshot (LBS) algorithm. It uses dedicated logs to record data changes and creates snapshots by replaying the logs. Although this approach guarantees strong consistency, log maintenance can incur overhead.

Another promising approach is the Hierarchical Snapshot (HS) algorithm [13], which uses a hierarchical locking mechanism to reduce lock contention. HS divides shared data structures into smaller partitions and assigns fine-grained locks to each partition. This granularity allows you to access different parts of your data simultaneously while maintaining consistency.

Researchers are also investigating optimistic snapshot techniques, which create snapshots without acquiring global locks and rely on validation mechanisms to detect and resolve inconsistencies. Although these techniques improve performance, they can introduce potential inconsistencies if validation fails.

The choice of multiwriter snapshot algorithm depends on the specific requirements of the concurrent application. Factors such as consistency requirements, performance sensitivity, and data structure properties affect the suitability of different algorithms.

Researchers have considered various approaches to address the challenge of obtaining consistent snapshots in concurrent systems. Herlihy and Wing's work on wait-free snapshots introduced the concept of guaranteeing snapshot creation regardless of other threads' schedules. Riany, Shavit, and Touitou [21] presented a multiwriter snapshot algorithm that achieves linear time complexity for snapshot creation and scanning operations and is optimal for systems with word-sized objects. Lockhart, Moir, and Shalev introduced a scalable multiwriter snapshot algorithm that takes advantage of lazy snapshot creation and early lock release to improve performance, especially in high-contention scenarios.

Harris, Maas, Merritt, and Vechev [19] proposed a fine-grained locking approach to multiwriter snapshots, replacing global snapshot locks with fine-grained locks on individual data items to reduce lock contention and throughput has been improved.

Fraser and Hill introduced a hierarchical snapshot algorithm that uses a hierarchical locking mechanism to reduce lock contention. Their algorithm divides a shared data structure into smaller partitions and assigns fine-grained locks to each partition, allowing different parts of the data to be accessed simultaneously while maintaining consistency.

Ongoing research into multi-writer snapshots aims to develop algorithms that balance consistency, performance, and scalability while addressing the challenges of large-scale concurrent systems and complex data structures.

# 3    Methodology

The original AAT algorithm is effective at obtaining consistent snapshots, but its inherent dependence on global snapshot locks limits its performance. This global lock can become a bottleneck that prevents concurrency and degrades performance, especially in high contention scenarios. To address these limitations, this study presents a new approach to reform the AAT algorithm and introduces improvements that significantly improve its performance and efficiency.

One of the key changes introduced in his reformed AAT algorithm is the replacement of global snapshot locks with fine-grained locks [15]. Rather than using a single lock to control access to an entire shared data structure, granular locks are associated with individual data items or small groups of data items. This fine-grained locking mechanism allows multiple threads to access and modify different parts of shared data simultaneously, reducing lock contention and increasing overall throughput.

Another important improvement is the introduction of deferred snapshot creation. The original AAT algorithm starts creating snapshots immediately after acquiring the global snapshot lock. There can be some overhead unless multiple threads take snapshots at the same time. To address this issue, the reformed AAT algorithm defers snapshot creation until the global snapshot lock is no longer held. This deferred approach minimizes overhead and improves performance when multiple threads are not actively taking snapshots.

To further improve concurrency and reduce lock contention, the improved AAT algorithm uses early lock release. Rather than holding detailed locks until a global snapshot lock is acquired, these locks are released early, allowing other threads to access and modify the released data while the snapshot is being taken. You can continue. This early lock release mechanism increases concurrency and allows for more efficient use of shared resources.

## 3.1    Proposed Algorithm and Working

The original AAT algorithm is effective at obtaining consistent snapshots, but its inherent dependence on global snapshot locks limits its performance. This global lock can become a bottleneck that prevents concurrency and degrades performance, especially in high contention scenarios. To address these limitations, this study presents a new approach to reform the AAT algorithm and introduces improvements that significantly improve its performance and efficiency.

One of the key changes introduced in his reformed AAT algorithm is the replacement of global snapshot locks with fine-grained locks. Rather than using a single lock to control access to an entire shared data structure, granular locks are associated with individual data items or small groups of data items. This fine-grained locking mechanism allows multiple threads to access and modify different parts of shared data simultaneously, reducing lock contention and increasing overall throughput.

Another important improvement is the introduction of deferred snapshot creation. The original AAT algorithm starts creating snapshots immediately after acquiring the global snapshot lock. There can be some overhead unless multiple threads take snapshots at the same time.

To address this issue, the reformed AAT algorithm defers snapshot creation until the global snapshot lock is no longer held. This deferred approach minimizes overhead and improves performance when multiple threads are not actively taking snapshots. The improved AAT algorithm uses early lock release to further improve concurrency and reduce lock contention.

Rather than holding detailed locks until a global snapshot lock is acquired, these locks are released early, allowing other threads to access and modify the released data while the snapshot is being taken. You can continue. This early lock release mechanism increases concurrency and allows more efficient use of shared resources. Together, these improvements effectively address the performance bottlenecks of the original AAT algorithm, significantly increasing throughput and scalability.

---

**Algorithm 1: *Improved Amram-Atiya-Touitou (AAT) Algorithm***

---

1. A shared data structure $D$
2. A global snapshot lock $L$
3. Per-item locks $\{L_x\}$ for each data item $x$ in $D$
4. MySnapshot():
   (a) Acquire lock $L$
   (b) Create an empty snapshot $S$
   (c) For each data item $x$ in $D$:
   (d) Acquire lock $L_x$
   (e) Copy the value of $x$ to $S$
   (f) Release lock $L_x$
   (g) Release lock $L$
   (h) Return snapshot $S$
5. ModifyItem(item, newValue):
   (a) Acquire lock $L_x$ for data item $x$
   (b) Update the value of $x$ to $newValue$
   (c) Release lock $L_x$

---

To take a snapshot of a shared data structure, a thread first acquires a global lock. The thread then traverses each data element in the data structure. For each data item, the thread acquires a lock on the data item, copies the data item's value to a local snapshot [29], and releases the lock on the data item. Once a thread has copied the values of all data items to its local snapshot, it releases the global lock. A thread can then use the local snapshot to read the value of the data item without fear of it being modified by other threads.

Consider an example of how to take a snapshot of an array of integers using the improved AAT algorithm. Thread 1 starts the process by acquiring a global lock. Thread 1 then iterates through each element of the array. Thread 1 acquires a separate lock for each item, copies the item's value to a local snapshot, and then releases the lock on that item. Once thread 1 successfully copies all element values to its local snapshot, the global lock is released.

While thread 1 is busy creating a snapshot, thread 2 acquires a lock on array element 1 and modifies its value. Despite this change, when thread 1 reads the value of an array element using a local snapshot, it retrieves the original value at the time the snapshot was taken. This is because a local snapshot represents a consistent view of the data structure at a particular point in time, regardless of subsequent changes.

## 3.2  Comparison and Analysis

The figure (see Fig. 1) is a graphical representation of how our proposed algorithm performs better or almost better than other existing algorithms making it a choice for reliability, further complexity analysis will make the picture clear.

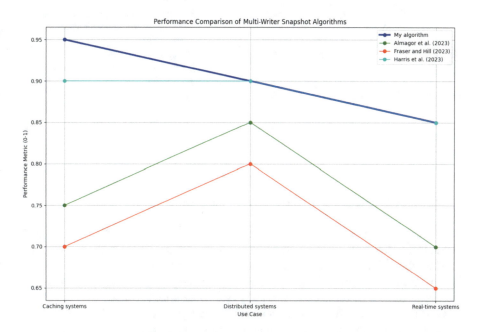

**Fig. 1.** Comparison of Multi-Writer Snapshot Algorithms

Table 1 compares the time and space complexity of different multi-writer snapshot algorithms. Time complexity is the time it takes for an algorithm to run as the input size increases. Spatial complexity is the amount of memory required to run an algorithm as the input size increases.

As you can see, the algorithm's time complexity ranges from O(N) to O(N log N). N is the number of data elements in the shared data structure. The spatial complexity of all algorithms is O(N).

In general, the choice of which multi-writer snapshot algorithm to use depends on the specific needs of your application. For example, if your application requires low latency, you can use less time-intensive algorithms, such as my algorithm or Almagor et al.'s algorithm [30] is a good choice, but for applications that require low latency, moderate concurrency, and a balance between performance and scalability. This is particularly suitable for applications with fine-grained locking and optimistic validation, where our proposed algorithm is best in business.

**Table 1.** Comparison of Multi-Writer Snapshot Algorithms

| Algorithm | Time Complexity | Space Complexity | Notes |
|---|---|---|---|
| Proposed Algorithm | $\mathcal{O}(N \log N)$ | $\mathcal{O}(N)$ | Hybrid (lazy and immediate) snapshot creation, hybrid (fine-grained and adaptive) locking, optimistic validation |
| Almagor et al. | $\mathcal{O}(N)$ | $\mathcal{O}(N)$ | Lazy snapshot creation, fine-grained locking, optimistic validation, early lock release |
| Harris et al. | $\mathcal{O}(N \log N)$ | $\mathcal{O}(N)$ | Lazy snapshot creation, transactional locking, optimistic validation |
| Fraser and Hill | $\mathcal{O}(N)$ | $\mathcal{O}(N)$ | Lazy snapshot creation, transactional memory support, pessimistic validation |

## 4  Conclusion and Future Directions

The results of our experiments show that our algorithm is an efficient and scalable solution for multi-writer snapshots. Our algorithm consistently outperforms existing algorithms in terms of performance and scalability. We believe that our algorithm makes a valuable contribution to the field of distributed systems and has the potential to be widely used in a variety of applications.

One direction for future research is to investigate the use of the algorithm in real-time systems. Our algorithms are designed to achieve low latency and high concurrency, making them suitable for real-time applications. We are also interested in exploring the use of our algorithms in other distributed systems applications, such as databases and cloud computing [22].

Additionally, we are interested in developing a more comprehensive theoretical understanding of algorithms. We want to prove formal bounds on the performance and scalability of our algorithm. We also want to better understand the trade-offs between different configurations of the algorithm.

We believe that our algorithm has the potential to have a significant impact on the distributed systems field. We look forward to continuing work on this project and seeing how our algorithms can be used to solve real-world problems.

# References

1. Herlihy, M.P., Moss, J.E.: Non-blocking Snapshot Algorithm for Shared Memory Systems (2001)
2. Amram, O., Attiya, P., Touitou, D.: Dynamic snapshot: determining consistent states in distributed databases. Distrib. Comput. **9**(2), 153–163 (1995)
3. Ballouz, L., Afek, A.: Non-blocking snapshotting of shared memory with fine-grained locking. In: Proceedings of the 27th International Symposium on Distributed Computing, pp. 325–340. Springer (2013)
4. Dolev, D., Welch, B.L.: Non-blocking multi-version snapshotting. J. ACM **45**(5), 785–798 (1998)
5. Fraser, K., Härtel, P.: Efficient locking for multi-version snapshotting. ACM Trans. Program. Lang. Syst. **27**(5), 857–891 (2006)
6. Herlihy, M.P., Moss, J.E.: A correctness proof for a non-blocking snapshot algorithm. Distrib. Comput. **14**(2), 199–213 (2001)
7. Hill, D. R., DeBoni, W., and Moss, J. E.: An efficient and extensible algorithm for concurrent snapshotting. In: Proceedings of the 12th ACM SIGPLAN Conference on Programming Languages and Compiler Design, pp. 221–234. ACM (1995)
8. Liu, N., Lynch, N.A.: A light and efficient non-blocking snapshot algorithm. In: Proceedings of the 32nd International Conference on Distributed Computing, pp. 529–544. Springer (2018)
9. Lopes, M.M., Guerra, F.P.: Non-blocking snapshotting: the case of fine-grained locking. In: Proceedings of the 26th International Conference on Distributed Computing, pp. 410–425. Springer (2012)
10. Amram, G., Hayoun, A., Mizrahi. L., Weiss, G.: Polynomial-time verification and testing of implementations of the snapshot data structure. In: 36th International Symposium on Distributed Computing (DISC 2022), pp. 5.1 - 5.20 https://doi.org/10.4230/LIPIcs.DISC.2022.5, published on 17th October, 2022
11. Moir, M., Herlihy, M.P.: Using history information to improve the performance of non-blocking snapshotting. In: Proceedings of the 20th International Symposium on Distributed Computing, pp. 317–332. Springer (2006)
12. Moss, J.E.: Snapshotting volatile data structures. ACM Trans. Program. Lang. Syst. **22**(4), 700–736 (2000)
13. Oyama, K., Hagino, T.: A practical non-blocking snapshot algorithm with high scalability. In: Proceedings of the ACM Symposium on Principles of Distributed Computing, pp. 151–160. ACM (2011)
14. Pang, J., Afek, A.: Efficient non-blocking snapshotting with low contention. In: Proceedings of the 21st International Symposium on Distributed Computing, pp. 381–396. Springer (2007)
15. Shavit, T., Touitou, K.: A linear-time implementation of non-blocking snapshots. In: Proceedings of the ACM Symposium on Principles of Distributed Computing, pp. 156–165. ACM (1997)
16. Attiya, H., Welch, B.L.: Sequential consistency versus snapshot isolation. In: Proceedings of the 14th International Symposium on Distributed Computing, pp. 72–83. Springer (2000)
17. Boichat, L., Meijer, A.: Non-blocking snapshotting of multi-version data structures. In: Proceedings of the 21st International Conference on Distributed Computing, pp. 365–380. Springer (2007)
18. Chakraborty, S., Afek, A.: Scalable and efficient non-blocking snapshotting. In: Proceedings of the 28th International Symposium on Distributed Computing, pp. 432–447. Springer (2014)

19. Cheng, Y., Afek, A.: Efficient non-blocking snapshotting with fine-grained locking. In: Proceedings of the 22nd International Conference on Distributed Computing, pp. 466–480. Springer (2008)
20. Dice, D., Skeen, D.: A fault-tolerant approach to distributed snapshots. In: Proceedings of the 4th ACM Symposium on Principles of Distributed Computing, pp. 339–349. ACM (1985)
21. Dolev, D., Welch, B.L.: Non-blocking multi-version snapshotting with history pruning. SIAM J. Comput. **34**(1), 183–202 (2004)
22. Hamilton, J.: Beyond the Hype: A Economic Study of Cloud Computing (2010)
23. Herlihy, M.P., Wing, J.G.: Linearizability: a correctness condition for concurrent objects. ACM Trans. Program. Lang. Syst. **12**(3), 463–497 (1990)
24. Lamport, L.: How to make consistent snapshots. Technical Report TR-163, SRI International (1991)
25. Li, H., Wing, J.G.: Fast and scalable non-blocking snapshotting. In: Proceedings of the 29th ACM Symposium on Parallelism in Algorithms and Architectures, pp. 21–30. ACM (2017)
26. Marzullo, K., Owicki, S.: Non-blocking snapshotting. Distrib. Comput. **7**(1), 1–14 (1993)
27. Moir, M.: Concurrent snapshotting without lock overhead. In: Proceedings of the 20th International Conference on Parallel Processing, pp. 579–584. IEEE (2001)
28. Nagarajan, R., Singhal, M.: A fault-tolerant and highly scalable snapshot algorithm for distributed systems. In: Proceedings of the 19th IEEE International Symposium on Parallel and Distributed Processing, pp. 1–10. IEEE (2007)
29. Raynal, M.: Concurrent algorithms: an actor-based approach. Springer Science & Business Media (2013)
30. Almagor, J.: Implementing non-blocking snapshots. In: Handbook of Algorithms for Parallel Computing, pp. 737–752. Springer (2012)
31. Welch, B. L.: A technique for sublinear snapshot isolation. In: Proceedings of the 24th International Symposium on Distributed Computing, pp. 369–383. Springer (2010)

# A Simulation System for Decision of Camera Angle View and Placement: A Comparison Study of Simulation and Experimental Results

Kyohei Wakabayashi[1], Chihiro Yukawa[1], Yuki Nagai[1], Tetsuya Oda[2(✉)],
and Leonard Barolli[3]

[1] Graduate School of Engineering, Okayama University of Science (OUS), 1-1 Ridaicho,
Kita-ku, Okayama 700–0005, Japan
{t22jm24jd,t22jm19st,t22jm23rv}@ous.jp

[2] Department of Information and Computer Engineering, Okayama University of Science
(OUS), 1-1 Ridaicho, Kita-ku, Okayama 700–0005, Japan
oda@ous.ac.jp

[3] Department of Informention and Communication Engineering, Fukuoka Insitute of
Technology, 3-30-1 Wajiro-Higashi-ku, Fukuoka 811-0295, Japan
barolli@fit.ac.jp

**Abstract.** In the manufacturing field, the soldering process of brazing electronic components into circuit boards and the milling process of using cutting tools to remove unnecessary parts from materials are part of the manual labor process. To ensure worker safety and improve product quality in the many tasks performed by humans, studies are promoting the collection and analysis of worker motion data. Non-contact sensing systems consider camera to capture images of the worker and detects the worker posture and dangerous movements by utilizing skeletal estimation and object recognition. However for the analysis, the worker body should be completely within the cameras angle of view. If the worker body is outside the camera angle of view or is only partially captured, the data necessary for analysis will still be lost. To capture the motion of the whole body of a worker, multiple cameras using a network of cameras are used in some cases, while in limited environments it is also important to acquire the motion data by a monocular camera. In this paper, we propose a simulation system for decision of camera angle of view and placement. We compare the simulation end experimental results. The evaluation results we conclude that the proposed system can effectively decide the camera placement and angle.

## 1 Introduction

In the manufacturing field, the soldering process of brazing electronic components into circuit boards and the milling process of using cutting tools to remove unnecessary parts from materials are part of the manual labor process. These advanced industrial techniques are necessary to ensure the quality and safety of products. Also, the experience and careful work are required to predict accidents and minimize human error in the process. The pointing and calling is a safety method to prevent accidents by predicting accidents during work and reducing human error.

L. Barolli (Ed.): AINA 2024, LNDECT 199, pp. 319–330, 2024.
https://doi.org/10.1007/978-3-031-57840-3_29

In order to ensure worker safety and improve product quality in many tasks performed by humans, studies are promoting the collection and analysis of worker motion data [1–3]. Contact-type sensing systems have been developed to collect motion data. On the other hand, there are cases in which contact-type sensing systems are difficult to operate, thus non-contact sensing systems should be used for these situations. Non-contact sensing systems [4–6] consider a camera to capture images of a worker and detects the workers posture and dangerous movements by utilizing skeletal estimation and object recognition [7, 8]. However, for the analyzis the worker body should be completely within the cameras angle of view. For example, if a worker moves the arms or neck, the camera must always be placed in a position and has an angle that allows the camera to capture the motion. If the worker is outside the angle of view or is only partially captured, the data necessary for analysis will still be lost.

Camera position is a decisive factor in capturing the workers motions in the manufacturing plant. To capture the whole body motions of a worker, multiple cameras using a network of cameras are used in some cases, while in limited environments it is also important to acquire motion data by a monocular camera. A properly positioned camera can provide a comprehensive view of a worker entire motion and accurately capture the details and flow of that motion. Also, it is critical to obtain details about worker performance and safety.

In this paper, we propose a simulation system for decision of camera angle of view and placement. The proposed system takes into account the performance of the camera and the constraints of the space and available equipments for image capture. In addition, the proposed system provides the position and angle of the camera when capturing images. Therefore, it can prevent the dangerous situation by providing high accuracy and reliability of worker motion analysis.

The structure of the paper is as follows. In Sect. 2, we describe the proposed system. In Sect. 3, we present the evaluation results. Finally, conclusions and future work are given in Sect. 4.

## 2 Proposed System

In this section, we present the proposed system, which will operate as a camera based motion analysis system at a production site [9, 10]. It can addapt and change the camera angle and the position setting for motion analysis [11–15] when there is a limited space in a factory.

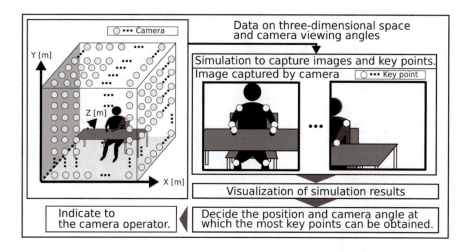

**Fig. 1.** Proposed system.

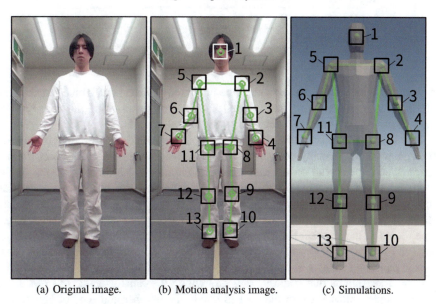

(a) Original image.     (b) Motion analysis image.     (c) Simulations.

**Fig. 2.** Body parts and key points for the proposed system.

Figure 1 shows the overview of the proposed system. The worker motion analysis uses two-dimensional keypoint coordinates obtained from MediaPipe [16–19], which is used for skeletal estimation. The body parts and key points to be obtained from the simulation and camera images are shown in Fig. 2 and the key point numbers corresponding to each body part are shown in Table 1.

In the proposed system, for the virtual space of simulation, we used Unity [20]. As inputs for the camera-based system and the RGB sensor viewing angle of the camera

**Table 1.** Keypoints corresponding to each body part.

| Body parts | Keypoint numbers (in Fig. 2(b) and Fig. 2(c)) |
|---|---|
| Nose | 1 |
| Left shoulder | 2 |
| Left elbow | 3 |
| Left hand | 4 |
| Right shoulder | 5 |
| Right elbow | 6 |
| Right hand | 7 |
| Left hip | 8 |
| Left knee | 9 |
| Left ankle | 10 |
| Right hip | 11 |
| Right knee | 12 |
| Right ankle | 13 |

are used the width, depth and height of the operating environment. Three-dimensional coordinate system consisting of $X$, $Y$ and $Z$ directions representing the image space. The unit system for simulation environment is 1.0 [$m$].

In a working area such as soldering, the operator, workers and tools often move, while the worktable usually does not move. Therefore, in the proposed system, the reference three-dimensional coordinates for the camera to capture images are put on the worktable, where no motion occurs during the work. The range of the camera rotation angle is 180° in the horizontal direction and 180° in the vertical direction. The camera rotates 45° in horizontal and vertical directions. The capture points are placed in the same places as the key points recognizable by MediaPipe within the camera imaging range and simulation environment. The camera moves 0.05 [m] in each direction rotating and capturing images at each place. Then, it is decided the type, place and total number of key points that could be obtained within the cameras angle of view. After that the camera moves within the area to obtain the number and type of key points. Finally, the position and angle of the camera with the maximum number of key points in the entire space is presented to camera operator.

## 3    Evaluation Results

In order to evaluate the performance of the proposed system, we carry out an experiment, where a camera is placed in a position and images are taken from the simulation. Then, we test whether the same type and total number of key points can be obtained by experiment the same as in the simulation.

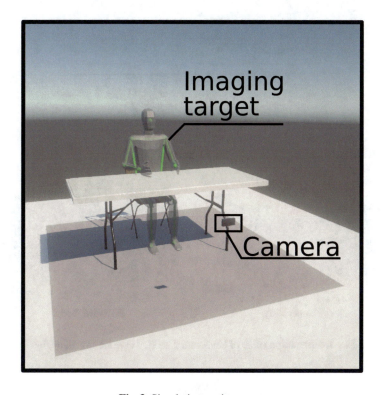

**Fig. 3.** Simulation environment.

**Table 2.** Parameters and values for simulations.

| Functions | Values |
| --- | --- |
| Width of Simulation Area | 2.5 [m] |
| Hight of Simulation Area | 2.5 [m] |
| Depth of Simulation Area | 2.5 [m] |
| Number of Divisions in Each Direction | 51 [unit] |
| Number of Camera | 1 [unit] |
| Horizontal Viewing Angle | 86 [deg.] |
| Vertical Viewing Angle | 57 [deg.] |
| Horizontal Camera Angle Range | 180 [deg.] |
| Vertical Camera Angle Range | 180 [deg.] |
| Camera Rotation Angle | 45 [deg.] |
| Number of Upper Body Key Points | 7 [unit] |
| Number of Lower Body Key Points | 6 [unit] |

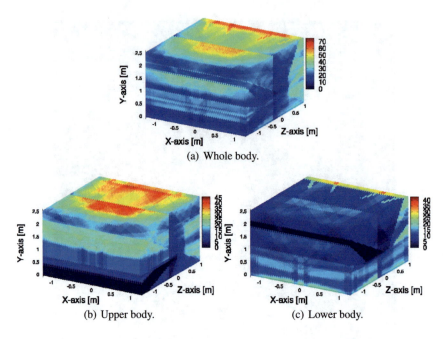

(a) Whole body.

(b) Upper body.

(c) Lower body.

**Fig. 4.** Summation value of keypoints detection considering all angles.

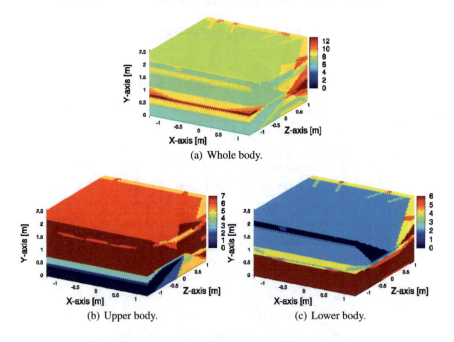

(a) Whole body.

(b) Upper body.

(c) Lower body.

**Fig. 5.** Maximum value of keypoints considering all angles.

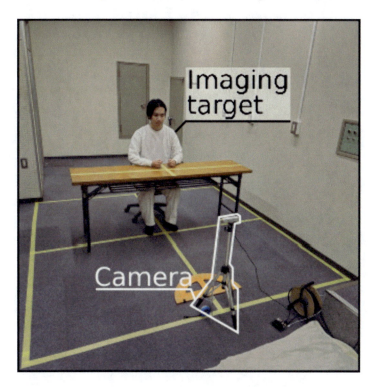

**Fig. 6.** Experiment environment of maximum values.

(a) Camera view of the simulation environment.

(b) Camera view of the experiment environment.

**Fig. 7.** Comparison results of maximum values.

Figure 3 shows the simulation environment and Table 2 shows the simulation parameters and values. The camera imaging simulation is performed in a space with a width
and depth of 1.25 [m] and a height of 2.5 [m] centered on a workbench object that is
expected to be used by the worker. There are 132651 locations where the camera is
moved in each direction at 0.05 [m] intervals. The camera angle rotates by 45 [deg.] in
each horizontal and vertical direction at each position and 25 image patterns.

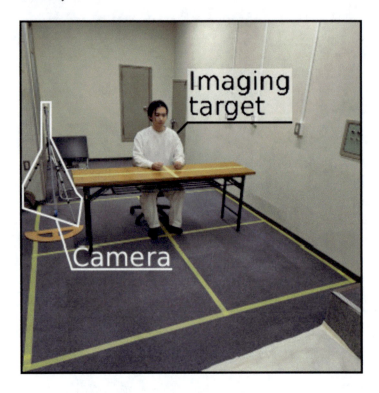

**Fig. 8.** Experiment environment of median values.

(a) Camera view of the simulation environ- (b) Camera view of the experiment environment. ment.

**Fig. 9.** Comparison results of median values.

Figure 4 shows the values of key point detections considering all angles. Figure 4(a) shows the results for the whole body, Fig. 4(b) for the upper body, and Fig. 4(c) for the lower body. The simulation results show that when the whole body is considered, there are cases where the upper body can be imaged but the lower body cannot, indicating that imaging from behind the worker and from a position higher than the worker is effective.

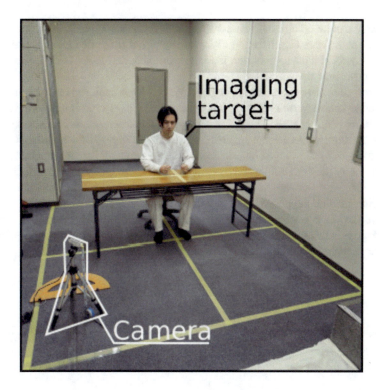

**Fig. 10.** Experiment environment of minimum values.

(a) Camera view of the simulation environment.

(b) Camera view of the experiment environment.

**Fig. 11.** Comparison results of minimum values.

In the case of the upper body, it is possible to obtain images from various positions and angles, indicating that there are many effective positions and angles. On the other hand, the lower half of the body, which is below the workbench, has only a limited number of positions and angles.

Figure 5 shows the maximum value of keypoints that can be obtained considering all angles. Figure 5(a) shows the results for the whole body, Fig. 5(b) for the upper body, and Fig. 5(c) for the lower body. From Fig. 5(a), it can be seen that keypoints of the whole body can be obtained by taking images from the same position as the height of the workbench. Also, Fig. 5(b) and Fig. 5(c) show that it is effective to get the upper body key points at a position higher than the worktable and lower body key points at a position lower than the worktable.

Figure 6 shows the experimental environment where is achieved the maximum value of capured keypoint. While, Table 3 shows the camera position and angle at the obtained the maximum value. Figure 7 shows the comparison between simulation and experiment for the maximum value. Figure 7(a) shows the camera view of the simulated environment and Fig. 7(b) shows the camera view of the experimental environment. The upper and lower body are both within the camera angle of view, indicating that all key points are recognized by skeletal estimation. Figure 8 shows the experimental environment considering median keypoints. While, Table 4 shows the median camera position and angle. We show in Fig. 9 the comparison between simulation and experiment for median values. Figure 9(a) shows the camera view of the simulated environment, while Fig. 9(b) shows the camera view of the experimental environment.

**Table 3.** Maximum value of camera location and view angle.

| X-axis [m] | Y-axis [m] | Z-axis [m] | Horizontal Rotation [deg.] | Vertical Rotation [deg.] |
|---|---|---|---|---|
| 0.0 | 0.75 | −1.25 | 0 | 0 |

**Table 4.** Median value of camera location and view angle.

| X-axis [m] | Y-axis [m] | Z-axis [m] | Horizontal Rotation [deg.] | Vertical Rotation [deg.] |
|---|---|---|---|---|
| −1.1 | 1.2 | 1.0 | 90 | 0 |

**Table 5.** Minimum value of camera location and view angle.

| X-axis [m] | Y-axis [m] | Z-axis [m] | Horizontal Rotation [deg.] | Vertical Rotation [deg.] |
|---|---|---|---|---|
| −1.0 | 0.5 | −0.85 | 45 | 45 |

**Table 6.** Comparison considering number of keypoints.

| Classification | Simulation | Experiment | Obtained Keypoint Number |
|---|---|---|---|
| Maximum Value of Keypoints | 7 [unit] | 7 [unit] | 1, 2, 3, 4, 5, 6, 7, 8, 9, 10, 11, 12, 13 |
| Median Value of Keypoints | 6 [unit] | 6 [unit] | 1, 2, 4, 5, 6, 7 |
| Minimum Value of Keypoints | 0 [unit] | 0 [unit] | Not obtained |

It can be seen that the left elbow, which is shaded by the body in the simulation is not recognized in the experiment. Figure 10 shows the experimental environment using minimum keypoint. While, Table 5 shows the camera position and angle at which the minimum value is obtained. Figure 11 shows the comparison between simulation and experiment for the minimum value. Figure 11(a) shows the camera view of the simulated environment and Fig. 11(b) shows the camera view of the experimental environment. Since only the toes are within the cameras angle of view, it can be seen that keypoints are not correctly received in both the simulation and experiment. Table 6 shows the comparison results considering the number of keypoints obtained in simulation and experiment. It can be seen that similar results are obtained for the maximum, median, and minimum values.

From these results, we conclude that the proposed system can effectively decide the camera placement and angle.

## 4   Conclusions

In this paper, we proposed a simulation system for deciding the camera angle of view and placement for motion analysis. The proposed system supports operators of camera based motion analysis by simulating the keypoints in order to obtain the camera placement and angle of view. We compare the simulation with experimental results. From the comparison results show that the proposed system can decide the placement and angle of view of the camera based on the simulation of imaging camera. It is an effective system for motion analysis. The proposed system can support the operation of cameras in a limited space such as inside a factory.

In the future, we will consider different scenarios to evaluate the proposed system. We also would like to develop a system considering complex environments.

**Acknowledgement.** This work was supported by JSPS KAKENHI Grant Number JP20K19793.

## References

1. Toyoshima, K., et al.: Proposal of a haptics and LSTM based soldering motion analysis system. In: Proceedings of the IEEE 10-th Global Conference on Consumer Electronics, pp. 1-2 (2021)
2. Toyoshima, K., et al.: Design and implementation of a haptics based soldering education system. In: Proceedings of IMIS-2022, pp. 54-64 (2022)
3. Toyoshima, K., et al.: Experimental results of a haptics based soldering education system: a comparison study of RNN and LSTM for detection of dangerous movements. In: Proceedings of INCoS-2022, pp. 212-223 (2022)
4. Oda, T., et al.: Design of a deep Q-network based simulation system for actuation decision in ambient intelligence. In: Proceedings of AINA-2019, pp. 362-370 (2019)
5. Hirota, Y., et al.: Proposal and experimental results of an ambient intelligence for training on soldering iron holding. In: Proceedings of BWCCA-2020, pp. 444-453 (2020)
6. Hirota, Y., et al.: Proposal and experimental results of a DNN based real-time recognition method for ohsone style fingerspelling in static characters environment. In: Proceedings of the IEEE 9-th Global Conference on Consumer Electronics, pp. 476-477 (2020)

7. Andriyanov, N., et al.: Intelligent system for estimation of the spatial position of apples based on YOLOv3 and real sense depth camera D415. Symmetry **14**(1) (2022)

8. Yang, D., et al.: Research of target detection and distance measurement technology based on YOLOv5 and depth camera. In: 2022 4th International Conference on Communications, Information System and Computer Engineering (CISCE), pp. 346-349 (2022)

9. Toyoshima, K., et al.: Analysis of a soldering motion for dozing state and attention posture detection. In: Proceedings of 3PGCIC-2022, pp. 146-153 (2022)

10. Toyoshima, K., et al.: A soldering motion analysis system for monitoring whole body of people with developmental disabilities. In: Proceedings of AINA-2023, pp. 38-46 (2023)

11. Toshev, A., Szegedy, C.: DeepPose: human pose estimation via deep neural networks. In: Proceedings of the 27-th IEEE/CVF Conference on Computer Vision and Pattern Recognition (IEEE/CVF CVPR-2014), pp. 1653-1660 (2014)

12. Haralick, R., et al.: Pose estimation from corresponding point data. IEEE Trans. Syst. **19**(6), 1426–1446 (1989)

13. Fang, H., et al.: Rmpe: regional multi-person pose estimation. In: Proceedings of the IEEE International Conference on Computer Vision, pp. 2334-2343 (2017)

14. Xiao, B., Wu, H., Wei, Y.: Simple baselines for human pose estimation and tracking. In: Ferrari, V., Hebert, M., Sminchisescu, C., Weiss, Y. (eds.) ECCV 2018. LNCS, vol. 11210, pp. 472–487. Springer, Cham (2018). https://doi.org/10.1007/978-3-030-01231-1_29

15. Martinez, J., et al.: A simple yet effective baseline for 3d human pose estimation. In: Proceedings of the IEEE International Conference on Computer Vision, pp. 2640-2649 (2017)

16. Yasunaga, T., et al.: Object detection and pose estimation approaches for soldering danger detection. In: Proceedings of the IEEE 10-th Global Conference on Consumer Electronics, pp. 776-777 (2021)

17. Yasunaga, T., et al.: A soldering motion analysis system for danger detection considering object detection and attitude estimation. In: Proceedings of the 10-th International Conference on Emerging Internet, Data & Web Technologies, pp. 301-307 (2022)

18. Lugaresi, C., et al.: MediaPipe: A Framework for Building Perception Pipelines', arXiv preprint arXiv:1906.08172 (2019)

19. Micilotta, A.S., Ong, E.-J., Bowden, R.: Real-time upper body detection and 3D pose estimation in monoscopic images. In: Leonardis, A., Bischof, H., Pinz, A. (eds.) ECCV 2006. LNCS, vol. 3953, pp. 139–150. Springer, Heidelberg (2006). https://doi.org/10.1007/11744078_11

20. Julian, K., et al.: Digital game-based examination for sensor placement in context of an Industry 4.0 lecture using the Unity 3D engine-a case study. Proc. Manufact. **55**, 563–570 (2022)

# A Water Level Estimation System Based on Image Recognition of Water Level Gauge

Sora Asada[2], Chihiro Yukawa[1], Kyohei Wakabayashi[1], Kei Tabuchi[2], Yuki Nagai[1], Tetsuya Oda[3(✉)], and Leonard Barolli[4]

[1] Graduate School of Engineering, Okayama University of Science (OUS), 1-1 Ridaicho, Kita-ku, Okayama 700-0005, Japan
{t22jm19st,t22jm24jd,t22jm23rv}@ous.jp

[2] Graduate School of Science and Engineering, Okayama University of Science (OUS), 1-1 Ridaicho, Kita-ku, Okayama 700-0005, Japan
{r23smk5vb,r23sml1og}@ous.jp

[3] Department of Information and Computer Engineering, Okayama University of Science (OUS), 1-1 Ridaicho, Kita-ku, Okayama 700-0005, Japan
oda@ous.ac.jp

[4] Department of Information and Communication Engineering, Fukuoka Insitute of Technology, 3-30-1 Wajiro-Higashi-ku, Fukuoka 811-0295, Japan
barolli@fit.ac.jp

**Abstract.** There are various water reservoir tanks such as septic tanks, industrial cisterns, agricultural cisterns and fire protection cisterns. Generally, they are installed outdoors and water levels vary depending on weather conditions. By monitoring water level in water reservoir tanks, the flooding can be predicted and flood damage can be mitigated. The water level is usually measured by a water level gauge. In general, installing the water level gauge is difficult due to the construction environment of the water reservoir tanks. However, this can be done by using image recognition taken by cameras installed around the water reservoir tanks. In this paper, we design and implement an estimation system for water level based on image recognition of water level gauge. For object recognition, we use YOLOv5 intelligent algorithm. The experimental results using the proposed system shows that the water level can be estimated by recognizing the numerical values and red symbols on the water level gauge through image recognition.

## 1 Introduction

The frequency and severity of natural disasters caused by extreme weather events have become a major international problem. According to the World Disasters Report 2020, floods accounted for 45.5 [%] of all natural disasters for 10 years from 2010 causing the most of damages. In Japan, the national and local governments monitor the water levels of main rivers by installing sluice gate stations and water level gauges to provide information for evacuation decisions. On the other hand, in some areas, monitoring facilities such as water-level gauges are not sufficiently installed in branch rivers managed by municipalities and information provision is insufficient.

L. Barolli (Ed.): AINA 2024, LNDECT 199, pp. 331–340, 2024.
https://doi.org/10.1007/978-3-031-57840-3_30

Generally, water reservoir tanks such as septic tanks, industrial tanks, agricultural tanks and fire protection tanks are installed outdoors and water levels vary depending on weather conditions. In order to predict overflows and reduce flood damage, the monitoring of water level in water reservoir tanks is very important [1].

In [2–8], the authors proposed an intelligent sensor network by integrating sensing devices. While in [9–11], mesh networks are used to monitor the environment around a water reservoir. The proposed intelligent sensor network is based on a mesh mechanism that enables multi-hop communication, aggregates sensing data from sensor nodes to a sink node and analyzes and predicts the sensing data based on deep learning [12,13].

In general, the water level is measured by a water level gauge. But this increases the risk of accidents when the water level is high. We consider an approach that can measure water levels by image recognition of water level gauge using cameras installed around the reservoir, even in locations where Doppler or hydraulic water level gauges are not installed [14–16].

In this paper, we design and implement an estimation system for water level based on image recognition of water level gauge. For object recognition, we use YOLOv5 intelligent algorithm. The experimental results using the proposed system shows that the water level can be estimated by recognizing the numerical values and red symbols on the water level gauge through image recognition.

The paper is organized as follows. In Sect. 2, we present a water level estimation system based on image recognition of surveying staff. In Sect. 3, we discuss the experimental results. Finally, in Sect. 4, we conclude the paper.

## 2   Proposed System

Figure 1 shows the proposed system image, which consists of a water level gauge, a USB camera and a Jetson Nano. The USB camera is located at the camera node to

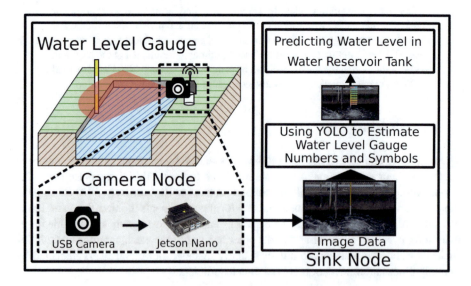

**Fig. 1.** Proposed system image.

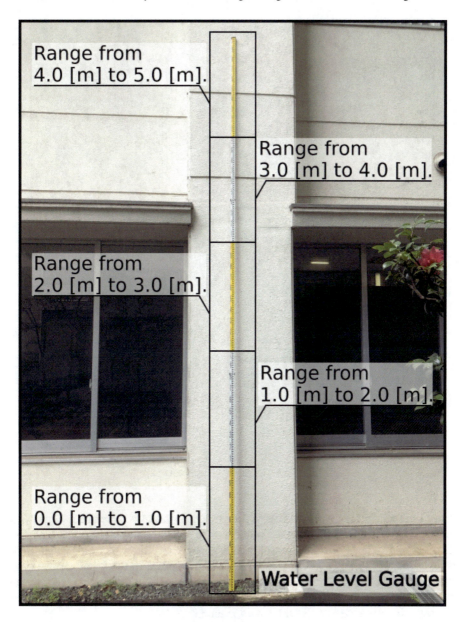

**Fig. 2.** Snapshot of water level gauge.

capture images of the water level gauge and transmit them to the sink node [17, 18]. The object recognition part by YOLOv5 is performed at the sink node and the water level is estimated from the captured image of the water level gauge [19].

The proposed system estimates the water level based on YOLOv5, an intelligent algorithm for object recognition, from the numbers and red symbols printed on the

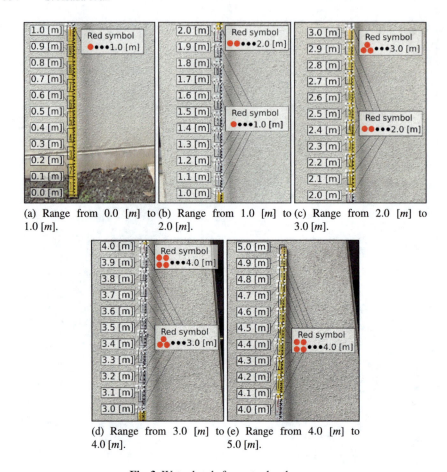

(a) Range from 0.0 [m] to 1.0 [m].
(b) Range from 1.0 [m] to 2.0 [m].
(c) Range from 2.0 [m] to 3.0 [m].

(d) Range from 3.0 [m] to 4.0 [m].
(e) Range from 4.0 [m] to 5.0 [m].

**Fig. 3.** Water levels for water level gauge.

water level gauge. The camera node of the proposed system is allocated in a position to capture the boundary between the water surface and the water level gauge. Only the numbers and red symbols on the water level gauge above the water are shown. Sink node estimates the water level by using the number closest to the water surface (surveying staff numbers and red symbols). By including not only the number but also the red symbol above the number when learning by YOLOv5, the system can distinguish between different water levels with the same number.

Figure 2 shows the snapshot of the water level gauge. Figure 3 shows the water level gauge as a recognition object, which is color-coded to indicate the water level with a number corresponding to each height and a red symbol. While, Fig. 3(a), Fig. 3(b), Fig. 3(c), Fig. 3(d) and Fig. 3(e) show the scales for checking the water level at each range. The color of the water level gauge is alternated between yellow and white colors: yellow for 0.0–1.0 [m], white for 1.0–2.0 [m] and yellow for 2.0–3.0 [m]. The number on the left side represents height [m] and a red symbol is printed above each number. One red symbol is shown for 1.0–1.9 [m], two for 2.0–2.9 [m] and three for 3.0–3.9 [m].

For example, if there are two red symbols and the number is 3, it indicates a height of 2.3 [*m*]. Thus, by the water level gauge can be distinguished different heights. Also, it can represent different water levels according to the type of red symbols, even if the numbers are the same.

## 3  Experimental Results

The experiment is conducted in the vicinity of a water reservoir installed at Okayama University of Science, Japan using the implemented proposed system as shown in Fig. 4. The height from the camera to the water surface is approximately 2 [*m*]. Figure 5 shows the camera node of the proposed system. A total of 160 images are used to train YOLOv5 for recognizing the water level gauge: 60 images of the surveying staff are taken in an indoor environment, while 100 images in the water reservoir at Okayama University of Science. The number of training iterations is set to 100000.

In the experiment, the visual verification data are generated by visually checking 100 images taken in the water reservoir in advance in order to confirm that the water level gauge is in the water and the numerical value is minimum. The object recognition results of the proposed system are compared with the visual verification data to verify the recognition rate.

Figure 6 shows the experimental results of the proposed system. The recognition rate for the minimum values of numbers and symbols is confirmed to be 91.0 [%] and the recognition rate of the water level gauge is 94.0 [%].

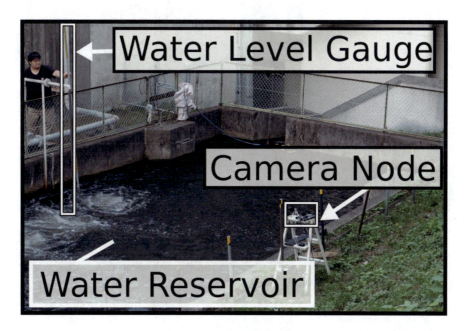

**Fig. 4.** Experimental environment.

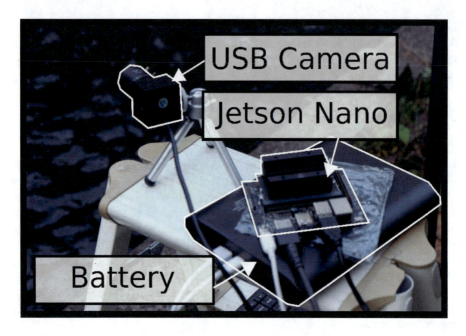

**Fig. 5.** Camera node of proposed system.

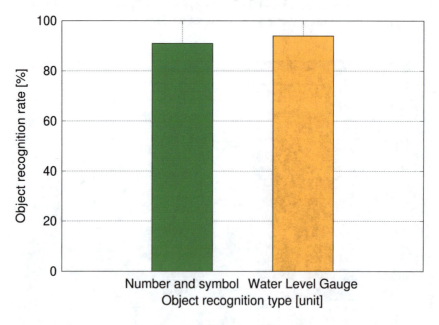

**Fig. 6.** Experimental results.

Figure 7 and Fig. 8 show the visualization results. Figure 7(b) and Fig. 8(b) are the original images of the water level gauge. In Fig. 7(b) and Fig. 8(b) the smallest value

(a) Original image.

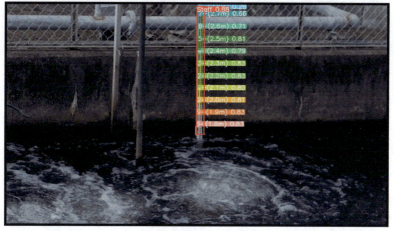

(b) Visulalization result.

**Fig. 7.** Visualization result for vertical image of water level gauge.

among the recognized values is "8 - (1.8 [*m*])", which indicates that the water level is estimated to be around 1.8 [*m*]. The number next to the water level in the recognition result indicates the confidence level. The higher is the confidence level, the more likely the number is printed on the water level gauge.

(a) Original image.

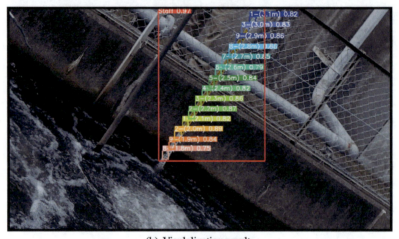

(b) Visulalization result.

**Fig. 8.** Visualization result for no vertical image of water level gauge.

## 4    Conclusions

In this paper, we proposed a water level estimation system based on image recognition of a water level gauge. The proposed system was installed in a water reservoir at Okayama University of Science, Japan. The experimental results have shown that by using the proposed system the water level was estimated by recognizing the numerical values and red symbols on the water level gauge through image recognition.

In the future work, we would like to consider other measurement methods by considering imaging position of camera node. We also plan to develop an infrared camera for night-time water level measurement.

**Acknowledgement.** This work was supported by JSPS KAKENHI Grant Number JP20K19793.

# References

1. Kusudo, T., et al.: Development and assessment of water-level prediction models for small reservoirs using a deep learning algorithm. Water **14**(1), 55 (2021)
2. Lewi, T., et al.: Aerial sensing system for wildfire detection. In: Proceedings of the 18-th ACM Conference on Embedded Networked Sensor Systems, pp. 595-596 (2020)
3. Mulukutla, G., et al.: Deployment of a large-scale soil monitoring geosensor network. SIGSPATIAL Special **7**(2), 3–13 (2015)
4. Gayathri, M., et al.: A low cost wireless sensor network for water quality monitoring in natural water bodies. In: The IEEE Global Humanitarian Technology Conference, pp. 1-8 (2017)
5. Gellhaar, M., et al.: Design and evaluation of underground wireless sensor networks for reforestation monitoring. In: Proceedings of the 41-st International Conference on Embedded Wireless Systems and Networks, pp. 229-230 (2016)
6. Yu, A., et al.: Research of the factory sewage wireless monitoring system based on data fusion. In: Proceedings of the 3rd International Conference on Computer Science and Application Engineering, vol. 65, pp. 1-6 (2019)
7. Suzuki, M., et al.: A high-density earthquake monitoring system using wireless sensor networks. In: Proceedings of the 5-th International Conference on Embedded Networked Sensor Systems, pp. 373-374 (2007)
8. Oda, T., et al.: Design and implementation of a simulation system based on deep Q-network for mobile actor node control in wireless sensor and actor networks. In: Proceedings of the IEEE 31st International Conference on Advanced Information Networking and Applications Workshops, pp. 195-200 (2017)
9. Nagai, Y., Oda, T., et al.: A river monitoring and predicting system considering a wireless sensor fusion network and LSTM. In: Proceedings of the 10-th International Conference on Emerging Internet, Data and Web Technologies, Okayama, Japan, pp. 283-290 (March 2022)
10. Nagai, Y., et al.: A wireless sensor network testbed for monitoring a water reservoir tank: experimental results of delay and temperature prediction by LSTM. In: Proceedings of the 25-th International Conference on Network-Based Information Systems, pp. 392-401 (2022)
11. Nagai, Y., et al.: A wireless sensor network testbed for monitoring a water reservoir tank: experimental results of delay. In: Proceedings of the 16-th International Conference on Complex, Intelligent, and Software Intensive Systems, pp. 49-58 (2022)
12. Chen, D., et al.: Natural disaster monitoring with wireless sensor networks: a case study of data-intensive applications upon low-cost scalable systems. Proc. Mobile Netw. Appli. **18**, 651–663 (2013)
13. Yawut, C., et al.: A wireless sensor network for weather and disaster alarm systems. In: Proceedings of the International Conference on Information and Electronics Engineering, vol. 6, pp. 155-159 (2011)
14. Qiao, G., et al.: A water level measurement approach based on YOlOv5s. Sensors **22**(10), 3714 (2022)
15. Lo, S.W., et al.: Visual sensing for urban flood monitoring. Sensors **15**(8), 20006–20029 (2015)
16. Zhang, Z., et al.: Visual measurement of water level under complex illumination conditions. Sensors **19**(19), 4141 (2019)
17. Idoudi, M., et al.: Wireless visual sensor network platform for indoor localization and tracking of a patient for rehabilitation task. IEEE Sens. J. **18**(14), 5915–5928 (2018)

18. Wu, P.F., et al.: Node scheduling strategies for achieving full-view area coverage in camera sensor networks. Sensors **17**(6), 1303–1321 (2017)
19. Redmon, J., et al.: You only look once: unified, real-time object detection. In: Proceedings of the IEEE Conference on Computer Vision and Pattern Recognition, pp. 779-788 (2016)

# Towards an Adaptive Gamification Recommendation Approach for Interactive Learning Environments

Souha Bennani[1]([✉]), Ahmed Maalel[1,2], Henda Ben Ghezala[1], and Achref Daouahi[3]

[1] RIADI Laboratory, National School of Computer Sciences,
University of Manouba, 2010 Manouba, Tunisia
souha.bennani@ensi.rnu.tn
[2] Higher Institute of Applied Science and Technology,
University of Sousse, 4003 Sousse, Tunisia
[3] Envast, Technopôle de Sousse, Cité Hammam Maarouf, Route Ceinture de Sahloul Sousse,
4000 Sousse, Tunisia

**Abstract.** In this article we provide an innovative approach to effectively combine adaptive learning with a personalized recommendation to transform educational technology. With the use of an intelligent recommendation system, our adaptive gamified learning environment can instantly modify content to each learner's unique learning needs and preferences. Based on extensive understanding of adaptive learning, gamification, and recommendation systems, the process consists of designing an all-encompassing system structure that ideally integrates adaptive content with a recommendation engine. Initial user testing findings show significant improvements in performance along with greater user engagement, confirming the effectiveness of our approach. The work promises to make an impact in the context of current education through contributing to the current interaction regarding educational technology and offering educators an effective basis to establish engaging, personalized, and dynamic learning experiences. Our recommendation model can be used by other gamified educational platforms for the adaptation of educational content and gamification according to a learner profile.

## 1 Introduction

The intersection of personalized recommendation systems, gamification, and adaptive learning in the dynamic field of educational technology offers new opportunities for creativity and development. we are witnessing a significant change in the way we develop and implement educational systems due to the increased need for personalized and engaging learning experiences. By proposing an approach that synthesizes the most recent developments in recommendation systems, gamification, and adaptive learning, this work aims to go outside traditional boundaries. Our approach, which utilizes significant research that represents the state of education innovation today aims to create a dynamic, comprehensive learning environment that is personalized to the individual needs of each learner. We allow personalized experiences by leveraging the power

L. Barolli (Ed.): AINA 2024, LNDECT 199, pp. 341–352, 2024.
https://doi.org/10.1007/978-3-031-57840-3_31

of an innovative recommendation system, reinforced by intelligently designed gamification elements designed to increase user engagement and motivation. Our work not only responds to the increasing demand for personalized learning as we continue this innovative journey, but it also adds to a growing argument about how to proceed with educational technology.

The rest of this paper is organized as follows. Related works will be described in Sect. 2, our proposed approach in Sect. 3 followed by experiments in Sect. 4 and evaluation in Sect. 5. Finally, Sect. 6 will conclude the paper.

## 2   Related Works

Recent studies have facilitated the integration of gamification, personalized recommendation systems, and adaptive learning in the continually evolving field of educational technology. Innovative gamification technologies and tailored adaptation methods are significant advancements which established an atmosphere for an interactive and adaptive learning environment [1, 2].

### 2.1   Gamification in Education

Recent studies have improved our understanding of how gamification affects student learning results. The systematic review by Koivisto & Hamari [1] clarifies the convoluted relationship between gamification and improved education. The integration of artificial intelligence (AI) and machine learning algorithms for personalizing gamification components to specific learner preferences is one notable advancement in this field. Furthermore, it has been shown that gamifying educational experiences using interactive technologies may enhance student retention and commitment of knowledge. These most recent advancements provide helpful advice regarding how to incorporate advanced gamification elements into our interactive learning environment. The primary objective of recent developments in interactive learning environments has evolved to using innovative technologies to improve the learning experience. Innovations in interactive online environments have become powerful tools to foster educational experiences [1, 2, 4].

### 2.2   Adaptive and Interactive Learning Environments

In the rapidly evolving field of adaptive learning, recent advancements underscore the significance of personalized adaptation to cater to individual learner needs. Notable advancements include the integration of machine learning algorithms capable of adaptation based on user interactions, allowing for a more dynamic and responsive learning experience [2]. Furthermore, the exploration of adaptive learning analytics, as seen in the work [3], contributes to refining adaptability by leveraging data-driven insights into learner behavior and performance.

## 2.3 Recommendation Systems in Education

According to the work of Ekstrand et al. [5], advances in personalized recommendation systems have led to a paradigm shift toward sophisticated algorithms and user-centric approaches that promotes transparency and precision. The incorporation of artificial intelligence (AI) architectures in recommendation are one of the recent advancements [6]. Recommendation systems are essential for improving personalized learning in the field of education. According to Khanal et al. [7], content-based recommendation systems use learners' past preferences and the specific characteristics of educational resources to offer individualized recommendations. For example, the system can suggest more advanced subject tasks if a student has done previous levels in the past. According to Wang and Fu's study [8], collaborative filtering makes recommendations based on the group preferences of learners who share characteristics like others. This might include providing recommendations of educational resources in a learning environment based on the preferences and actions of classmates who share the same academic interests. Hybrid approaches are also gaining popularity; they combine aspects from collaborative and content-based filtering techniques. In this work we propose the first steps validation of our recommendation approach based on collaborative filtering.

## 3  Proposed Approach

In this section we will present the Adaptive Gamification Recommendation Approach for Interactive Learning Environments called "AG-Rise". We also detail the connections between all phases of data collection, the classification and recommendation approach, and use of each method/technique to clearly explain the usefulness and nature in relation to our study context (Fig. 1).

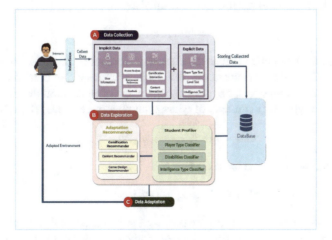

**Fig. 1.** Proposed approach: AG-Rise

### 3.1 Data Collection and Classification Phase

- Collecting data. We start our procedure by collecting both static and dynamic data from the learners.
- Classification phase: The student profiler. Dynamically collecting information about how students engage with e-learning activities, such as their profile and interactions with the learning environment [10].

The learner profiling based on interaction data and test results as follows:

- *Player Type Test:* A Hexad questionnaire is provided. Data related to gamification covers things like player types and preferred game elements [11].
- *The intelligence type Test* is a self-developed questionnaire based on Gardner's Multiple Intelligence Theory and Gardner's original test to determine what type of intelligence learners possess and whether they have any limitations.
- *Level Test:* The taxonomy developed by Bloom, which is used in Tunisian institutions, is used. It is a classification scheme for cognitive learning that goes from fundamental concepts and standards to highly developed learning levels.

### 3.2 Recommendation Phase

The learning scenario recommendation problem can be considered as a special case of a recommendation problem for which the factorization matrix presents a solution [9]. In our case, the factorization matrix characterizes the learners and the elements of the learning scenario (game elements, activities, questions). In our approach, we work on 3 classifiers characterizing learning profiles corresponding to 3 types which are the type of player, type of intelligence and type of learning disability. For each type, we identify m subtypes which construct the learner profile. For each of its subtypes, we identify the n elements to recommend or also called the adaptation elements. The AE set of n adaptation elements *ae* to recommend is defined as:

$$AE = \{ae_1, ae_2, ..., ae_n\} \tag{1}$$

It is important to notice that each adaptation element corresponds to one or more learner types. Thus, the ith element, $ae_i$, is associated with a vector of motivation indices $GM_i$, , where each component, $m_i$, , is the percentage of motivation that it can provoke in one of the **m** types $t_j$ (m number of types, n number of elements) (Fig. 2)

$$GM_i = (m_1, m_2, ..., m_n) \forall i = 1...n \tag{2}$$

Therefore, we define the matrix, M as

$$M = GM_{i1 \leq i \leq n} = (mij)_{1 \leq i \leq n, 1 \leq j \leq m} \tag{3}$$

where the rows of this matrix are indexed by the elements, and the columns by the types. Thus, $m_{i,j}$ represents the percentage of motivation (motivation index) that the i-th element produces for the j-th type (Fig. 3).

n selected adaptation elements, $ae_i$, enumerated by i = 1, ..., n. Each column represents a type of learner, $t_j$, , (j = 1, ..., m). For player types, 12 selected game elements,

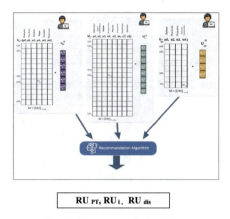

**Fig. 2.** Recommendation Phase

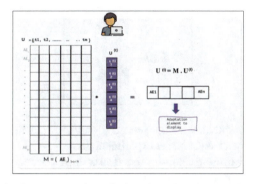

**Fig. 3.** Motivation vector and matrix for adaptation elements for each model.

$GE_i$, enumerated by $i = 1,\dots, 12$. Each column represents a type of learner, $t_j$, , ($j = 1, \dots, 6$), for intelligence types, 16 selected activities, $AM_i$, enumerated by $i = 1, \dots, 16$. Each column represents a type of intelligence, $t_j$, , ($j = 1, \dots, 8$), and for disability types, 8 selected activities, $QM_i$, enumerated by $i = 1, \dots, 8$. Each column represents a type of disability, $t_j$, , ($j = 1,\dots, 4$). Naturally, the learner's U type is dynamic over the course of the learning experience. Thus, each type at the beginning of the learning experience, can be approximated by an initial learner type, $U^{(0)}$, This type also varies depending on the interactions as well as their feedback on the recommended items, which allows them to complete their true types, $U^{(t)}$. Thus, we iteratively update the learner's profile at time t, $U^{(t)}$, and then calculate the utility of showing that element to the learner, $ae^{(t)}$. Each iteration consists of three steps:

[1] **Step 1**: Calculate the new type classification.

$$U^{(t+1)} = (1 - \varepsilon)U^{(t)} + \varepsilon(M^+ * S^{(t)}) \qquad (6)$$

[2] **Step 2**: Calculate the utility of showing an element to a learner at a specific time t+ 1 (denoted U(t+1)); And

[3] **Step 3**: Select the item to recommend based on the assigned scores.

The Fig. 4 shows our recommendation process. The green circles represent the constant data, and all other elements have dynamic values which change over time, and which are:

- $U^{(t)}$: the type of learner in each model at time t.
- $S^{(t)}$: interaction indices.

$$S_{AE}(t) \quad \begin{pmatrix} S^{(t)}AE_1 \\ \cdot \\ \cdot \\ \cdot \\ S^{(t)}AE_n \end{pmatrix}$$

- $\varepsilon$: to avoid extreme fluctuations between $U^{(t)}$ and $U^{(t+1)}$, where $0 < \varepsilon < 1$— The value of this parameter must be set experimentally.
- $M+$: the Moore-Penrose pseudo-inverse matrix of $M^1$, necessary for the interpretation of $S^{(t)}$ and $U^{(t)}$ in the same space.

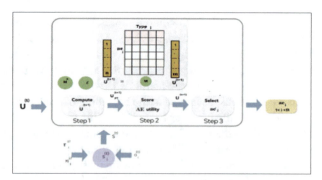

**Fig. 4.** Steps to calculate the utility of showing $ae_i$ associated with a type of user at time $t + 1$

Thus, our recommendation algorithm is presented below:

---

[1] M+: is the pseudo-inverse of M, which is formed by replacing each non-zero diagonal entry with its inverse and transposing the resulting matrix.

```
Recommendation algorithm
Input: Reading (learner. Profile)
Output: Write (GE Element, AM Element, QM Element)
Variables
MPT: Matrix (12:6)
MDIS: Matrix (8:4)
MI : Matrix (16 : 8)
Begin
Calculate U PT (t)
Calculate U DIS(t)
Calculate U i (t)
For each iteration do
 Calculate UPT (t+1)
 Calculate RUPT
 GE Element Selection
 Calculate U DIS (t+1)
 Calculate RUDIS
 AM Element Selection
 Calculate Ui (t+1)
 Calculate RUI,
 QM Element Selection
End for
End
```

## 4   Experiments

In this section, we take the example of a real-life learner from the Tunisian Class Quiz[2] learning application. Our learner begins the learning process. Our application collects data from learner as shown in Fig. 5.

**Fig. 5.**  Learner data collection via Class Quiz

### 4.1   Profiling and Recommendation Phases

Throughout our process, we consider the first iteration for a new learner's iteration to be 10 min. At t = 5 min, we measure each indicator to consider it as the moment t - After each iteration is done after 15 min and t−1 refers to the previous iteration. For this part, we set the value of ε equal to 0.01.

---

[2] https://www.classquiz.tn/.

**Fig. 6.** Example of unlockable display

**Fig. 7.** Example of an Egg easter displayed.

**(a) Classification and Recommendation for Dynamic PlayerType.** The learner's player types were identified at the first testing ($t^0$) as follows: 20% disruptor, 10% free spirit, 30% achiever, 30% player, 20% philanthropist, and 0% socializer. Challenge, Easter egg, Point, and Level are the four game elements that have been chosen for the learning process. After interacting with those elements at the second interaction ($t^{0+1}$), the learner's classification changed to 20% disruptor, 50% free spirit, 50% achiever, 50% player, 0% socializer, and 0% philanthropist. There have become six game elements: Challenge, Level, Point, Easter Egg, Unlockable, and Gift. This was made possible by the addition of two additional game elements (Unlockable and Gift) that better suit players' preferences. To recalculate the type at each iteration $t + 1$, we consider the current $UPT^{(t)}$ as well as the interactions that learners made with the different elements and his feedback. we iteratively update the learner's profile at time t, UPT(t), and thus we calculate the usefulness of showing the game elements to the learner, $ge^{(t)}$ following the three steps Each iteration consists of the three steps decribed in *Sect. 3.2 (We note that these three steps are also applied for section b and section c).* Game items are numbered from 1 to 6. We calculate the interaction index for each game element displayed. To determine which game item is selected, weighted random selection would use the vector components as weights, so game items with higher values would be more likely to be selected. The result is 2 game items randomly chosen based on these probabilities. So, we display the unlockable and Easter Egg shown in Figs. 6 and 7.

**(b) Classification and Recommendation for Dynamic Learning Disability Type.** This learner has 40% Dyslexia, 70% Dyscalculia, 40% Dysgraphia, 40% attention/concentration. For the start of the application, we take dyscalculia as the majority disorder. We offer the learner the AM2 and AM5 activities for dyscalculia. The list of activities for a 7-year-old was proposed by an expert for each type of learning disability. Learning disability activities are numbered from 1 to 8. We calculate the interaction index for each activity displayed. The result is 2 activities randomly selected based on these probabilities, which are AM2 and AM5. The activities they need and are shown in Figs. 8 and 9.

**(c) Classification and Recommendation for Dynamic Intelligence Type.** This learner's intelligence types are 20% linguistic, 0% logical-mathematical, 20% spatial, 10% intrapersonal, 50% musical, 30% interpersonal, 10% kinesthetic, 30% naturalistic.

**Fig. 8.** Counting Red Cars in the Photo.

**Fig. 9.** Find Number 7 in Buildings

The result is 2 randomly selected activities QM based on these probabilities, which are QM9 and QM10 as shown in Figs. 10 and 11.

**Fig. 10.** QM10: Playing a Musical Instrument.

**Fig. 11.** QM9: Sing with Quizo

The recommendation results for Iteration 1 are shown in Fig. 12.

**Fig. 12.** The different recommendation results for Aziz

## 4.2 Classification and Recommendation Update

As detailed below, a change to the learner percentages for each player type, intelligence type, and disorder type has been passed. This resulted in the recommended elements being modified from one iteration to the next as follows:

**-Player type-based recommender:** Our learner went from 20% disruptor, 50% free spirit, 50% achiever, 50% player, 0% socializer, 0% philanthropist to 19% disruptor, 49% free spirit, 49% achiever, 49% player, 0% socializer, 0% philanthropist in iteration 1, we recommended the *unlockable and Easter Egg* to move to 18% disruptor, 48% free spirit, 48% achiever, 48% player, 0% socializer, 2% philanthropist at iteration 2 and recommend the *Point and Leaderboard*.

**-Intelligence-based Recommender:** Our learner's intelligences were at the beginning 20% linguistic, 0% Logico-mathematical, 20% spatial, 10% intrapersonal, 50% musical, 30% interpersonal, 10% kinesthetic, 30% naturalistic, at iteration 1, the learner became 19% linguistic, 0% logical-mathematical, 19% spatial, 9% intrapersonal, 49% musical, 29% interpersonal, 9% kinesthetic, 29% naturalistic, we recommended *QM9 and QM10* activities. By iteration 2, he became 18% linguistic, 0% logical-mathematical, 18% spatial, 8% intrapersonal, 48% musical, 28% interpersonal, 8% kinesthetic, 28% naturalistic and we recommended *QM14 and QM5 activity*.

**-Learning Disability-based Recommender:** At the start, our learner had 40% Dyslexia, 70% Dyscalculia, 40% Dysgraphia, 40% attention/concentration. At iteration 1, it became with 38% dyslexia, 68% dyscalculia, 39% dysgraphia, 39% attention/concentration and the recommended activities were *AM2* and *AM5*. Subsequently, in iteration 2, the learner became with 38% dyslexia, 67% dyscalculia, 38% dysgraphia, 38% attention/concentration and the recommended activities were *AM8* activity and *AM4* activity. These results show the dynamism of each type reflected in the change in the percentages and indicate the effects of the recommended game elements on the change of player type percentages, and the recommended activities on the percentages of intelligence types. Also, the learning disabilities percentages have decreased due to the learning disability recommended activities, which improved how effectively students are able to deal with their difficulties.

## 5    Evaluation

In our case, to evaluate the results of our recommendation, we opted for an offline evaluation. The dataset used in this assessment was extracted from Class Quiz after almost 3 months of collecting feedback from learners' parents. Since the Class Quiz app is dedicated to children, the feedback survey was specifically targeted to parents to assess the impact of the recommendations. The dataset contains 500 parents. Parents generated 480 upvotes which is equal to 96%. We assessed the accuracy, Mean Absolute Error (MAE) and Mean Square Error of our recommendation system using feedback from learners and parents as an indicator for the performance of our recommendation. We obtained the following results shown in Fig. 13.

```
Accuracy Score :      85.92
Mean Absolute Error:  0.0518
Mean  Squard Error:   0.1297
```

**Fig. 13.** Evaluation of the recommender system

Our recommendation system has an accuracy of 0.86, which means that 86% of the predictions are correct. The MAE is 0.05, which means that the average difference between the predictions and the actual values is 0.05. The MSE is 0.13, which means that the average squared errors are 0.13.

## 6 Conclusion

Our work focuses mainly on proposing an Adaptive Gamification Recommendation Approach for Interactive Learning Environments. To test the effectiveness of our approach, we applied it to the educational platform "Class Quiz". We collected the information, interactions, and experiences of the app's learners. Our learner profiler is made up of a combination of three classifiers which are the Player type classifier, Intelligence type Classifier, and Disability Classifier. After the classification of the learners, we test the performance of our recommendation system and to have a recommendation of a learning scenario composed of recommendations of game elements and activities for a learner's profile in relation to his needs and preferences of the learners identified from his profile. Having a good evaluation of out actual recommendation, we plan in the future, to work on our learner profile ontology [12], the ontology-based recommender will consolidate our recommender so that it is hybrid. This type of recommender will consolidate the recommendation results and present an improved prototype of our approach.

## References

1. Koivisto, J., Hamari, J.: The rise of motivational information systems: a review of gamification research. Int. J. Inf. Manag. **45**, 191–210 (2019)
2. Bennani, S., Maalel, A., Ben Ghezala, H.: Adaptive gamification in E-learning: a literature review and future challenges. Comput. Appl. Eng. Educ. **30**(2), 628–642 (2022)
3. Gligorea, I., Cioca, M., Oancea, R., Gorski, A.T., Gorski, H., Tudorache, P.: Adaptive learning using artificial intelligence in e-learning: a literature review. Educ. Sci. **13**(12), 1216 (2023)
4. Suresh Babu, S., Dhakshina Moorthy, A.: Application of artificial intelligence in adaptation of gamification in education: a literature review. Comput. Appl. Eng. Educ. e22683 (2023)
5. Ekstrand, M.D., Carterette, B., Diaz, F.: Distributionally-informed recommender system evaluation. ACM Trans. Recommender Syst. (2023)
6. Wu, S., Sun, F., Zhang, W., Xie, X., Cui, B.: Graph neural networks in recommender systems: a survey. ACM Comput. Surv. **55**(5), 1–37 (2022)
7. Khanal, S.S., Prasad, P.W.C., Alsadoon, A., Maag, A.: A systematic review: machine learning based recommendation systems for e-learning. Educ. Inf. Technol. **25**, 2635–2664 (2020)
8. Wang, H., Fu, W.: Personalized learning resource recommendation method based on dynamic collaborative filtering. Mob. Netw. Appl. **26**, 473–487 (2021)
9. Meder, M., Jain, B.J.: The gamification design problem. arXiv preprint arXiv:1407.0843 (2014)
10. Bennani, S., Maalel, A., Ben Ghezala, H., Daouahi, A.: Integrating machine learning into learner profiling for adaptive and gamified learning system. In: Nguyen, N.T., Manolopoulos, Y., Chbeir, R., Kozierkiewicz, A., Trawiński, B. (eds.) ICCCI 2022. LNCS, vol. 13501, pp. 65–71. Springer, Cham (2022). https://doi.org/10.1007/978-3-031-16014-1_6

11. Tondello, G.F., Mora, A., Marczewski, A., Nacke, L.E.: Empirical validation of the gamification user types hexad scale in English and Spanish. Int. J. Hum. Comput. Stud. **127**, 95–111 (2019)
12. Bennani, S., Maalel, A., Ghezala, H.B.: AGE-Learn: ontology-based representation of personalized gamification in E-learning. Procedia Comput. Sci. **176**, 1005–1014 (2020)

# Image Quality Distortion Classification Using Vision Transformer

Nay Chi Lynn[✉] and Tetsuya Shimamura

Saitama University, Saitama, Japan
nay.c.l.566@ms.saitama-u.ac.jp, shima@mail.saitama-u.ac.jp

**Abstract.** In this paper, we propose a method for classifying image quality distortions to identify common types of distortions typically present in images, utilizing a vision transformer. The method aims to enhance quality-related image processing approaches by identifying specific distortions as the initial step in distortion-based blind image quality assessment (BIQA). This simplifies the quality reconstruction process by tailoring it to the prior knowledge of distortion types, thereby aiding in improving image classification and potentially reducing biases caused by certain distortions. The proposed method is experimented on common benchmark image quality assessment (IQA) databases, including LIVE2008, TID2013, and KADID-10k. To generalize the performance with a larger database, we distorted images using four general distortion types: Gaussian noise, Gaussian blur, JPEG compression, and contrast degradation, applied to the ImageNet-1k database. The experimental results demonstrate that the proposed method outperforms other solutions in terms of accuracy

## 1 Introduction

The quality of digital images can be affected by various distortions at different stages, including acquisition, transmission, and any image processing system. Two major trends in image quality research are image reconstruction techniques and image quality assessment (IQA).

The first trend, image reconstruction, aims to produce high-quality images from degraded ones [1]. Many image reconstruction algorithms are distortion-specific, addressing denoising, deblurring, quality restoration from compression, and contrast enhancement from contrast degradation, among others. Incorporating prior knowledge about the distortion is crucial for the restoration process.

The second trend, IQA, involves assessing the perceptual quality of an image, aligning with human subjective opinions. IQA databases provide mean opinion score (MOS) or differential MOS (DMOS) values, derived from human observer opinions, considering visually significant attributes influencing the image. Blind IQA (BIQA) methods are more popular than full-reference IQA (FR-IQA) methods because BIQA methods use only distorted images as input, while FR-IQA methods require both the distorted and pristine reference images, which might

© The Author(s), under exclusive license to Springer Nature Switzerland AG 2024
L. Barolli (Ed.): AINA 2024, LNDECT 199, pp. 353–361, 2024.
https://doi.org/10.1007/978-3-031-57840-3_32

not always be available together in the real world. Distortion-specific BIQA, a type of BIQA method predicting quality scores under a specified distortion type, therefore it needs to detect the degradation cause or distortion type first.

Moreover, these quality distortions significantly impact the performance of image classification models. Having prior knowledge about the distortion type allows the classifier to focus on features more resilient to that particular distortion, improving classification reliability [4,5]. Knowing the distortion type also facilitates applying appropriate preprocessing techniques to mitigate distortion effects before classification. Therefore, precise distortion type detection is critical.

To achieve this, we propose a classification method capable of identifying distortion types affecting an image. In this context, we conducted experiments with pretrained neural networks and transformers. The performance accuracies are then assessed and compared using publicly available benchmark IQA databases, including LIVE2008 [6], TID2013 [7], KADID-10k [8]. Additionally, experiments were conducted on a relatively large image database, ImageNet [9], to improve generalization performance. The experimental results indicate that the proposed method is an effective and practical approach for distortion classification for certain common distortions.

The paper is organized as follows. In Sect. 2, we describe the architecture of the proposed method designed for distortion classification. In Sect. 3, we explain the IQA benchmark databases and the large-scale external database experimented with the proposed method. Section 4 discusses the implementation and experimental results, followed by concluding remarks in Sect. 5.

## 2   Proposed Method

The proposed method is based on the idea of transfer learning using pretrained networks. There are many types of pretrained network, here we utilize the vision transformer (VIT) architecture. The network architecture using self-attention mechanism in the transformer model was originated in natural language processing (NLP), pioneered by Vaswani et al. [10]. Adapting the successful self-attention mechanism from NLP to the realm of image-related applications, Dosovitskiy et al. [11] introduced a VIT architecture. In line with this inspiration, the proposed method utilizes a pretrained VIT incorporating valuable features and patterns acquired from a vast pretrained database.

In this paper, we compare two architectures by applying the pretrained VIT network on benchmark IQA databases and our generated database. The architecture is comprised of several layers. The initial layer utilizes a VIT b16 pretrained head with a patch size of 5, serving as the backbone for feature extraction. A flatten layer follows to facilitate compatibility with subsequent fully connected layers. A synchronized batch normalization layer is then incorporated for effective normalization and regularization. We also add a dense layer with the Gaussian error linear unit (GELU) activation function to enhance the representation power of the neural network. Subsequently, two more dense layers are added,

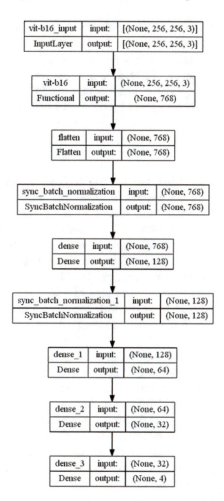

| vit-b16_input | input: | [(None, 256, 256, 3)] |
|---|---|---|
| InputLayer | output: | [(None, 256, 256, 3)] |

| vit-b16 | input: | (None, 256, 256, 3) |
|---|---|---|
| Functional | output: | (None, 768) |

| flatten | input: | (None, 768) |
|---|---|---|
| Flatten | output: | (None, 768) |

| sync_batch_normalization | input: | (None, 768) |
|---|---|---|
| SyncBatchNormalization | output: | (None, 768) |

| dense | input: | (None, 768) |
|---|---|---|
| Dense | output: | (None, 128) |

| sync_batch_normalization_1 | input: | (None, 128) |
|---|---|---|
| SyncBatchNormalization | output: | (None, 128) |

| dense_1 | input: | (None, 128) |
|---|---|---|
| Dense | output: | (None, 64) |

| dense_2 | input: | (None, 64) |
|---|---|---|
| Dense | output: | (None, 32) |

| dense_3 | input: | (None, 32) |
|---|---|---|
| Dense | output: | (None, 4) |

**Fig. 1.** Architecture of the proposed system using pretrained vision transformer

applying L2 regularization for additional regularization. The final one is a dense layer with a Softmax activation function, indicating the model's classification task with four distortion classes.

In this experiment, we designed two similar network architectures both following the same sequential layer order. However, a significant distinction lies in the number of filters employed within each layer. The first architecture, the layers are configured with 64, 32, and 16 filters, respectively. In contrast, the second architecture mirrors the layer order of the first one but employs a more robust configuration with 128, 64, and 32 filters in the respective layers. The filter sizes are tailored to the requirements of the task, aiming to enhance the model's capacity to capture insightful patterns and potentially leading to improve performance.

To visually describe the structural differences between the two architectures, Fig. 1 is illustrated. This figure serves as a clear visual demonstration, highlighting the specific filter sizes associated with each layer in the second network architecture. The adjustment in filter sizes in the second architecture is intended as a strategic decision to contribute in the proposed method as it enhances the model's performance when compared to the first architecture.

## 3   Databases

The proposed method is subjected to two distinct experimental approaches. In the first approach, we conducted tests on the benchmark IQA databases, including LIVE2008, TID2013, and KADID-10k. Those databases contain 29, 29, and 23 distorted images with 5, 5, 5 distortion levels respectively. Each of the databases contains the following common distortions: additive Gaussian noise, Gaussian blur, JPEG compression and JPEG2000 compression. In total, there are 2928 distorted images from the three integrated database. Each distorted image is standardized to the size of $256 \times 256 \times 3$. Figure 2 presents sample distorted images regarding the four common distortions. The content of all databases varies from landscapes through nature, city, people, to portrait photography.

In the second approach, we downloaded ImageNet database in which 100,000 images in total. We generated four distortions usually occurred in image distortions namely additive Gaussian noise, Gaussian blur, JPEG compression, and contrast degradation with four distortion levels on each ImageNet image as shown in Fig. 3. This variety of distortion levels on a large-scale database provided the proposed method to conduct a sufficient training of distorted images. Hereafter, we clarify the process of creating distortions applied to ImageNet images, aimed at the generation of a comprehensive distortion database.

Noise can be caused by a number of factors from camera settings to environmental conditions such as low lighting, rapid shutter speeds, heat-up camera sensors, etc. For the generation of additive Gaussian noise, we employ Scikit library, specifying variance of ranges 0.01, 0.03, 0.07, 0.10 as four distortion levels with a constant zero mean.

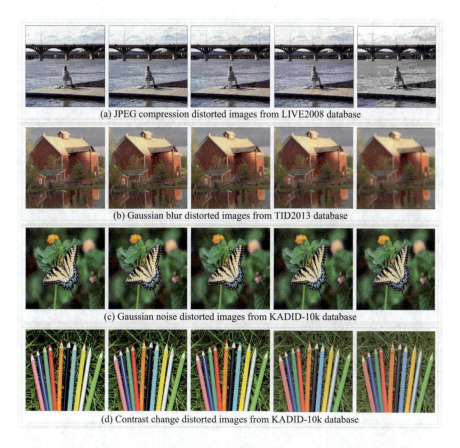

(a) JPEG compression distorted images from LIVE2008 database

(b) Gaussian blur distorted images from TID2013 database

(c) Gaussian noise distorted images from KADID-10k database

(d) Contrast change distorted images from KADID-10k database

**Fig. 2.** Distorted images with five distortion levels of JPEG compression, Gaussian blur, Gaussian noise and contrast changes from LIVE2008, TID2013 and KADID-10k databases

Blur can occur intentionally or unintentionally due to technical limitations, camera shake, low resolutions, or scaling issues, etc. To evaluate the impact of Gaussian blur, we apply the OpenCV filter2D functionby means of the horizontal kernel with sizes 3, 5, 7, 9.

JPEG compression, a significant distortion affecting image quality, occurs by eliminating redundant data and discarding some of the less noticeable color information within the image with different compression levels. To simulate quality degradation by JPEG compression, we reproduce 4 stages of JPEG factors using the Python Imaging Library (PIL) library by adjusting the quality of the image with (75, 50, 25, 5) parameter factors.

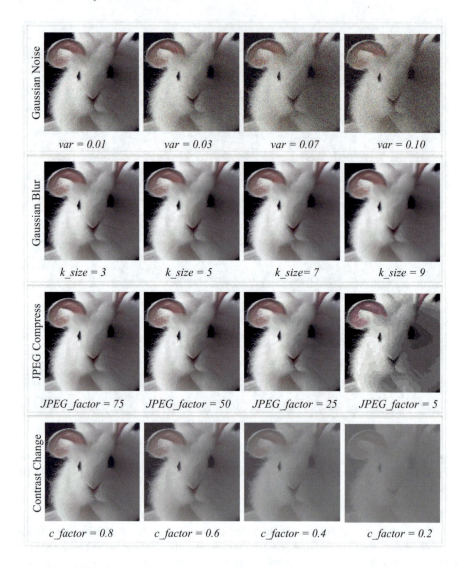

**Fig. 3.** ImageNet images after distorted with four distortion levels of Gaussian noise, Gaussian blur, JPEG compression, and contrast degradation

In addition to noise and blurring, contrast degradation is another influential distortion affecting image quality because it makes difficult to distinguish details and features of the distorted image. To evaluate the influence of contrast degradation, we utilized the Pillow library on images with lower contrast factor values decreasing from 0.8 to 0.2 in steps of 0.2.

Figure 3 provides examples of the resulting distorted images.

**Table 1.** Experimental results on the LIVE2008, TID2013, and KADID-19k integrated database

| Architecture | Accuracy |
|---|---|
| Architecture 1 | 94.368% |
| Architecture 2 | 95.712% |

**Table 2.** Experimental results on the ImageNet generated database

| Architecture | Accuracy |
|---|---|
| Architecture 1 | 97.336% |
| Architecture 2 | 98.153% |

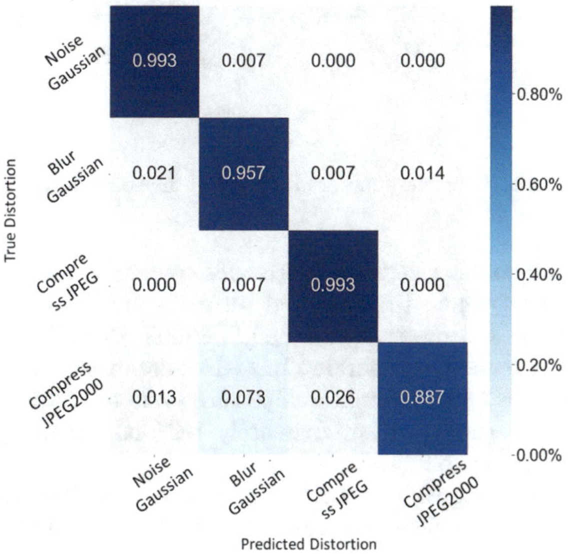

**Fig. 4.** Confusion matrix on the combination of LIVE2008, TID2013, and KADID-10k databases for the proposed method

## 4   Implementation and Experimental Analysis

The experiment is implemented by dividing into separate groups, specifically training data (80%) and validation data (20%), for both benchmark databases and the generated database. We set up this setting as it is a common practice used in machine learning algorithms to mitigate the risk of overfitting. Experiments on both databases are conducted using Python, and the implementation use the TensorFlow framework for VIT integration. We set the same number of epochs for training, with monitoring based on a metric aimed at minimizing the loss by checking at the end of each single epoch.

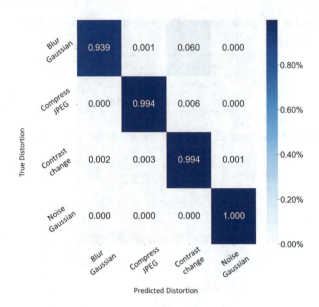

**Fig. 5.** Confusion matrix on the generated ImageNet database for the proposed method

Tables 1 and 2 summarize the performance comparison achieved by the two model architectures where the first and second architectures are denoted by Architectures 1 and 2, respectively on the LIVE2008, TID2013, and KADID-10k integrated database and the generated ImageNet database, respectively. We can see that architecture 2 outperforms architecture 1 on both database experiments exhibiting an improvement of approximately 1.4% on the integrated database and 0.8% on the generated database.

To gain deeper insights into performance and potential improvement, we report the confusion matrices. Figure 4 presents the resulting confusion matrix on the combination of the three integrated databases. It reveals that the proposed method performed less effectively on class 4 (JPEG2000 compression), possibly due to visual characteristics similar to other distortions such as Gaussian blur and JPEG compression, making it challenging to distinguish. This could lead to its inability to accurately recognize the distinctive features.

Figure 5 presents the confusion matrix on the generated database. It indicates the proposed distortion classifier's efficiency in detecting distortion classes on the large training database. In Fig. 5, the Gaussian blur class label is the weakest one in which some are misclassified with contrast degradation class label. For the Gaussian noise class label, a 100% correct classification rate is notably achieved.

## 5   Conclusion

In this paper, we proposed a transformer-based classification method for identifying distortions types of images. Training and validation in the proposed method are experimented not only on prevalent distortions present in three benchmark IQA databases-LIVE2008, TID2013, and KADID10k, but also on manually distorted and generated images sourced from the large-scale ImageNet database. The experimental results demonstrate the robustness of the proposed method in accurately distinguishing the types of image quality distortions. As future work, we aim at expanding the capabilities of our classifier model to predict additional distortion classes and conducting more extensive testing on images derived from real-world scenarios.

# References

1. Demoment, G.: Image reconstruction and restoration. overview of common estimation structures and problems. IEEE Trans. Acoust. Speech, Signal Process. **37**(12), pp.2024-2036 (1989)
2. Mittal, A., Moorthy, A.K., Bovik, A,C.: No-reference image quality assessment in the spatial domain. IEEE Trans. Image Process. **21**(12), 4695-4708. (2012)
3. Lynn, N.C., Sugiura, Y., Shimamura, T.: Blind noisy image quality assessment using spatial, frequency and wavelet statistical features. J. Signal Process. **28**(1) 19–27 (2024)
4. Algan, G., Ulusoy, I.: Image classification with deep learning in the presence of noisy labels: a survey. Knowl.-Based Syst. **215**, 106771 (2021)
5. da Costa, G.B.P., Contato, W.A., Nazare, T.S., Neto, J.E., Ponti, M.: An empirical study on the effects of different types of noise in image classification tasks. arXiv preprint arXiv:1609.02781. (2016)
6. Sheikh, H.R., Wang, Z., Cormack, L., Bovik, A.C.: LIVE image quality assessment database release 2. http://live.ece.utexas.edu/research/quality
7. Ponomarenko, N., et al.: A New Color Image Database TID2013: Innovations and Results, Proceedings of ACIVS, pp.402-413 (2013)
8. Lin, H., Hosu, V., Saupe, D.: KADID-10k: a large-scale artificially distorted IQA database. In: Proceedings of 11th IEEE International Conference on Quality of Multimedia Experience (QoMEX), pp.1-3. (2019)
9. Imagenet. https://www.image-net.org/
10. Vaswani, A., et al.: Attention is all you need. In: Proceedings of 31st Conference on Neural Information Processing Systems (2017)
11. Dosovitskiy, A., et al.: An image is worth 16x16 words: Transformers for image recognition at scale. arXiv preprint arXiv:2010.11929. (2021)

# Towards Ecological Tunisian Sign Language Recognition System

Sana Fakhfakh[(✉)] and Yousra Ben Jemaa

L3S Laboratory , ENIT, El Manar University, Tunis, Tunisia
{sana.fakhfakh,yousra.benjemaa}@enis.tn

**Abstract.** The Deaf and Hard of Hearing community relies heavily on sign language as a mode of communication since it connects them to the hearing world. The distance between these two communities is being closed, thanks in large part to the use of autonomous sign language recognition devices.

This paper investigates innovative techniques to offer an autonomous and ecological Tunisian sign language recognition system (AETSLRS). This system is carefully constructed to function well in real-world circumstances based on efficient vision-based techniques that are proven compared to existing works. The potential advantages and difficulties of this work are examined in this study to pave the way for a more sustainable future.

## 1 Introduction

Sign languages (SL) represent unique visual-gestural modes of communication utilized by deaf populations worldwide. While many SL recognition systems have been developed for American, French, German, Arabic, Indonesian [1–6], and so on, not many have been developed for Tunisian Sign Language (TSL) [7]. This is unexpected since there are a significant number of deaf people in Tunisia-approximately 53,000 of the 11 million people who live there are deaf[1] Therefore, improving communication among Tunisia's deaf people is the first goal of this work.

Existing works, particularly in isolated word fields, often ignore the deaf brain's interpretation while presenting sign motions [7–10]. Furthermore, these approaches sometimes impose environmental and signer variance constraints without considering ecological and environmental aspects in an effort to simplify the feature extraction process [1,6–11]. Moreover, deep learning methods have gained popularity recently as a useful remedy despite their complicated

---

[1] African sign languages resource center Homepage: https://africansignlanguagesre sourcecenter.com/tunisia/.

© The Author(s), under exclusive license to Springer Nature Switzerland AG 2024
L. Barolli (Ed.): AINA 2024, LNDECT 199, pp. 362–371, 2024.
https://doi.org/10.1007/978-3-031-57840-3_33

algorithms and large data requirements [12,13]. So, to create a sustainable and environmentally friendly future, it is necessary to shift the focus to environmentally friendly methods [10] and fight climate change [12].

In this context, this paper highlights as its second goal a new approach towards a sustainable environment [10] and introduces the first autonomous and ecological Tunisian sign language recognition system (AETSLRS). This presents a revolutionary tool that allows Deaf people in Tunisia to interact easier with each other.

The remaining part of this paper is structured as follows: In part Sect. 2, the proposed system is presented. Section 3 displays the validation of the new autonomous and ecological approach. In the end, our work is accomplished in Sect. 4.

## 2   Proposed System

Developing an ecological SL recognition system with minimal energy consumption has become an essential endeavor. This approach requires the exploration of various concepts, such as neural deafening systems and advanced computer vision methods [9,10].

Our research strategy, established since 2017 [7–11], focuses on two proven steps: the first step involves Key points sign motion extraction [8], while the second step involves its shape analysis [7,9,10].

Figure 1 depicts the framework design linked to the adopted strategy to offer the first AETSLRS, which employs a sophisticated solution entailing decreased data and energy consumption.

This framework, primarily ignoring all deep learning solutions as methods that investigate more data and more energy consumption [12] and incorporates two proven techniques for the sign motion shape analysis stage:

- Shape descriptors: this approach takes advantage of HOG as a shape descriptor technique in sign motion shape analysis step [9].
- Distance analysis: This approach takes advantage of geodesic distance as an elastic distance metric in sign motion shape analysis step [10].

These methods were recently highlighted by American sign language through RWTH-BOSTON [14] and SIGNUM corpora [15] and proved to enhance energy efficiency and processing time compared to deep solutions [10]. Algorithm 1 presents all AETSLRS pipeline steps.

**Algorithm 1.** AETSLRS pipeline:

**Input system**: video stream.
**Step 1:** Key points sign motion extraction
**Step 2:** Key points sign motion shape analysis
**Step 3:** Classification : KNN classifier
**Output system:** word text.

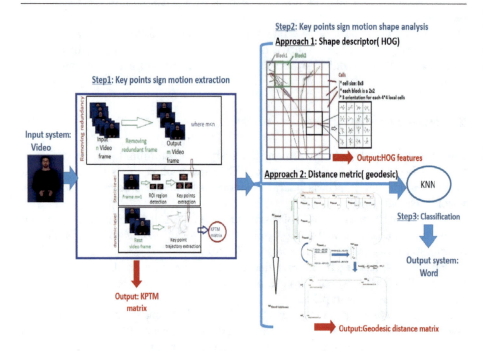

**Fig. 1.** AETSLRS framework design

## 2.1  Key Points Sign Motion Extraction

During this phase, the suggested key-point trajectory matrix(KPTM) [8] is utilized to articulate the motion and shape features of head and hand movements.

To offer a system that is both natural and self-sufficient, the initial phase involves identifying crucial locations based on the brain's operation, especially focusing on finger flexion. These locations encompass:

- Head and hand gravity centers,
- Ten finger point positions,
- Four wrist line extremities.

Then, 17 trajectories are generated using a particular filter for all 17 proposed points.

To enhance ecological efficiency, two key stages are introduced during the construction of the KPTM matrix:

- Real-time processing and data redundancy reduction: we explore the Removal of Redundant Frame (RRF) stage [8],
- Clothing condition invariance: we introduce a wrist detection stage as initial step [11].

The performance of the KPTM using international and public databases have been widely proven [7–10]. All KPTM construction steps are illustrated in Algorithm 2 and detailed in the work [8]. Figure 2 illustrates some extracted KPTM taken by different Tunisian signers for the same sign word (Mother sign).

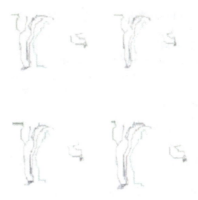

**Fig. 2.** Examples of KPTM images related to Mother sign

## 2.2 Key Points Sign Motion Shape Analysis

This section underscores the significance of shape trajectory analysis [7,9,10] in understanding the cognitive functioning of the deaf brain.

The objective of this stage is to interpret all sign features related to shape and motion (represented by KPTM) as sign paths, reflecting cognitive brain functions.

Consequently, issues related to sign dynamics transform into objectives centered around shape description in the field of computer vision.

Only proven ecological solutions, such as shape descriptors [9] and distance techniques [10] need to be considered in this paper to provide our proposed AETSLRS.

**Algorithm 2.** Key points sign motion extraction pipeline [8]:

**Input system**: video stream.
**Step 1:** Key points extraction:
Frame=1
**Input:**color image.
1.Pre-processing stage.
1.1 Skin color's detection: YCbCr color space.
1.2. Morphology operation application.
1.3 ROI detection: Only 3 biggest regions extraction
1.4. Hands and Face identification
1.5. Wrist line detection
1.6 Hands and Face regions bounding.
2.Key points' extraction stage
2.1 Face gravity center extraction
2.2 Extraction of the gravity centers of both hand
2.3 Extraction of 5 finger point positions, and 2 wrist line extremities of both hand.
**Output:** Vector V(x,y) of key extracted points: 17 points
**Step 2:** KPTM matrix construction:
Frame=2: nombre of frames
1. RRF stage
2. Vector V normalization.
3. KPTM initialized matrix by V vector.
Initialize 17 PF.
4. For each PF and in each frame Normalize and update KPTM position matrix with each updated position by each PF
**Output system:** KPTM.

### 2.2.1  Shape Descriptors Solution(HOG [9])

As presented in Fig. 3 a HOG descriptor is directly applied to all extracted KPTM and used as a feature vector for the KNN classifier.

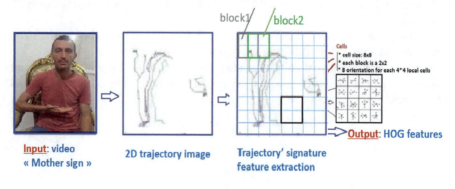

**Fig. 3.** Sign motion shape analysis: shape descriptor solution (HOG) [9]

All details are examined in the work [9].

### 2.2.2 Elastic Distance Metric(geodesic [10])

This approach involves the direct projection of all obtained trajectories(KPTM) into the Riemannian shape space, where a comparison is made using the geodesic metric by the square root of the velocity (SRV:Square Root Velocity) function [16] as detailed in the work [10].

This process allows an intrinsic shape trajectory analysis. Consequently, a trajectory path within the space of shape analysis is regarded as an object within a metric space and takes advantage of:

- Geometric transformations invariant: translation, scaling, and rotation,
- Characteristics regularity,
- Geometrical shape' structure preservation,
- Coherent coordinate transformation,
- All trajectory parts connections are maintained.

Thus offering a big advantage related to data consumption in which only a simple distance calculation is involved.

All key points motion shape analysis steps are presented in Algorithm 3:

---

**Algorithm 3.** Key points sign motion shape analysis pipeline [10]:

---

**Step 1:** The geodesic distance matrix calculation: $M_{Geod}$
**Input system :** KPTM matrix: size N*M.
i=1:n
j=1:m
$M_{Geod}$[i,j]=SRV(KPTM[i,j])
**Output :** $M_{Geod}$ matrix:
**Step 2:** Optimizing $M_{Geod}$ matrix
**Input :** $M_{Geod}$ matrix: size N*M.
i=1:n
j=1:m
For each $D_{Geod_{i,j}}$ between training word $MA_i$ and testing word $MT_j$ in $M_{Geod}$
1/Calculate the minimum distance $d_c$ for each line $D_{Line_c}$ in $D_{Geod_{i,j}}$.
$Vec_{dist}[k] = d_{c_k}$ as k from 1 to 17
2/ Summation of all obtained distance $d_c$.
$Val_{Dist-intrinsec_{MA_i,MT_j}} = \sum_{k=1}^{17} Vec_{dist}[k]$
3/ Each $D_{Geod_{i,j}}$ between training word $MA_i$ and testing word $MT_j$ in $M_{Geod}$ was be replaced by $Val_{Dist_{MA_i,MT_j}}$
**Output system:** $M_{Geod-intrinsec}$ matrix: size n*m.

---

## 3 Performance Analysis

### 3.1 Proposed Tunisian Dataset

To emphasize the superiority of the proposed approach [9,10] on the one hand and introduce the inaugural AETSLRS on the other, we propose to evaluate it with the new complex TunSigns database [7]. The TunSigns dataset was developed as a first step to provide academics with the opportunity to suggest an effective TSL recognition system in collaboration with the Tunisian Association for Aid to the Deaf (ATAS). It is hoped that it might even be recognized as the primary dataset for the SL recognition system under natural conditions with regards to offering a green and ecological recognition system [12,13], as the key studies scheduled for 2040 have identified[2].

This database comprises 25 gestures collected in diverse and complex environments as presented in Fig. 4, involving nine different signers as illustrated in Fig. 5 across varying ages and genders [7].

**Fig. 4.** TunSigns database: (a) Grandfather word sign, (b) Mother word sign

**Fig. 5.** TunSigns databases' signers [7]

## 3.2   Results

A recognition rate of 97% is achieved using the geodesic distance approach and 97.07% with the HOG approach, as presented in Table 1.

---

[2] European commission Homepage: https://cordis.europa.eu/project/id/956090.

**Table 1.** TunSigns dataset results

| Approaches | Accuracy |
|---|---|
| HOG [9] | 97.07% |
| Elastic distance [10] | 97% |

These significant levels of accuracy demonstrate how well the proposed methods work with a sophisticated database collected in the natural scenario (TunSigns).

Hence, they could serve as a promising alternative for fostering a more eco-friendly environment by just introducing a simple similarity metric.

Thus, it highlights the first AETSLRS without any environmental or acquisition constraints.

### 3.3   Comparative Studies

To draw attention to the proposed strategy based principally on the shape trajectory analysis step in order to offer a more eco-friendly environmental solution with minimum carbon footprint consumption, we firstly evaluate it using different databases (TunSigns, SIGNUM, and RWTH-BOSTON databases), taking into consideration different sign languages: Tunisian and American and then we compare it to sign language recognition system using deep learning approach [7].

This approach investigates, for the learning step, as presented in Fig. 6, three convolutions, three pooling, and one fully connected layer [7].

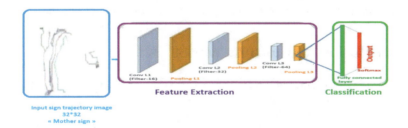

**Fig. 6.** Sign motion shape analysis: deep solution [7]

Thus, it can give an idea of the quantity of energy, time processing, and data consumption [12] compared to other proposed approaches.

As presented in Table 2, the sign motion shape stage is proved with all proposed databases. Also, the proposed approach always gives good accuracy with the Tunisian database. An amelioration of only 1% for all datasets is shown when using CNN/deep learning technique compared to geodesic approach.

**Table 2.** TunSigns,SIGNUM and RWTH-BOSTON databases results

| Approaches | TunSigns | SIGNUM | RWTH-Boston |
|---|---|---|---|
| CNN [7] | 98% | 98.21% | 95.83% |
| HOG [9] | 97.07% | 96.93% | 92.27% |
| Elastic distance [10] | 97% | 97.87% | 94.71% |

This signifies that the implementation of HOG [9] and elastic distance [10] techniques not only accomplishes the objective of establishing the first AETSLRS but also turns the vision of an environmentally isolated word sign language recognition system into reality, with a thoughtful consideration of the following crucial factors:

- Signer Variation: the variety of ways deaf people can express themselves through signing, taking into consideration the differences in signers' styles, speeds, and nuances.
- Geometric Variation: the structural and morphological disparities that exist between signals, taking into account the variances in hand shape and location.
- Environmental Variation: the dynamic nature of environments where signing takes place, encompassing diverse lighting, backgrounds, clothes, conditions,etc.
- Data Consumption: the volume of data needed for precise identification, taking into account storage constraints and optimizing data utilization without sacrificing speed.
- Time Processing: the recognition system runs in a reasonable amount of time, cutting down on processing delays to deliver feedback in real-time or almost real-time and maximizing the recognition process's overall efficiency.

## 4    Conclusion and Future Works

We make a big step toward a society that is more technologically sophisticated, environmentally conscious, and inclusive with our proposed AETSLRS which is both autonomous and environmentally friendly.

Our solution, compared to existing ones, gives more precision, less carbon footprint, encourages a cleaner environment, and works toward a more sustainable future.

In the future, adding more words to the TunSigns database will be required to propose a final and complete system. This global edition must fully address all biometric data protection issues.

# References

1. Joksimoski, B., et al.: Technological solutions for sign language recognition: a scoping review of research trends, challenges, and opportunities. IEEE Access. **10**, 40979–40998 (2022)
2. Bilge, Y., Cinbis, R., Ikizler-Cinbis, N.: Towards zero-shot sign language recognition. IEEE Trans. Patt. Anal. Mach. Intell. **45**, 1217–1232 (2023)
3. Li, D., Yu, X., Xu, C., Petersson, L., Li, H.: Transferring cross-domain knowledge for video sign language recognition. In: 2020 IEEE/CVF Conference on Computer Vision and Pattern Recognition (CVPR), pp. 6204–6213 (2020)
4. Tunga, A., Nuthalapati, S., Wachs, J.: Pose-based sign language recognition using GCN and BERT. In: 2021 IEEE Winter Conference on Applications of Computer Vision Workshops (WACVW), pp. 31–40 (2021)
5. Hu, H., Zhao, W., Zhou, W., Li, H.: SignBERT+: hand-model-aware self-supervised pre-training for sign language understanding. In: IEEE Transactions on Pattern Analysis and Machine Intelligence. **45**, 11221–11239 (2023)
6. Razieh Rastgoo, K., Escalera, S.: Sign language recognition: a deep survey. Expert Syst. Appl. **164**, 113794 (2021)
7. Fakhfakh, S., Jemaa, Y.: Deep learning shape trajectories for isolated word sign language recognition. Int. Arab J. Inf. Technol. **19**, 660–666 (2022)
8. Fakhfakh, S., Jemaa, Y.: Gesture recognition system for isolated word sign language based on key-point trajectory matrix. Computación Y Sistemas. **22**, 1415–1430 (2018)
9. Fakhfakh, S., Ben Jemaa, Y.: Shape trajectory analysis based on hog descriptor for isolated word sign language recognition. In: International Conference on Advanced Information Networking and Applications, pp. 46–54 (2022)
10. Fakhfakh, S., Jemaa, Y.: A green isolated word sign language recognition system based on geodesic metric space. In: International Conference on Cyberworlds, CW 2023, Sousse, Tunisia, October 3-5, 2023, pp. 274-281 (2023)
11. Fakhfakh, S., Jemaa, Y.: Hand and wrist localization approach for features extraction in Arabic sign language recognition. In: 2017 IEEE/ACS 14th International Conference on Computer Systems and Applications (AICCSA), pp. 774–780 (2017)
12. Savazzi, S., Rampa, V., Kianoush, S., Bennis, M.: An energy and carbon footprint analysis of distributed and federated learning. In: IEEE Transactions on Green Communications and Networking. **7**, 248–264 (2023,3). https://doi.org/10.1109
13. Bosio, A., et al.: Design, verification, test and in-field implications of approximate computing systems. In: 2020 IEEE European Test Symposium (ETS), pp. 1–10 (2020)
14. Zahedi, M., Keysers, D., Deselaers, T., Ney, H.: Combination of tangent distance and an image distortion model for appearance-based sign language recognition. In: Pattern Recognition: 27th DAGM Symposium, Vienna, Austria, August 31-September 2, 2005. Proceedings 27, pp. 401-408 (2005)
15. Agris, U., Kraiss, K.: Signum database: video corpus for signer-independent continuous sign language recognition. In: Sign-lang@ LREC 2010, pp. 243–246 (2010)
16. Joshi, S., Klassen, E., Srivastava, A., Jermyn, I.: A in Rn. In: 2007 IEEE Conference on Computer Vision and Pattern Recognition, pp. 1–7 (2007)

# Enabling Intelligent Data Exchange in the Brazilian Energy Sector: A Context-Aware Ontological Approach

Matheus B. Jenevain[1](✉) , Milena F. Pinto[2] , Mario A. R. Dantas[1] ,
Regina M. M. B. Villela[1] , Jose M. N. David[1] , and Victor S. A. Menezes[1]

[1] Federal University of Juiz de Fora (UFJF), Juiz de Fora, Brazil
matheus.jenevain@estudante.ufjf.br
[2] Federal Center for Technological Education Celso Suckow da Fonseca
(CEFET/RJ), Rio de Janeiro, Brazil

**Abstract.** The ever-expanding requirement for interoperability between systems, coupled with the growing nu mber of participants in the energy sector, creates a scenario where diverse actors, each with unique realities and specificities, must exchange data and knowledge. This exchange often occurs among significant differences in context. These disparities can lead to many difficulties and misunderstandings, as the information exchanged is susceptible to misinterpretation based on the sender's and receiver's contexts. Therefore, this work addresses this issue by proposing an extension to the Open Energy Ontology (OEO) [1] that focuses on context. It investigates how an actor's understanding is shaped by their context, the methods for inferring this context, and strategies to enhance interoperability. The results obtained demonstrate the potential of the approach proposed by this work.

## 1 Introduction

Since the release of the Smart Grid Report in 2015[2], many projects have been undertaken, primarily focusing on interoperability and how it could be done using standards and protocols [3,4]. For example, ENTSO-e [5], through its TDX-Assist project [6], aimed to find better ways to interoperate between different transmission system operators across Europe using the Common Information Model [7]. IRENA [8], on its World Energy Transitions Outlook 2023 [9], heavily uses smart grids, smart meters, and IoT.

When it comes to protocols, there is no universal consensus. ENTSO-e, for example, uses the CIM [7] protocol, defined by standard 61970-301, to perform interoperability within TDX-Assist [4]. AEMO [10], on the other hand, has chosen to use its format [11], specifically created for the Australian scenario. Regardless of the selected model, it is maintained that communication is carried out through the exchange of structured data according to the chosen standard with established fields and data types. Per the specifications above, information is passed as is, with nothing beyond the raw values. An example is the ENTSO-e, which includes various European countries [12], each with specificities. Bringing

L. Barolli (Ed.): AINA 2024, LNDECT 199, pp. 372–381, 2024.
https://doi.org/10.1007/978-3-031-57840-3_34

it to the local reality, we have several Brazilian states, especially when comparing the north and south, which have their own climatic and energy generation realities. Therefore, it is relevant to consider that there is a loss in meaning and a potential loss of information due to this "decontextualization" of data. The literature suggests that patterns are of greater value when used to enable the extension of models for building a profile that meets the specific requirements of each scenario. The Open Energy Platform [1] constructs and refines its standard ontology based on various sources' insights. As Costa et al. [13] noted, ontologies are powerful tools to analyze data and infer strategic information about what is being exchanged. For this work, interoperability will be defined as two systems, heterogeneous or not, that can exchange data securely and auditable [14].

As such, the contribution of this work lies in the modeling and proposition of an extension to Open Energy Ontology (OEO), grounded in the theory of context and perception applied to collaborative systems and the knowledge of domain business rules. OEO was chosen based on its communal construction and work, especially for use in the energy sector. The main contributions are as follows:

- Proposal of extending the existing Open Energy Ontology (OEO) to encompass energy-related data and information related to perception and context, aligning with distributed systems theories.
- Understanding and adaptation of OEO, transferring the knowledge acquired in the previous stages to the resulting classes and subclasses. This will be done first by modeling an individual ontology and then importing it into the main OEO file. The ontology already accomplishes this with some components, as seen in the files in the 'imports' folder within the official OEO release.

With this, the intention is to enhance OEO to allow the transmission of richer data with greater potential for generating information and value for the parties involved. This study aims to address this gap by enriching the ontology through developing a dedicated module capable of synthesizing information and knowledge. This enhancement involves integrating and merging concepts developed by various authors with theories derived from energy and software domains.

## 2    Background and Related Works

Modernizing power systems has led to the emergence of critical entities, such as Renewable Energy Distributors (REDs), tasked with distributing renewable energy locally. Traditional information systems have employed a shared data representation and vocabulary in the context of information communication technology (ICT) infrastructure. However, this conventional approach has given rise to interoperability challenges [15]. Therefore, a need for standardized information exchange becomes apparent in this dynamic energy scenario. This is where ontologies come into play. Ontology is intended to serve rich and complex knowledge about subjects and the relations between them. Gruber [16] first cited the term ontology. Since then, ontologies became more significant in computer science concerning knowledge representation and processing. Concerning smart grids and energy systems, ontologies formally represent knowledge, describing relationships between concepts within a specific domain. They serve as a structured framework for representing and exchanging information, ensuring seamless communication among diverse entities involved in the energy sector.

Extending the Open Energy Ontology (OEO) becomes particularly significant in this context. Kuccuk et al. [17] highlight the limited examples of domain ontologies specific to the energy domain. The proposed extensions to OEO address this gap by enhancing the ontology with a focus on context and perception. These extensions aim to enable entities like DERs and DSOs to exchange data enriched with contextual information, ultimately contributing to more effective and meaningful communication within the energy domain. The collaborative and open nature of OEO's development, as emphasized by Booshehri et al. [1], adds value to its utility in distributed development, aligning with the collaborative nature of the broader energy community. Table 1 provides a chronological overview of a few contributions from various authors in ontologies, focusing on their applications in the energy sector. This methodology introduces an innovative approach to expanding the OEO, specifically focusing on integrating perception and context.

**Table 1.** Contributions in Ontologies for the Energy Sector.

| Year | Author | Contribution |
|------|--------|--------------|
| 1993 | Gruber [16] | Introduction to the term ontology and highlight their growing importance in representation and knowledge processing. |
| 1999 | Chandrasekaran et al. [18] | Presents a comprehensive view of the role of ontologies in object-oriented software design. |
| 2014 | Santodomingo et al. [19] | Describes an ontology matching system for the smart grid domain, aiming to mitigate interoperability challenges. |
| 2015 | Lopez et al.[20] | Proposes an ontology explicitly addressing energy efficiency in smart grid neighborhoods. |
| 2021 | Booshehri et al. [1] | Offers an overview of the evolution of energy system modeling and highlights the role of the Open Energy Ontology (OEO). |
| 2023 | Proposed Methodology | Insertion of Perception and Context into OEO |

## 3   Proposed Methodology

As mentioned before, the context in which a particular operator is situated can lead to a misguided interpretation, even if the data adheres to the agreed-upon protocol. To mitigate such occurrences, this work seeks to introduce a "contextualization" into processes, conveying semantically richer data. Figure 1 illustrates a procedure in which three DERs exchange information with three DSOs, using an ontology to be developed in the subsequent sections without losing context.

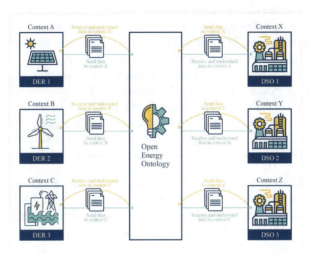

**Fig. 1.** Overall idea of the proposed contribution.

In this scenario, three REDs and three DSOs exchange data using the Open Energy Ontology as a communication standard. However, by implementing the contributions proposed in this work, they can both send data with information from their contexts and understand those of the recipients. This is the main difference between what OEO offers now and what is being proposed. As seen in Santos et al. [21], an ontology based on context requires information related to the actor, such as who or what it is, its location, and so on. However, it is important to notice that the energy domain requires some formula changes to convey the data more semantically, generating knowledge according to each context. There are many ways in which this can be achieved. Paneque et al. [22] suggests seven steps that range from defining the domain to modeling the classes. They are, from first to last:

1. *Determine domain and scope for the ontology* - The scope was built around the concept of context, aiming to convey information about the meaning of a given term from the sender to the receiver to ensure mutual understanding. As noted, the domain revolves around the energy sector.
2. *Consider reusing existing ontologies* - As defined in Sect. 1-A, the OEO [1] will act as the foundation for the proposed contribution. In this case, it will be extended to convey context.
3. *Enumerate important terms in the ontology* - In this instance, the terms are relevant to the proposed extension and will play a key role in defining the

class properties. Examples include *actorContext, term,* and *termMeaningBy-Context.*

4. *Define classes and class hierarchy* - Using the most significant terms, a series of ontology classes have been established. As shown in Figs. 2a and 2b, these classes will be created to work alongside those present in OEO, enabling a more direct interlinkage in the future.
5. *Define the properties of classes* - These properties were formulated based on the requirements to convey the context in which the sender operates. Examples include *meaning, actor, context,* and *term.*
6. *Define facets for the slots* - This stage involves establishing cardinality and a set of rules governing the ontology. As mentioned earlier, these will be created using Semantic Web Rule Language (SWRL) [23].
7. *Create Instances* - Instances, also known as individuals, correspond to specific terms and data that vary based on the context of the sender or receiver. They represent the focal point of the issue being discussed in this work.

The following classes were proposed, using OEO as a reference, to provide context to the terms being exchanged with the assistance of the proposed ontology. Figure 2 (a) depicts the organization of classes and subclasses, and Fig. 2 (b) depicts the object properties linking each one of them.

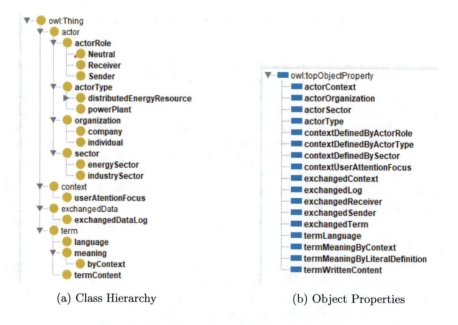

(a) Class Hierarchy                    (b) Object Properties

**Fig. 2.** Class Hierarchy and Object Properties.

# 4    Results and Discussion

The ontology presented in previous sections will be applied to a hypothetical use case for validation purposes. This application will demonstrate how ontology aids when two actors in the communication exchange process operate within different contexts. The example draws from the interoperability analysis between actors with different contexts, albeit on a smaller scale, presented in Donohoe et al. [24].

## 4.1    Actors

In this case, two actors are proposed: DERHousehold and EnergyCompany. To simplify comprehension, DERHousehold represents an individual producing energy via solar panels on their house, selling surplus energy to EnergyCompany. These actors operate within distinct contexts, altering the interpretation of specific terms exchanged between them. For this exercise, DERHousehold acts as the sender, while EnergyCompany acts as the receiver.

## 4.2    Context

In this context, DERHousehold functions as a small-scale distributed energy resource, symbolizing a house that generates energy via solar panels and sells the excess production. All energy produced by DERHousehold is directed to an EnergyCompany responsible for managing the grid and all associated processes. Even though they operate within the same region and domain, several terms can differ based on the sender and receiver. Three terms will be explored for this case, as shown in Table 2.

**Table 2.** Terms and Their Respective Meanings Based on Each Actor's Context

| Term | Sender's Meaning | Receiver's Meaning |
|---|---|---|
| Net Metering | Surplus energy generated, credited, or compensated by the energy company | The act of measuring surplus energy supplied to the grid by households, determining compensation or credits |
| Storage | Batteries or storage systems to store surplus energy for later use | Larger-scale energy storage solutions implemented by the energy company |
| Grid Connection | Physical linkage of household's renewable sources to the grid | Integration of distributed energy resources into the grid, impacting grid stability and management |

## 4.3    Object and Data Properties

For the sake of simplicity, the following table presents the critical object and data properties necessary for conceptualizing and representing the context in which an actor is situated. These properties work alongside the classes used to model the ontology, particularly those dealing with actors, terms, meaning, and context. Each table will display the properties of each critical actor (i.e.,

**Table 3.** Actor Class

| Name | NameDomain and Range | Explanation |
|---|---|---|
| actorContext | actor and context | Connects an instance of an actor to its respective context |
| actorOrganization | actor and organization | Connects an instance of an actor to its respective organization |
| actorSector | actor and sector | Connects an instance of an actor to its respective sector |
| actorOrganization | actor and actorType | Connects an instance of an actor to its respective type |

**Table 4.** Context Class

| Name | NameDomain and Range | Explanation |
|---|---|---|
| contextDefinedByActorRole | context and actorRole | Represents that a context is defined by the role of its actor, whether Sender or Receiver |
| contextDefinedByActorType | context and actorType | Represents that a context is defined by the type of its actor |
| contextDefinedBySector | context and sector | Represents that a context is defined by the sector of its actor |

Table 3). These object properties connect the actor to an instance that defines its belonging within a context. In this case study, actorType serves as the defining connection.

The displayed object properties link actors to their context, aiding the ontology in inferring term meanings from user context. Future additions to connections and context definitions could involve creating new classes or using existing ones in OEO. ExchangedData signifies data exchange between sender and receiver. In the scenario, 'exchangedContext' defines the sender's context. When received by EnergyCompany, the conveyed meaning aligns with the sender's context, enabling EnergyCompany to interpret it akin to DERHousehold, as seen in Table 2, and terms can hold multiple meanings based on context.

**Table 5.** ExchangedData Class

| Name | NameDomain and Range | Explanation |
|---|---|---|
| exchangedContext | exchangedData and context | Connects an instance of exchanged data to the context to which it belongs |
| exchangedReceiver | exchangedData and receiver | Connects an instance of exchanged data to the its receiver |
| exchangedSender | exchangedData and sender | Connects an instance of exchanged data to the its receiver |
| exchangedTerm | exchangedData and term | Connects an instance of exchanged data to the term being exchanged |

**Table 6.** Term Class.

| Name | NameDomain and Range | Explanation |
|---|---|---|
| termMeaningByContext | term and by context | Connects a term to the definition according to the context to which it belongs |
| termWrittenContent | term and by context | Connects a term to its content, the word that represents it. |

## 4.4   Data Exchange Scenario

In the described scenario, the sender communicates net metering, storage, and grid connection to the receiver through an ontology, as seen in Table 2. This structure, through the connections established on Tables 4 and 3, ensures that the exchanged terms, as per Tables 6 and 5 align with the sender's intended meaning, aiding accurate interpretation by the receiver. The sender's context, depicted as a circle, guides the receiver in understanding the information correctly, minimizing potential misunderstandings across operational levels or areas (as illustrated in Fig. 3).

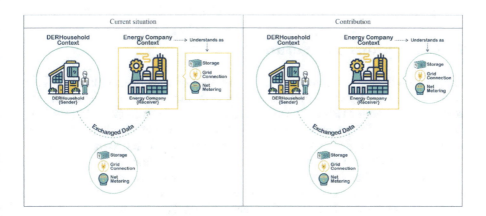

**Fig. 3.** Experimental Scenario.

Figure 4 presents a graph illustrating how the classes interact and collaborate. In this case, only the most relevant classes pertaining to this scenario were displayed for ease of comprehension. Within this framework, actors and context exhibit the most connections with other classes, highlighting how the actor plays a pivotal role in defining the context and the meanings of the transmitted data.

The ontology proposed, through the utilization of object properties, data properties, classes, and presented rules, can infer relations and information. It generates knowledge and aids in comprehending the context and its connections to the exchanged data in a manner that, in some cases, was not previously executed. In the end, the proposed solution ensures that, in a scenario where the

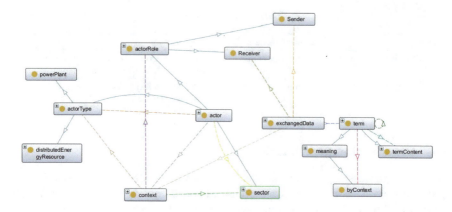

**Fig. 4.** Ontology Graph.

terms presented in Table 6 are exchanged, the receiver comprehends them within the sender's context. For instance, 'Storage' would be interpreted as 'batteries or storage systems intended to store surplus energy for later use,' rather than 'larger-scale energy storage solutions implemented by the energy company'.

## 5   Conclusions and Future Work

The proposed work addressed semantic interoperability challenges in smart grids by extending the OEO to incorporate context and perception elements. This extension allowed for richer data transmission in energy systems, acknowledging their unique conditions and meanings. A key challenge involved creating a test scenario rooted in industry realities, as most existing research focused on theory rather than practical applicability. The constructed fictitious case demonstrated one potential application, laying the groundwork for future exploration. Future work aims to expand context definition by integrating spatial data and exploring actor-centered ramifications. It also includes investigating alternative methods of extending OEO for data inference, utilizing actor characteristics for context-awareness, exploring applicability in diverse industries, and employing semantics for richer context, among other research avenues.

**Acknowledgment.** The authors thank Brazilian research agencies CNPq, CAPES, INESC P&D Brasil, CEFET-RJ, INERGE, and FAPERJ for supporting this work.

## References

1. Booshehri, M., et al.: Introducing the open energy ontology: enhancing data interpretation and interfacing in energy systems analysis. Energy and AI **5**, 100074 (2021)
2. IEEE Smart Grid, IEEE smart grid annual report (2015)

3. Anisie, A., Navarro, J.P.J., Antic, T., Pasimeni, F., Blanco, H.: Innovation landscape for smart electrification: Decarbonising end-use sectors with renewable power. In: International Renewable Energy Agency June (2023)
4. Suljanović, N., et al.: Design of interoperable communication architecture for tsodso data exchange. In: 2019 IEEE Milan PowerTech, pp. 1–6, IEEE (2019)
5. ENTSO-E - european network of transmission system operators for electricity. https://www.entsoe.eu/. Accessed: 15 Dec 2023
6. Commission, E.: Cordis - project id 774500 results (2023). Accessed 9 March 2024
7. ENTSO-E, Common information model (cim) (2023). Accessed 9 March 2024
8. I. R. E. A. (IRENA), "World energy transitions outlook 2023," 2023. Accessed 9 March 2024
9. International Renewable Energy Agency (IRENA), World energy transitions outlook 2023 (2023)
10. Australian Energy Market Operator — aemo.com.au. https://aemo.com.au/. Accessed 15 Dec 2023
11. A. E. M. O. (AEMO), "Aemo - asexml standards and schemas," 2023. Accessed 9 March 2024
12. ENTSO-e, "Entso-e members," 2023. Accessed 10 Nov 2023
13. Costa, G.C., Werner, C.M., Braga, R., Dalpra, H.L., Araújo, M.A.P., Ströele, V.: Deriving strategic information for software development processes using provenance data and ontology techniques. Int. J. Bus. Process. Integr. Manag. 9(3), 170 (2019)
14. M. de Minas e Energia, "Brasil registra maior produção de energia limpa dos Últimos 12 anos," 2023. Accessed 9 March 2024
15. Motiwalla, L.: V dan jeff thompson.(2009). enterprise systems for management.
16. Gruber, T.R.: A translation approach to portable ontology specifications. Knowl. Acquis. 5(2), 199–220 (1993)
17. Küçük, D., Küçük, D.: Ontowind: An improved and extended wind energy ontology arXiv:1803.02808 (2018)
18. Chandrasekaran, B., Josephson, J.R., Benjamins, V.R.: What are ontologies, and why do we need them? IEEE Intell. Syst. Appl. 14(1), 20–26 (1999)
19. Santodomingo, R., Rohjans, S., Uslar, M., Rodríguez-Mondéjar, J., Sanz-Bobi, M.A.: Ontology matching system for future energy smart grids. Eng. Appl. Artif. Intell. 32, 242–257 (2014)
20. López, G., Custodio, V., Moreno, J.I., Sikora, M., Moura, P., Fernández, N.: Modeling smart grid neighborhoods with the enersip ontology. Comput. Ind. 70, 168–182 (2015)
21. dos Santos, V.V., Tedesco, P., Salgado, A.C.: Percepçao e contexto. In: Sistemas Colaborativos (2012)
22. Paneque, M., del Mar Roldán-García, m., García-Nieto, J.: e-lion: Data integration semantic model to enhance predictive analytics in e-learning. Expert Syst. Appl. 213, 118892 (2023)
23. World Wide Web Consortium, SWRL: A semantic web rule language combining owl and ruleml (2004)
24. Donohoe, M., Jennings, B., Balasubramaniam, S.: Context-awareness and the smart grid: requirements and challenges. Comput. Netw. 79, 263–282 (2015)

# RoHR: Regionalized Hierarchical Routing for Vehicular Networks

Cairo Aparecido Campos[1], João Carlos Giacomin[1], and Tales Heimfarth[2(✉)]

[1] Computer Science Department, Universidade Federal de Lavras, Lavras, Brazil
cairoapcampos@gmail.com, giacomin@ufla.br
[2] Applied Computing Department, Universidade Federal de Lavras, Lavras, Brazil
tales@ufla.br

**Abstract.** Vehicular ad hoc networks (vanet) are composed by mobile nodes that are vehicles. They face difficulties in maintaining communications due to their dynamic topology, where routes are frequently broken as communicating nodes move apart. Geographic routing protocols, particularly those based on clusters are frequently employed to mitigate this problem. However, they are not suited to the special characteristics of urban environments. Aiming at a better performance for this kind of environment, this work proposes the Regionalized Hierarchical Routing (RoHR) protocol. In this protocol, the network topology control is based on clusters, composed by sets of vehicles located within delimited regions, denominated sectors. The position of each cluster-head (CH) is fixed within each sector. Each vehicle is associated with a cluster, depending on its position. Each CH knows the positions of its members. The organization of clusters is hierarchical, since a set of adjacent clusters forms a new top-level cluster. Data transmissions over long distances are carried out by multiple hops, using greedy geographic routing. CHs are used to query the position of the vehicle to which the communication is addressed. All the communications are V2V (vehicle to vehicle), RoHR do not use infrastructure. Each vehicle are aware of its own position, the position of the CH to which it is associated and the map of sectors. RoHR was compared with the canonical AODV (Ad-hoc On-demand Distance Vector) protocol, presenting a superior performance in terms of end-to-end latency.

## 1 Introduction

The evolution of wireless communication technologies has made direct communication between mobile equipment possible, without the need for infrastructure. This standard is known as mobile ad hoc communication (Mobile Ad Hoc NETwork - MANET). Among the various applications of MANETs, vehicular networks (Vehicular Ad Hoc NETworks - VANETs) present themselves as a central element of an intelligent transport system designed to increase safety on the roads [2]. Vehicular networks allow applications to exchange information to predict and prevent accidents, increase traffic efficiency, improve passenger comfort and even disseminate commercial advertisements [4]. Communication in VANET networks can be classified as V2V (vehicle-to-vehicle), V2I (vehicle-to-infrastructure) in which vehicles communicate with Roadside Units (RSU) and V2X, which is a combination of the previous two [10]. Although different applications for this type of network have emerged, routing in VANET is still a major challenge due to the high mobility and dynamic topology. To provide more efficient routing, some algorithms were developed or adapted, considering the limitations and characteristics of VANETs. However, routing protocols designed for MANETs, such as AODV [8], are also used in vehicular networks since they are efficient in certain communication scenarios. Among the classes of protocols developed, those based on position and clustering allow, in many cases, to achieve lower latency and lower overhead in routing.

In position-based (geographic) routing, the position of neighboring nodes and the destination is an important information, which is provided by a location service [3]. In cluster-based protocols, virtual groups are created between vehicles. In each cluster, one node is elected as cluster-head (CH). This node is responsible for managing intra-cluster and inter-cluster communication [1]. The cluster topology can be divided into two categories: single-hop and multi-hop cluster. In the first, the cluster members are in the radio range of the CH, in the second, communication is carried out using message relays between member nodes. Recent work has given preference to multi-hop communication, which results in greater cluster stability [10].

In VANETs, the lifetime of a cluster may be short due to high vehicle speeds and direction changes. A vehicle grouping technique for the road environment is presented in [9], which uses, to form clusters, information on the location and direction of the vehicle and the difference in speeds between vehicles. Although this technique increases the stability and lifetime of the clusters, the same advantages are not expected in an urban environment, as it has its own characteristics, such as lower speeds and constant changes of direction.

In order to overcome the inefficiencies of reactive routing protocols, in this work, RoHR is proposed for urban environments. It fuses geographic routing with cluster-based routing protocol. Each cluster is a sector delimited by a set of streets. The routing of control and data messages is done using greedy geographic routing algorithm. Each sector has a cluster-head (CH) that stores the position of the vehicles in its own sector. Our approach is hierarchical, as a group of clusters can form a new higher-level cluster. The CH at the lowest level knows the positions of vehicles within their sector. A higher-level CH communicates with lower-level CHs, maintaining position tables of all associated members. In each cluster, the cluster-head role is assumed by the node closest to a predetermined position. Employing geographic positions to delimit each cluster and using a hierarchy of cluster levels allows RoHR to route packets efficiently in urban environments. Our protocol also maintains the scalability, working with different sizes of regions. The contribution of this work is a new routing protocol based on multi-level clustering. Clusters maintain tables of positions in order to support the multi-hop routing process.

## 2   Related Work

In this section, related work on cluster-based routing algorithms will be presented. In vehicular networks, each clustering algorithm has its own characteristics. Some consider only vehicle-to-vehicle (V2V) communications, others include vehicle-to-infrastructure (V2I) communications. The position of the target vehicle can be known or unknown. Some algorithms are dedicated to urban environments, others to highways. Next, some cluster-based routing protocols are presented.

CBLR (Cluster Based Location Routing) [12] is a hierarchical cluster-based protocol. Communications are of the V2V type, without the use of infrastructure. Member nodes of a cluster communicate only with the cluster-head (CH), which forwards the message to the destination node. A CH can communicate with another adjacent CH through special nodes called bridges. When the destination node does not belong to the same cluster as the message source node, the CH sends position request messages to adjacent CHs. Upon receiving the destination's position information, the source node sends the message to the CH, which forwards it via geographic routing through the other CHs.

LORA-CBF (Location Routing Algorithm with Cluster Based Flooding) [11] is a hierarchical cluster-based protocol, similar to CBLR. Every position request from destination nodes belonging to different clusters is made via multiple hops, using bridge nodes. Unlike CBLR, data message transmission is done independently of CHs. When the source receives the location of the destination node, it simple forwards the message using geographic routing.

The HCB (Hierarchical Cluster Based) protocol [18], initially proposed for a heterogeneous MANET network, has a two-layer communication hierarchy and considers two types of nodes in the network. Nodes do not have position information. Type 1 nodes have only a short-range interface (<100 m) and form clusters at layer 1. Type 2 nodes, also called super nodes, have a short-range wireless interface and a long-range interface. Super nodes act as cluster-heads at layer 1 and communicate, at layer 2, with other super nodes. Layer 2 communication can be direct or through a base station. Communication, at layer 1, between a cluster-head and a member of its cluster is done via multiple hops. The cluster-heads periodically send HEARTBEAT messages, which have a counter that is incremented at each hop. This way, each member will know the shortest route to the CH. If the destination node is in a different cluster, the cluster-head forwards the packet to the cluster-head with which the destination node is associated. Upon receiving the data packet, this CH forwards it to the destination node.

CBR (Cluster Based Routing) [7] is a position and cluster-based routing protocol. In CBR, when a source node wants to send a packet to a destination node, it inserts the location of the destination node into the data packet and sends it to the CH of its cluster. When the CH receives the data packet, it forwards it to the neighboring CH, selected according to the minimum angle criterion. The data packet is forwarded between the CHs, always with the premise of selecting the next ideal neighbor CH, until the packet reaches the CH of the cluster where the destination node is located.

In TIBCRPH (Traffic Infrastructure Based Cluster Routing Protocol with Handoff) [16], nodes know their geographic positions through some location service. This protocol also considers a location service that allows the source node to detect the position and speed of the destination node. In this protocol, the cluster-heads are fixed to the existing traffic infrastructure equipment. Each cluster has a CH and a group of members. The union of CHs forms a backbone network covering a large area with several streets. In a TIBCRPH cluster, members communicate directly with each other. When a data message needs to be sent from a source node to a destination node in another cluster, it is transmitted between the backbone CHs until it reaches the destination node's CH.

Differently from our proposal, which is focused in urban scenario, some protocols are proposed taking in consideration mainly highway applications. Examples are CBDRP (Cluster-Based Directional Routing Protocol) [14] and an algorithm based on Euclidean distance, proposed by [15].

From the reviewed literature, it is possible to notice four problems that can affect the implementation of a VANET network. Firstly, due to the constant changes in the trajectory and speed of vehicles, the lifetime of a cluster may be short, due to inclusion and exclusion of members. Cluster recomposition has a cost due to control messages. Unlike the others, TIBCRPH reduces this cost by using static infrastructure-based CHs and a technique that allows vehicles to find new CHs. Secondly, it is desired that vehicles communicate with each other when traveling in cities, which implies infrastructure costs if V2I communication is used. Third, communication between nodes in urban areas using a long-range signal may fail due to the presence of obstacles such as buildings and trees. For example, the HCB algorithm is susceptible to this problem due to direct communication between CHs, when using long range communication. Fourth, routing between clustered vehicles moving in the same direction on a highway prolongs the life of the cluster, but cannot be employed in urban environments due to the constant changes in traffic direction caused by mobility. Examples of directional routing protocols are CBDRP and the Euclidean Distance Based Algorithm.

The innovation of our work is to propose a new routing algorithm for urban areas with the characteristics:

- Clusters are delimited and defined by streets and not by direction and speed, which increases the lifetime of each cluster;
- Communications are always made with short-range radios, using greedy geographic routing, even when a data message needs to be sent to a vehicle located in a distant cluster;
- Routing is done in any direction of traffic, there is no preferred direction.

Different from other urban routing methodologies encountered in the state of the art, in our proposal, the source vehicle did not need to known the destination's position in order to initiate a transmission. Furthermore, a dedicated network infrastructure is not required in the city, as only V2V communication is used.

## 3   Regionalized Hierarchical Routing (RoHR)

Regionalized Hierarchical Routing (RoHR) is a protocol designed for the urban environment. It has an ad hoc nature, eliminating the need for infrastructure to exchange messages. It uses a multi-level architecture, giving it scalability, resilience and agility in exchanging messages. At the lower level, several clusters are formed by geographically close vehicles. The cluster topology is multi-hop, with vehicles communicating using short-range radios. All vehicles are aware of their geographic location. Each cluster at this level is composed of all vehicles located within a defined region in urban space, which is called sector. The leader node, called cluster-head level 1 (CH1), keeps location tables of member nodes (vehicles) updated periodically. A set of geographically close level-1 clusters are organized into a level-2 cluster, which is led by a level-2 cluster-head (CH2). The location tables of each member node are updated regularly, with intervals greater than those used to update the CH1 tables. As the number of network nodes increases, more levels will be used in our organization to ensure scalability. Clusters at level *n* are composed by adjacent clusters at the previous level (n - 1), except for clusters at level 1, which members are common vehicles.

The idea of organizing the network in this way aims to facilitate the communication of messages. Whenever a vehicle, called CAR-SOURCE, has a data message to be transmitted to a destination vehicle, called CAR-DEST, it will send a query to the CH1 of its sector through geographic routing asking for the location of the destination. Upon receiving the destination's location, CAR-SOURCE sends the data message to CAR-DEST, also using greedy geographic routing, propagating the message by multihop through intermediate vehicles. Geographic routing is used to take the message to a position close to the CAR-DEST, as it may be moving. The message is delivered using the destination identifier. The routing method is explained later.

Figure 1 illustrates the exchange of control messages between CAR-SOURCE and CH1 and the sending of data messages from CAR-SOURCE to CAR-DEST. If CAR-DEST is not in the same sector as CAR-SOURCE, CH1 must forward the destination location request to CH2. Figure 2 illustrates the exchange of messages involving a level 2 cluster-head. If this CH does not have the requested information, higher-level cluster-heads may be involved in this process.

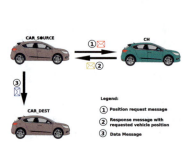

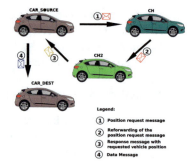

**Fig. 1.** Exchange of control and data messages employing solely CH1.

**Fig. 2.** Exchange of control and data messages using CH2.

## 3.1  Cluster Management

In the present work, unlike others, the information on the sectors and location of the CHs is previously generated and each vehicle, upon arriving at a region, already has this information in its navigation system. Therefore, the maintenance of clusters, as shown in this section, is done in order to save control messages, using this information.

The objective of the CH election is to select a vehicle that has little movement and is close to the previously specified geographic position, as shown above. Furthermore, all nodes in a cluster (sector) already know how to communicate with the CH: simply sending a message to the beforehand known position of the CH. A vehicle positioned close to this position will receive the message and fulfill the role of CH. The election of the CH is made according to necessity. Vehicles that are at a certain maximum distance from the chosen position for the CH are natural candidates. The choice is made spontaneously: if a request arrives in the CH region and does not generate a response from it, a suitable node in the region will assume the role of CH.

To achieve this, all nodes that are within a radius of the desired position for the CH configure their radios to listen to messages opportunistically. When a request to the CH is detected, a timer is activated. If, for a vehicle within the radius, the time interval expires and no CH response is heard by then, the vehicle assumes the CH role and responds. The time set for each timer is inversely proportional to how suitable a vehicle is for the CH role. Vehicles with low movement and closer to the predefined position will wait less time, assuming the role of CH before others that are less suitable.

Cluster members do not need to be informed about the CH election. When a request is needed, simply send a message to the known CH position. Furthermore, as each sector has its predefined geographic location, each vehicle has information about its cluster and also automatically detects the change of sector based on location information. When leaving one sector and entering another, a vehicle sends a HELLO message to the CH of the new sector to notify of the new situation. Furthermore, to maintain tables in CHs, moving vehicles send HELLO messages containing position and location periodically. When a vehicle is parked, this message is sent during the procedure. Eventually, parked vehicles can send updates to the CH, if the algorithm is configured to do so. In this manner, the network traffic is reduced.

## 3.2 Adapted Geographic Routing

RoHR uses greedy geographic routing to route its control messages and data. A node trying to communicate will request to the CH the position of the destination node and uses geographic routing to achieve its destination.

During the routing procedure, some obstacles may occur. Since the vehicles will only be positioned along the streets, the occurrence of local minimum points, known as dead ends, will be frequent. In order to increase the message delivery rate, an adaptation to geographic routing was proposed in the present work. Three conditions are observed in this new geographic routing. For the sake of clarity, some terms are defined: *target position* is the position to which the message should be forwarded; *transmitter* is the vehicle that has a message to be sent; *Neighbors* are all vehicles that can communicate directly with the transmitter. In this case, the first transmitter is the vehicle in which the message is generated (the CAR-SOURCE), the last target position is the known position of the vehicle for which the message is intended (the CAR-DEST position).

The three condition are the following:

1. If the transmitter is on the same street as CAR-DEST, simple geographic routing will be adopted, the message is transmitted to the neighboring vehicle that is closest to CAR-DEST.

2. If the transmitter is not on the same street as the CAR-DEST, the message must be routed in the direction of the target position, defined as the corner of the street segment where the CAR-DEST is located, always choosing the corner closest to the transmitter.
3. In the event of a local minimum (dead end), the message must be forwarded to a new target position, defined as the corner closest to the transmitting vehicle. When crossing the corner, conditions 1 and 2 must be observed again.

Figure 3 presents examples of message routing following the adapted greedy routing algorithm described here. Three square blocks are considered. Four corners are identified, e1, e2, e3 and e4. CS1, CS2 and CS3 have the role of CAR-SOURCE, which send messages to CAR-DEST, CD. CS1 is in the first condition, therefore its message is transmitted via direct route to the CD vehicle. CS2 is in the second condition, therefore its message is initially transmitted towards corner e1, passing through corner e3. At corner e1, condition 1 is reached, therefore, the transmitter located at that corner will use a direct route to CAR-DEST. The CS3 is in the second condition, so its message is initially transmitted in the direction of corner e2. Corners e1 and e2 delimit the street segment where CAR-DEST is located, with e2 being the closest to CS3. At point *m* the local minimum condition is reached, there is no other vehicle close to the transmitter (located at point *m*) capable of shortening the distance to the CD. In this case, condition 3 is observed, and the message must be forwarded in the direction of corner e4, which is closest to the relay node that identified the dead end. The first vehicle beyond corner e4 that receives the message will identify condition 2 and will transmit the message towards corner e2. At corner e2, condition 1 is reached.

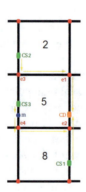

**Fig. 3.** Message routing according to adapted greedy algorithm

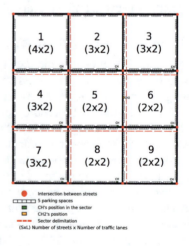

**Fig. 4.** Urban scenario considered in communications

## 4   Experimental Results

This section presents the results and discussions of the evaluation of the protocol proposed in this article. The results obtained were compared with the canonical AODV (Ad-Hoc On-Demand Distance Vector) protocol [8],widely used in ad hoc wireless networks and also in many vehicular network applications. AODV, as RoHR, is a routing protocol independent of infrastructure, since only V2V communications are used.

For the AODV protocol simulations, the GrubiX simulator (extension of ShoX [6]) for sensor networks was used. For the RoHR protocol, simulations were performed using the OMNeT++ [17] simulator, the SUMO [5] simulator and the Veins [13] framework. Veins enables the development of routing protocols by interconnecting the OMNeT++ event-driven simulator and the SUMO traffic simulator. It is important to highlight that the parameters were adjusted on both platforms and the movement model used was the same.

### 4.1   Simulation Setup

The traffic scenario used was based on the Manhattan mobility model generated by SUMO's netgenerate tool. The scenario is made up of avenues in a grid format, each block being considered a sector. All avenues are two-way with parking space. Each avenue has a total length of 600 m. Within each sector, the part of the avenue included in it is 200 m. This avenue segment contains, on each side, 4 parking areas, each with capacity for 5 vehicles. Figure 4 presents the scenario used for the simulations.

For the tests, 2 clustering levels were employed. It was defined that each sector makes up a level 1 cluster. The CH function of each cluster is performed by the node closest to the lower left intersection in the sector, as shown in the figure. These cars remained fixed throughout the simulation. To fill parking spaces, in the simulations, 600 cars were generated that park in the parking lots described above. In our simulation, the routing of intermediate messages was carried out only by parked cars. In addition to the 600 parked cars, 11 moving cars were used. All 9 sectors form a level 2 cluster. The position of the level 2 CH can be seen in Fig. 4.

Communications with the following message sizes were tested: 230, 1152 and 2304 bytes. The largest packet size is based on the maximum service unit size of the IEEE 802.11 MAC protocol. The average size chosen was half of the largest, and the smallest package size was approximately 10% of the largest. For communication between vehicles, the IEEE 802.11p protocol was used for RoHR with 9 Mbps and a range of 70m. In AODV, IEEE 802.11 with 11Mbps and 70 m of range was selected. In each simulation, two random vehicles (among those moving) were drawn and a data packet was sent between them. For each packet size, 45 simulations were performed.

In RoHR protocol, some control messages are sent periodically. In the present simulation, it was configured that every 2 s the moving nodes send a control message indicating their position to the CH.

### 4.2   Results

Experiments were carried out to measure end-to-end latency. This metric is characterized by the total travel time of a data packet from CAR-SOURCE to CAR-DEST.

Figure 5 presents the result of the average end-to-end latency for the different packet sizes. As expected, for each protocol, an increase in latency was observed with increasing message size, although this increase is not proportional to message sizes. This is because many control packets are required to route any size data packet. It was also observed that the RoHR protocol performed better in all cases. The latency obtained with AODV was at least 20% higher than the latency of the RoHR protocol.

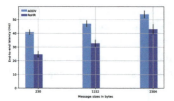

**Fig. 5.** Evaluation of the end-to-end latency of the AODV and RoHR protocols for the presented scenarios.

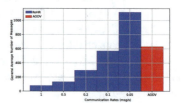

**Fig. 6.** Evaluation of the average number of messages used for each reception of a data packet, for different network traffic conditions.

The lower latency of RoHR can be explained by a different nature of the two protocols: in RoHR, the position of the destination is queried from the CH(s), and after this phase, the data can be sent to the destination, using geographic routing. In the case of AODV, a flooding of RREQ messages occurs to discover the path, until the destination is reached, when the RREP message is transmitted on the reverse path to the source node, in order to inform the route that should be used. It was noticed during the simulations that there was a congestion of RREQ packets, considerably increasing the transit time of the RREP message, as the forwarder needed to wait for the channel to be available to send the message.

After determining the communication latency of the two protocols, the total number of messages exchanged in each protocol was evaluated. As the total number of messages was similar for all message sizes, Fig. 6 presents the results for a data packet of size 2304 bytes. As AODV is a reactive protocol and RoHR is proactive, where CHs maintain an updated position table, we decided to evaluate our protocol with different communication rates, to check under which conditions it is more economical than AODV. For example, if only one message per hour is exchanged, the use of AODV is more advantageous, as periodic control messages are not necessary. In an environment with a lot of message transmission, a proactive protocol generally has superior performance.

For the experiment, we used different message exchange rates: 0.05; 0.10; 0.20; 0.50 and 1.00 messages per second. The total number of packets sent was counted for each data packet delivered, to calculate the protocol overhead. The result can be seen in Fig. 6, along with the total number of packets exchanged per data message delivered using AODV. It can be seen that even for scenarios with a very small traffic requirement (one data delivery every 10 s), far below the realistic urban environment scenario, the protocol presented in this article uses a smaller number of packets to accomplish the task. It can be noted that for realistic scenarios, with thousands of data packets being transmitted, this RoHR advantage tends to grow.

It can also be noted in the result that, for very small communication rates, the largest amount of packets exchanged with RoHR are control messages. For example, it was observed that, for rates of 0.10 and 0.05, the number of messages per data packet received practically doubled. This is because approximately the same number of messages passed through the network for both cases. However, it must be verified that in the first case $N1 = 100 s \cdot 0.10$ messages/s $= 10$ data messages were received, while in the second case $N2 = 100 s \cdot 0.05$ messages/s $= 5$ data messages were received. Thus, the average communication rate for each data message exchanged doubled for the lowest rate, since the denominator was half. The higher the communication rate, the lower the ratio of control packets to each data packet.

For the AODV protocol, the average number of hops that the data message needed to travel from CAR-SOURCE to CAR-DEST were approximately 9.7, 9.6 and 9.6 for messages of 230, 1152 and 2304 bytes respectively. The RoHR protocol spends, in general, a slightly smaller number of hops than AODV: approximately 8.9, 8.8 and 9.2 hops for the presented message sizes, respectively. This can be explained by the fact that geographic routing tends to use a straight path between the origin and destination. Path discovery by AODV is done through controlled flooding. The first RREQ packet to arrive at destination did not necessarily use the shortest path. It can also be seen that the data size does not appear to influence the metric.

The results found here showed that, for the urban environmental conditions tested, the proposed protocol presented better performance in all evaluated criteria.

## 5   Conclusion

In the present work, the RoHR (Regionalized Hierarchical Routing) protocol for vehicular routing in urban environments was presented. The protocol uses ad hoc communication between vehicles, eliminating the need for infrastructure, and uses a hierarchy of clusters to allow the determination of the position of the destination node, which makes the protocol scalable. Formation and maintenance of clusters are carried out in order to minimize overhead. When a given vehicle wants to transmit a message to a given destination, it queries the cluster structure for the location of the destination and upon receiving the answer uses a modified geo-routing algorithm for message delivery.

Simulations were performed of RoHR as well as AODV reference protocol to evaluate and compare the performances in an urban environment. The latency obtained with RoHR was at least 20% lower than that obtained with AODV, for all data packet sizes considered. Efficiency in data communication was measured in terms of total number of messages exchanged for each data packet delivered to destination. For data sending rates equal to or greater than 0.10 messages per second, RoHR required a smaller amount of packet exchange to perform delivery in our tests. For the scenarios and parameters considered, RoHR was a more effective alternative for establishing communication between vehicles in an urban environment. A mechanism to recover communication failures is a possible future improvement.

**Acknowledgments.** Cairo is grateful for the support received from IFSULDEMINAS - Campus Machado by releasing him from his duties to contribute to this research.

# References

1. Bengag, A., El Boukhari, M.: Classification and comparison of routing protocols in vanets. In: 2018 International Conference on Intelligent Systems and Computer Vision (ISCV), pp. 1–8, April (2018)
2. Boussoufa-Lahlah, S., Semchedine, F., Bouallouche-Medjkoune, L.: Geographic routing protocols for vehicular ad hoc networks (vanets): a survey. Veh. Commun. **11**, 20–31 (2018)
3. Cadger, F., Curran, K., Santos, J., Moffett, S.: A survey of geographical routing in wireless ad-hoc networks. IEEE Commun. Surv. Tutorials **15**(2), 621–653 (2013)
4. Eze, E.C., Zhang, S., Liu, E..: Vehicular ad hoc networks (vanets): Current state, challenges, potentials and way forward. In: 2014 20th International Conference on Automation and Computing, pp. 176–181, Sept 2014
5. Krajzewicz, D., Erdmann, J., Behrisch, M., Bieker, L.: Recent development and applications of SUMO - Simulation of Urban MObility. Int. J. Adv. Syst. Measure. **5**(3&4), 128–138 (2012)
6. Lessmann, J., Heimfarth, T., Janacik, P.: Shox: an easy to use simulation platform for wireless networks. In: Tenth International Conference on Computer Modeling and Simulation (uksim 2008), pp. 410–415 (2008)
7. Luo, Y., Zhang, W., Hu, Y.: A new cluster based routing protocol for vanet. In: 2010 Second International Conference on Networks Security, Wireless Communications and Trusted Computing, vol. 1, pp. 176–180, April (2010)
8. Perkins, C., Belding-Royer, E.: Ad-hoc on-demand distance vector routing. In: Proceedings WMCSA'99. Second IEEE Workshop on Mobile Computing Systems and Applications, pp. 90–100 (1999)
9. Rawashdeh, Z.Y., Mahmud, S.M.: A novel algorithm to form stable clusters in vehicular ad hoc networks on highways. In: EURASIP Journal on Wireless Communications and Networking, 2012, 12 (2012). https://doi.org/10.1186/1687-1499-2012-15
10. Ren, M., Zhang, J., Khoukhi, L., Labiod, H., Vèque, V.: A review of clustering algorithms in vanets. Ann. Telecommun. **76**, 02 (2021)
11. Santos, R.A., Edwards, R.M., Seed, L.N., Edwards, A.: A location-based routing algorithm for vehicle to vehicle communication. In: Proceedings. In: 13th International Conference on Computer Communications and Networks (IEEE Cat. No.04EX969), pages 221–226, Oct (2004)
12. Santos, R.A., Edwards, R.M., Seed, N.L.: Using the cluster-based location routing (cblr) algorithm for exchanging information on a motorway. In: 4th International Workshop on Mobile and Wireless Communications Network, pp. 212–216, Sept. (2002)
13. Sommer, C., German, R., Dressler, F.: Bidirectionally coupled network and road traffic simulation for improved IVC Analysis. IEEE Trans. Mob. Comput. **10**(1), 3–15 (2011)
14. Song, T., Xia, W., Song, T., Shen, L.: A cluster-based directional routing protocol in vanet. In: 2010 IEEE 12th International Conference on Communication Technology, pp. 1172–1175 Nov (2010)
15. DTian, D., Wang, Y., Lu, G., Yu, G.: A vanets routing algorithm based on euclidean distance clustering. In: 2010 2nd International Conference on Future Computer and Communication, volume 1, pages V1–183–V1–187, May (2010)

16. Wang, T., Wang, G.: Tibcrph: traffic infrastructure based cluster routing protocol with hand-off in vanet. In: The 19th Annual Wireless and Optical Communications Conference (WOCC 2010), pp. 1–5, May (2010)
17. Varga, A.: The omnet++ discrete event simulation system. In: ESM'01 (2001)
18. Xia, Y., Yeo, C.K., Lee, B.S.: Hierarchical cluster based routing for highly mobile heterogeneous manet. In: 2009 International Conference on Network and Service Security, pp. 1–6 June (2009)

# Development of Automated Satellite Image Processing System of JPSS-1 for Hotspot Monitoring Application

Nikorn Sutthisangiam[1(✉)], Suwatchai Kamonsantiroj[1],
and Phongsak Keeratiwintakorn[2]

[1] Department of Computer and Information Science, King Mongkut's University of Technology
North Bangkok, Bangkok 10800, Thailand
{nikorn.s,suwatchai.k}@sci.kmutnb.ac.th

[2] Department of Electrical and Computer Engineering, King Mongkut's University of
Technology North Bangkok, Bangkok 10800, Thailand
phongsak.k@eng.kmutnb.ac.th

**Abstract.** This research represents the creation of an automated processing platform for JPSS1 satellite imaging data, with a specific focus on hotspot monitoring applications using the VIIRS sensor. It leverages the CSPP processing program to compute data from satellite images and extract various parameters. The developed system has the capability to collect, process, analyze, store and visually present hotspot data through a user-friendly web application. It significantly reduces four times of human labour with optimizes operational time efficiency by threefold.

## 1 Introduction

Advancements in satellite technology have opened up opportunities for data collection and analysis, particularly in the field of meteorology. Satellite data has become a valuable resource for weather forecasting. Monitoring and analyzing weather conditions and tracking changes in the climate are crucial. Current weather patterns are dynamic and can cause damage to the environment and natural resources. Additionally, they can impact air quality by generating smoke and a significant amount of pollutants. These factors have implications for human health. For example, Thailand has been experiencing continuous issues with forest fires since 1992 to 2022. In 2022 alone, approximately 100,772.92 acres of land were burned, primarily in the northern region, accounting for 83% of the total burned area. The use of satellite technology in meteorology allows for a comprehensive understanding of weather systems on a global scale. It provides real-time data, enabling accurate weather predictions and timely response to natural disasters. Satellite imagery helps in monitoring cloud formations, tracking storm movements, and identifying areas prone to extreme weather events.

In Thailand, the organization which responsible of satellite technology usage is GISTDA (Geo-Informatics and Space Technology Development Agency). GISTDA is using JPSS-1 (Joint Polar Satellite System-1) for climate research and monitoring for whole country. The JPSS-1 is a polar-orbiting weather satellite launched by NASA

L. Barolli (Ed.): AINA 2024, LNDECT 199, pp. 394–401, 2024.
https://doi.org/10.1007/978-3-031-57840-3_36

(National Aeronautics and Space Administration) in partnership with NOAA (National Oceanic and Atmospheric Administration). It is part of a series of satellites designed to provide crucial data for weather forecasting, environmental monitoring, and climate research.

The main purposes of JPSS-1 are including

1. JPSS-1 is primarily designed for Earth observation, with a focus on collecting data related to weather patterns, climate, and environmental changes.
2. The satellite is equipped with advanced sensors and instruments to gather data about various Earth parameters. Some of the key instruments on JPSS-1 include the Advanced Technology Microwave Sounder (ATMS), Cross-track Infrared Sounder (CrIS), Visible Infrared Imaging Radiometer Suite (VIIRS) [1, 2] and the Ozone Mapping and Profiler Suite (OMPS). This research we especially focused on VIIRS data for hotspot monitoring application.
3. JPSS-1 follows a polar orbit, which means it passes over the Earth's poles on each orbit. This orbit allows it to capture global data coverage, making it particularly useful for monitoring changes in the Earth's atmosphere, oceans, and land surfaces.
4. The data collected by JPSS-1 is shared with meteorologists, climatologists, and researchers worldwide. This data is essential for improving weather models and gaining a better understanding of climate trends [3].

So JPSS-1 is a vital tool in our efforts to understand and predict weather and climate-related phenomena. It plays a critical role in enhancing our ability to respond to natural disasters, track climate change, and protect the environment.

For process satellite imaging data, this research has been used the CSPP (Community Satellite Processing Package) open-source software, it is a collection of software tools and packages designed for processing and analyzing data from weather and environmental satellites. CSPP is typically used by meteorologists, researchers, and government agencies to process and derive valuable information from satellite data for weather forecasting, climate monitoring, and environmental research.

In this paper, we present the automate processing data platform of JPSS-1 by VIIRS sensors. The CSPP has been use for processing the data then VIIRS active fire by hotspot prediction and processing to be displayed in a simple and efficient web application.

## 2   Related Theories and Principles

A. Forest fire detection
   Forest fire detection is a crucial aspect of managing forest fire incidents. The ability to detect forest fires quickly and accurately helps minimize the damage caused by these fires. Modern technologies play a significant role in forest fire detection. Satellite imagery, drone photography, and sensors installed in high-risk areas can support the detection of forest fires. These technologies allow for the rapid transmission of fire detection data to control centers.

   Detecting forest fires using satellite tools is an efficient and wide-ranging method for monitoring and controlling fires globally. This call hospot measuring. A hotspot, in the context of forest fires, refers to a specific location or area on the Earth's surface that is unusually warm or hot compared to its surroundings.

Generally, any hotspots can be detected using various technologies, including satellite-based remote sensing systems, such as VIIRS sensors from JPSS-1, which can detect the heat emitted from the Earth's surface.

B.  Visible Infrared Imaging Radiometer Suite

Satellite-based forest fire detection plays a significant role in early warning systems and fire management operations. By providing accurate and timely information about active fires, it helps authorities and emergency responders make informed decisions and allocate resources effectively.

VIIRS, or Visible Infrared Imaging Radiometer Suite, is an advanced Earth-observing instrument tool installed on NASA's Suomi NPP and JPSS-1 (NOAA-20) satellites [4]. It provides data on temperature, humidity, cloud coverage, light intensity, infrared radiation, and hotspots on the Earth's surface. The VIIRS system's hotspot detection works by measuring the heat emitted from infrared radiation at hotspots. It can detect hotspots even from a considerable distance.

The satellite captures images in both visible and infrared wavelengths, allowing for the identification of active fires. The thermal anomalies caused by the fires can be distinguished from the surrounding areas, aiding in their detection and monitoring. Additionally, satellite imagery allows for the mapping and tracking of fire progression over time. This information is crucial for assessing fire behavior, understanding fire spread patterns, and developing effective fire management strategies.

The VIIRS generated data stores in HDF5 (Hierarchical Data Format version 5) format. This file format is commonly used for storing and managing data from various scientific instruments and applications, including satellite data (Fig. 1).

| RNSCA-RVIRS_npp_ | RNSCA-RVIRS_npp_ | RNSCA-RVIRS_npp_ | RNSCA-RVIRS_npp_ |
|---|---|---|---|
| d20230131_ | d20230131_ | d20230131_ | d20230131_ |
| t0501485_ | t0635415_ | t1731280_ | t1909371_ |
| e0504392_b00001_ | e0649551_b00001_ | e1740000_b00001_ | e1922252_b00001_ |
| c2023013105051570 | c2023013106491296 | c2023013117401014 | c2023013119222822 |
| 7000_all-_dev.h5 | 0000_all-_dev.h5 | 7000_all-_dev.h5 | 1000_all-_dev.h5 |

**Fig. 1.** HDF5 file format from VIIRS sensor data.

The raw data file of VIIRS sensor in HDF5 file format keep in ftp server.

C.  Thermal point processing

The software that can process data obtained from VIIRS is called VIIRS Active Fire. This software is designed to detect forest fires in real-time. VIIRS Active Fire software works by analyzing the reflected light from the Earth's surface, particularly in the "infrared bands" frequency range. These bands are capable of detecting the heat emitted by forest fires. The algorithm examines each pixel of the image acquired from VIIRS and classifies whether that pixel contains a forest fire or not, using the following Table 1.

**Table 1.** VIIRS channel input to VAF algorithm.

| Channel | Wavelength (um) | Resolution (m) | Pixel ID | Purpose |
|---------|-----------------|----------------|----------|---------|
| I04 | 3.74 | 375 | BT4 | Fire detection (Primary) |
| I05 | 11.5 | 375 | BT5 | Fire detection & cloud classification |
| I01 | 0.64 | 375 | P1 | Cloud & Water classification (daytime only) |
| I02 | 0.86 | 375 | P2 | Cloud & Water classification (daytime only) |
| I03 | 1.61 | 375 | P3 | Water classification (daytime only) |
| m13 | 4.05 | 750 | BT13 | Fire Radiative Power (FRP) retrieval |

From Table 1, the I01, I02, and I03 bands are used to filter out solar radiation because the reflected solar energy can generate strong signals within the frequency range of I04, leading to false fire detections. The I04 band or 3.74 $\mu$m is the primary band for fire detection, while the I05 band or 11.5 $\mu$m is a secondary band for fire detection and the primary band for cloud classification. And the m13 band or 4.05 $\mu$m is used to calculate the Fire Radiative Power (FRP) because it has significantly higher temperatures than I04. The low temperature of I04 can potentially result in inaccurate data (Fig. 2).

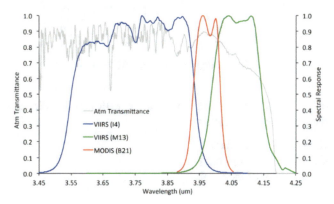

**Fig. 2.** Wavelength Detection from VIIRS.

The VIIRS Active Fire Algorithm is a system designed to detect and track fires and hotspots on the Earth's surface using data obtained from the sensors on the VIIRS satellite, which is installed on the Suomi NPP and JPSS satellites. The algorithm is complex, so in this research, we break it down into simpler components for better understanding.

Each pixel is evaluated and processed based on these components:

- Day/Night classification (Night = Solar Zenith Angle $\geq$ 90°)
- Cloud/Water pixel classification
- Fixed Threshold tests
- Potential background fires
- Filter Bright reflective targets
- Candidate fire pixels
- Contextual fire detection tests
- Secondary test
- Persistence tests
- Radiative Power retrieval (FRP)
- Persistent Anomaly flag

```
# Active Fires I-band EDR
#
# source: AFIMG_npp_d20230131_t0501496_e0503137_b58352_c20230424133222985506_cspp_dev.nc
# version: CSPP Active Fires version: 2.0.0
#
# column 1: latitude of fire pixel (degrees)
# column 2: longitude of fire pixel (degrees)
# column 3: I04 brightness temperature of fire pixel (K)
# column 4: Along-scan fire pixel resolution (km)
# column 5: Along-track fire pixel resolution (km)
# column 6: detection confidence ([7,8,9]->[lo,med,hi])
# column 7: fire radiative power (MW)
# column 8: Persistent Anomaly
#            0 - none
#            1 - oil/gas
#            2 - volcano
#            3 - solar panel
#            4 - urban (not in use)
#            5 - unclassified
#
# number of fire pixels: 12
#
   17.47797012,  120.92579651,  333.98773193,  0.375,  0.375,  8,  2.29858112,   0
   17.51397705,  120.81193542,  335.77362061,  0.375,  0.375,  8,  3.04649425,   0
   17.54649162,  120.92170715,  338.01324463,  0.375,  0.375,  8,  3.42852449,   0
   17.54579353,  120.91751862,  332.51486206,  0.375,  0.375,  8,  3.42852449,   0
   17.54509544,  120.91332245,  329.69772339,  0.375,  0.375,  8,  7.33991528,   0
   17.54983902,  120.81300354,  333.64559937,  0.375,  0.375,  8,  2.44141030,   0
   17.55331039,  120.81241608,  334.95590210,  0.375,  0.375,  8,  2.87747121,   0
```

**Fig. 3.** Output from VIIRS Active Fire

The figure in Fig. 3 represents the output of the VIIRS Active Fire software in text format. The data is arranged as follows:

- Date and time of collecting data
- Pixel position (latitude, longitude)
- Radiation temperature of the pixel
- Area of sensor scans
- Confidence in the accuracy of the data.
- Fire Radiative Power, the amount of energy that fire emits in the form of radiation.

As can be observed, the processing of hotspot result from start till the end are complex, manually and using variety tools. Currently, GISTDA employs four staff members to manually operate and spend six hours computing a single dataset. This research aims to develop an automate system, encompassing all operations.

# 3   Development of Automate Processing System

A. Operation of the automate processing system

The working process of the development system starts from downloading the JPSS-1 data to the process of displaying it through the browser. All steps can be divided into the following steps:

1. Downloading raw satellite data that has not gone through any process at all through ftp server.
2. Data-specific filtering required by the required data is level 0 data with the extension.hdf5 of the VIIRS sensor.
3. Data processing level 0 to level 1 with script file from CSPP program.
4. Choosing the data to be used in the next step or selecting the next level data. to process the data level.
5. Bringing level 2 data obtained from level data processing to convert data to have the extension *.csv.
6. Storing the final data into the database for use
7. Creating a backend service to bring the data in the database to be sent to the frontend side for use.

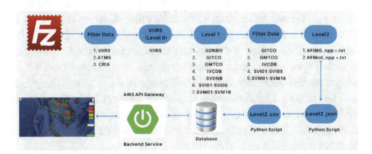

**Fig. 4.** Proposed Automate Processing Workflow of Hotspot Monitoring Application

Figure 4 shows the operation from the beginning to the end of hotspot process and store results in the database by CSV format.

B. Visualize on web representation

The rendering on the webpage will receive the hotspot data from the API and display it on the webpage in an easy-to-understand format. The data that will be displayed in the final step will be in JSON using tools like React-Leaflet In pairing with Leaflet plugins such as leaflet.heat, leaflet-control-geocoder.

Also the system has several features, it can be search by location or province, turn on or off the desired heat spot according to the severity, select the date of display, select the base map, turn on and off highlighting in Thailand and turn on and off the province name.

Figure 5 shows the part from the display of the obtained data with the following features.

**Fig. 5.** Visualization on web

## 4 Conclusions

This research presents the automate processing system of satellite imaging data for hotspot monitoring application. The focus of this research is on extracting temperature values from the VIIRS data sensor of the JPSS-1 satellite. It starts by downloading data from the FTP server and processing the obtained data. The downloaded data from the FTP server is in raw format and cannot be directly utilized. It needs to be processed and converted into level 1 data, which provides valuable information about the capabilities of the respective satellite. Level 1 data is further processed to obtain the desired level 2 data, such as hot spot information. The level 2 data is typically in a TXT file format, which is then converted into CSV format for storage in a database. The processed data is then sent to the frontend by API in JSON format for easy interpretation and display. It is also used to support further processing of other types of sensor data which CSPP compatible. As a result, the system can now be operated by a single staff member, with reducing the computing time to 2 h″.

**Acknowledgments.** The author would like to thank Geo-Informatics and Space Technology Development Agency for providing the research data.

## References

1. Schroeder, W., Giglio, L.: Visible Infrared Imaging Radiometer Suite (VIIRS) 750 m Active Fire Detection and Characterization Algorithm Theoretical Basis Document 1.0, pp.3–10, September 2017
2. Barnes, W., Molthan, M., Miller, S., et al.: NASA's earth observing system data and information system (EOSDIS): overview and future challenges. In: Earth Science Informatics, vol. 8, no. 2, pp. 236–252 (2015). Zhang, S., Zhu, C., Sin, J.K.O., Mok, P.K.T.: A novel ultrathin elevated channel low-temperature poly-Si TFT. IEEE Electron. Device Lett. **20**, 569–571 (1999)
3. Ryan, R.G.: Development of the visible infrared imaging radiometer suite (VIIRS) sensor data record (SDR). J. Atmos. Ocean. Tech. **30**(12), 2872–2884 (2013)

4. Ramapriyan, N., O'Kuinghttons, L.T., Wolfe, R.E., et al.: Land data products and services at the NASA National Snow and Ice Data Center Distributed Active Archive Center (NSIDC DAAC). Earth Sci. Inform. **4**(3), 135–147 (2011) (2002) The IEEE website. http://www.ieee.org/
5. Diner, D.J., Davis, A., Braswell, B.H., et al.: The Airborne Multiangle SpectroPolarimetric Imager (AirMSPI): a new tool for aerosol and cloud remote sensing. Atmos. Meas. Tech. **6**(9), 2007–2025 (2013)

# MotionInsights: Object Tracking in Streaming Video with Apache Flink

Dimitrios Banelas and Euripides G. M. Petrakis[✉]

Technical University of Crete (TUC), School of Electrical and Computer Engineering, Chania, Greece
dbanelas@tuc.g, petrakis@intelligence.tuc.gr

**Abstract.** MotionInsights facilitates object detection and tracking from multiple video streams in real-time. The system models video processing as a stream processing pipeline. Each video frame is split into smaller blocks, which are dispatched to be processed by a number of Flink operators. Each block undergoes background subtraction and component labeling. The connected components from each frame are merged into objects. In the last stage of the pipeline, all objects from each frame are concentrated to produce the trajectory of each object. The Flink application is deployed as a Kubernetes cluster in the Google Cloud Platform. Experimenting on a 7-machine Flink cluster revealed that MotionInsights achieves up to 6x speedup compared to a non-parallel implementation while providing accurate trajectory patterns. The highest (i.e., up to 6x) speedup was observed with the highest resolution video streams.

**Keywords:** Video processing · Background Subtraction · Object Detection · Object Tracking · Apache Flink · Apache Kafka · Kubernetes

## 1 Introduction

MotionInsights facilitates multiple object detection and tracking from multiple video sources. The system acquires real-time videos from the Web (i.e., streaming video), cameras, or video stored in files. Video processing and analysis is modeled as a data flow pipeline that is processed on Apache Flink's [3] distributed stream processing engine. Apache Kafka [9] facilitates the communication between the video production sources and Apache Flink. MotionInsights runs on a Kubernetes (K8s) cluster [8] in the Google Cloud Platform (GCP).

Each source video is split into frames and each frame is split into fixed sized blocks that enable rapid transfer and distribution to the partitions (i.e., queues) of the Kafka broker. To take full advantage of the available computing resources, MotionInsights enables the parallel (i.e., concurrent) acquisition of blocks from multiple video sources and ensures that these blocks are processed in the correct order on Apache Flink. Following background subtraction, each binary image block undergoes connected component labeling where the objects of each frame are detected. Consecutive video frames are aggregated and their trajectories are reported.

© The Author(s), under exclusive license to Springer Nature Switzerland AG 2024
L. Barolli (Ed.): AINA 2024, LNDECT 199, pp. 402–414, 2024.
https://doi.org/10.1007/978-3-031-57840-3_37

MotionInsights achieves up to 6x speedup compared to its sequential (monolithic) implementation. These speed-up results are achieved with common CPUs. This results in great cost savings compared to similar implementation using GPUs or GRID computers [1]. The system can process video up to 226 MB/sec input throughput but this can increase when Apache Flink uses more powerful (i.e., faster) machines. As with every Kubernetes deployment, it is easily reproducible and portable to all common computing infrastructures and the cloud.

Related work is discussed in Sect. 2. Architecture design and implementation issues are discussed in Sect. 3 followed by an analysis of its performance in Sect. 4. Conclusions and issues for future work are discussed in Sect. 5.

## 2   Related Work

The parallel implementation of fundamental image and video processing algorithms (e.g., background subtraction, and connected component labeling) are the basis of all real-time systems. However, most are tied to special purpose architectures such as CUDA-enabled GPUs [10], or GRID computers [7,11]. Antonakakis et al. [1] applied YOLOv5 to perform object detection in ultra-high resolution images on Nvidia Jetson AGX Xavier that includes GPU, CPU and RAM. They split each 8K image into K parts to delegate the processing of each part to a different grid of the GPU running YOLOv5 in a Docker container. The intermediate detections are merged at the end. This method achieves processing times <1 s per frame, matching the capturing frame rate of a high-resolution (i.e., 8K) camera.

Video2Flink (V2F) [6] handled video segmentation as a stream processing problem and can process real-time or batch video streams with the same efficiency. Aiming at video segmentation, its role is complementary and can be used to limit the scope of object detection and tracking to image sequences showing a continuous action. SIAT [14] is a distributed video processing framework on Apache Kafka and Apache Spark [12]. It achieves up to 3.5:1 times speed-up for multiple video analytics services such as video encoding, RGB to grayscale conversion, key-frame extraction shape extraction, compression, segmentation, classification, etc. MotionInsights achieved an even higher speed-up by decomposing the video into smaller chunks (i.e., pixel blocks) and by applying operations at an even lower granularity (i.e., pixel) level. RIDE [7] is a distributed multilayered video processing architecture with GPU and CPU worker machines. Coarse-grained parallelism is achieved by partitioning the video stream into smaller chunks (i.e., groups of frames) and by dispatching the chunks to different machines. Fine-grained parallelism is implemented on the GPUs of each worker. The authors reported up to 5:1 speed-up on 6 servers (Virtual Machines) in the cloud. RIDE is a custom software and hardware deployment and, is hardly replicable on common CPU architectures.

Several other attempts take advantage of the computational power of GRID architectures and reported significant speed-up compared to the processing of the entire video on a single machine. Parallel Horus [11] split the video into smaller chunks and video frames (one from each chunk) that are processed on different machines. This allows a high degree of parallelization with excellent (i.e., many times) speed-up for batched video. The authors reported 45:1 speed-up (compared to a sequential implementation) on a grid with 64 servers. Streaming Video Engine (SVE) [4] is a parallel video processing framework for Facebook. The videos are split into smaller chunks (i.e., 2 min at the most) that are processed separately on a large cluster of machines. The processing tasks include encoding, segmentation, and video track extraction. SVE achieved a speed-up ranging from 2x for short videos (i.e., less than 1 MB size) up to 9x for large videos (i.e., 1 GB size or more). Table 1 summarizes the above discussion. The comparison is mainly about functionality, architecture, and speedup.

**Table 1.** Comparison of video processing systems.

| System | Application | Processing | Hardware platform | Software platform | Type of processing | Maximum speedup |
|---|---|---|---|---|---|---|
| [4] | Video processing | Group of Pictures | Cloud | Custom | Batch | 9:1 (N/A) |
| [14] | Video analytics | Image frames | Cloud | Apache Kafka, Spark, Hadoop | Real-time | 3.5:1 (N/A) |
| [7] | Video analytics | Image frames | Cloud | Apache Kafka, Hadoop | Real-time | 5:1 (6) |
| [11] | Video processing library | Image frames | Grid | Custom | Batch | 45:1 (64) |
| [1] | Object detection, recognition | Image subframe | GPU | Nvidia Jetson AGX Xavier, Docker | Real-time | 1 FPS at 8K |
| [6] | Shot detection | Image blocks | Cloud | Apache Kafka, Flink | Batch, real-time | 7:1 (8) |
| *MotionInsights* | Object detection, tracking | Image blocks | Cloud | Apache Kafka, Flink | Batch, real-time | 6:1 (7) |

## 3   Architecture and Workflow

The videos are produced by the clients (i.e., online video sources or files). Each client splits the video into blocks that are easier to transfer to Apache Flink and process in parallel. The blocks are categorized into two topics (i.e., one to hold the blocks sent from the clients and one to hold the outputs). The input topic has $N$ partitions and the output topic has one. For each topic, the blocks are evenly distributed (i.e., in round-robin) to multiple Apache Kafka partitions. An equal number of parallel Apache Flink pipelines consume that data simultaneously. They process each block by applying a series of transformations (i.e., Flink operators). Figure 1 illustrates the system's architecture.

The client converts the video to raw RGB format, splits each frame into blocks, generates a unique key for each block, and broadcasts the block to Apache Kafka. Figure 2 illustrates this process. Each block contains a number of fully filled rows. If $H$ is the frame height and $W$ is the frame width, then the number $K$ of raws in each block must be such that $H \mod K = 0$. For RGB888 video the block size is $K \times W \times 3$ bytes. Splitting a video frame into blocks allows to parallelize the processing of a frame. Processing smaller block sizes allows the use of weaker Flink machines. If the Flink Cluster uses more powerful machines, a larger size can be set for each block.

The blocks are uniquely identified by keys generated by the client. The keys allow Flink to properly distribute blocks to partitions (without losing the identity of the frame

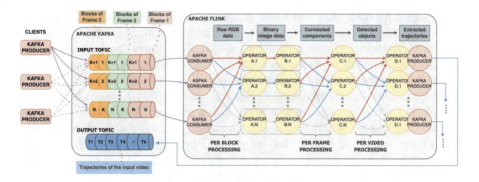

**Fig. 1.** MotionInsights architecture.

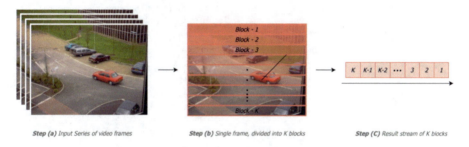

**Step (a)** Input Series of video frames    **Step (b)** Single frame, divided into K blocks    **Step (C)** Result stream of K blocks

**Fig. 2.** Creating a stream of blocks.

and video they come from) and later, to correctly group the blocks back to objects. A key contains the identifier of the video, the identifier of the frame it belongs, and its sequence within the frame. Additional information is the number of blocks per frame, video resolution, the number of rows contained in each block, and a boolean value denoting whether the block belongs to the last frame of the video. Figure 3 shows an example key. The *videoID* component is generated using the timestamp (i.e., the time the program started streaming the specific video in milliseconds). The last two digits are randomized to accommodate clients that start streaming a video at the exact same time.

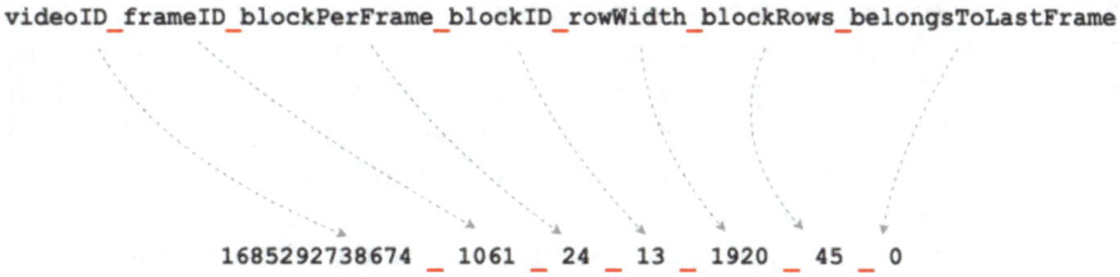

```
videoID_frameID_blockPerFrame_blockID_rowWidth_blockRows_belongsToLastFrame
```

```
1685292738674 _ 1061 _ 24 _ 13 _ 1920 _ 45 _ 0
```

**Fig. 3.** Example key of a block.

Blocks from more than one video source are distributed to $N$ Kafka partitions. Interleaving blocks from different videos in different partitions is not a problem. It is Flink's responsibility to reassemble information about blocks that form objects in a video, regardless of their processing order and video source. Figure 4 illustrates two clients sending frames from two videos to 4 Kafka partitions. Two frames are sent to Kafka from Client 1 (Red and Green blocks) and one frame is sent from Client 2 (Blue blocks). The potential omission of a block can significantly compromise the performance of the system overall (i.e., can cause awaiting the missing block and the risk of producing inaccurate results). It is the responsibility of Kafka to guarantee that all blocks are received. To deal with the issue, the clients are configured to require (from Kafka) an acknowledgment of receipt for each block sent.

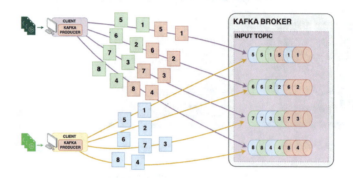

**Fig. 4.** Two clients distributing blocks of two videos to the 4 Kafka partitions.

### 3.1  Flink Workflow

MotionInsights defines 4 operators (excluding the source and sink operator) denoted as $A$, $B$, $C$, and $D$ respectively. The source operator is coupled with a Kafka consumer that reads blocks from a topic; operator $A$ executes background subtraction; operator $B$ detects connected components in image blocks; operator $C$ aggregates compound (i.e., whole) objects within video frames; operator $D$ computes object trajectories; the Sink operator outputs the results (i.e., objects and their trajectories) to a Kafka topic for further processing. The sequence of 4 operators defines a processing pipeline with parallelism 1. The pipeline should receive input from a single Kafka partition. Scaling the system up is realized by increasing the number of processing pipelines and the number of Kafka partitions accordingly. An application can run on $N$ parallel instances of the pipeline. Each pipeline includes instances of all operators, that is, each operator will have $N$ parallel instances.

All operators need to update past results (e.g., operator $A$ instance needs to update the background model of the previous block in a video sequence). These operators are stasteful and store data in a local key-value store, referred to as *state*. The coordination of parallel operators in time, is achieved using key information so that the stored states

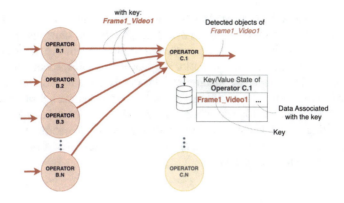

**Fig. 5.** Keys and keyed state example.

are identified unambiguously. For example, the outputs of operators $B$ $C$ and $D$ are transformed to keyed streams based on block, frame and video identifiers using the *keyBy()* function of Flink. Their state is then distributed to the operator instances that follow in the pipeline. Figure 5 illustrates this process. For example, to detect objects in a frame, the results of each operator $B$ instance (i.e., components in image blocks) are keyed (i.e., grouped) by frame and video identifier, and are dispatched to an instance of operator $C$. Similarly, the outputs of each operator $C$ instance (i.e., whole objects) are keyed by video identifier and sourced to an operator $D$ instance to update object trajectories.

Figure 1 is a view of the Flink pipeline with the intertwined data flows. The arrows between the operator instances indicate the redistribution of data among the pipelines during the processing. The output of each operator is shown at the top. The arrows between each pair of operators show the redistribution of the stream. The instances of operator $A$ can just forward blocks to instances of operator $B$ without any key information. Flink can chain instances of $A$ and $B$ operators together to execute them together on the same thread.

**Source Operator:** It retrieves video blocks from Kafka and transforms them to Java objects each comprising a key and the RGB pixel array of the block. The RGB array is passed as an argument to the background subtraction method on operator $A$ followed by noise reduction using morphological closing. The outputs are binary byte blocks.

**Operator $A$:** Each parallel instance of operator $A$ performs background subtraction on an image block. Background subtraction resorts to The Gaussian Mixture Model (GMM) method [13]. The GMM method is available in OpenCV *cv2.createBackgroundSubtractorMOG2()* and *cv2.createBackgroundSubtractorMOG()* functions. The method relies on learning and updating the statistical model about the distribution of pixel intensities in the background over time. Each pixel is modeled according to its counterparts in the previous frames and this makes GMM a stateful process. This introduces certain constraints related to which blocks can be sent to each operator *A* instance, and the order in which these blocks are processed. Blocks with the same *blockID* of the same video must be processed by the same instance of operator *A*. Moreover, these blocks must be processed in the order of the frames they come from. For blocks arriving in order, operator *A* updates their background model and forwards it to a parallel instance of operator *B*. Blocks arriving out of order are stored (in a HashMap) in the state of *A* awaiting for the block needed (i.e., a previous block). Figure 6 is an example of this process. Frame 5 arrived out of order and is stored in the state awaiting frame 4 (from the same video) to arrive. When the block of frame 4 arrives, its subsequent blocks (including block 5) are retrieved from the state and processed to update the background model.

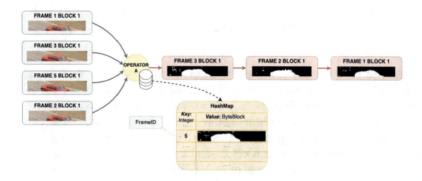

**Fig. 6.** Operator *A* functionality.

**Operator *B*:** It performs connected component labeling on the outputs of operator *A*. Each binary array is passed as an argument to *cv2.connectedComponentsWithStats()* method of OpenCV. This method returns the number of connected components in the block, along with their centroids, bounding boxes, and binary areas. This operator performs a one-to-many transformation without the use of state information. Figure 7 illustrates the input and output of an instance of operator *B*. Very small (i.e., noisy) regions are also detected. Components with an area less than 0.01% of the total block area are filtered out.

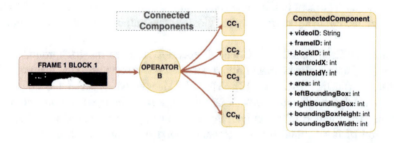

**Fig. 7.** Operator *B* (left) and representation of a connected component (right).

**Operator** *C*: The components detected in the previous stage correspond to image blocks rather than whole frames. Each parallel instance of operator *C* collects connected components of the same video frame and aggregates compound (i.e., whole) objects using state information. A counter keeps track of the number of components received. The process ends when all blocks of a frame have been received (i.e., the value of the counter matches the number of frame blocks) and no more components can be merged. The algorithm iterates over all components of a frame received to check whether any two of them can be merged. A component can be merged with other components from the same block, or components above or below it, based on adjacency or overlapping information of their bounding boxes [2]. A component can also be merged with other already merged components. Merging also updates the representation of the connected components by modifying their bounding boxes, centroids, and areas. Figure 8 illustrates an example of merging.

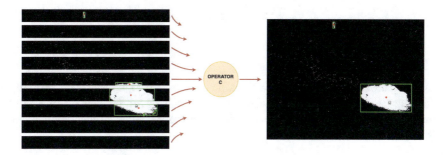

**Fig. 8.** Example of connected components merging.

**Operator** *D*: Each parallel instance of operator *D* receives a stream of information about objects (i.e., lists with bounding box, and centroid for each object in each frame) and computes their trajectories in the video. To differentiate among objects in different videos, the stream is keyed by video identifier (i.e., *VideoID*). Each parallel instance must process the frames of a video in order. Similar to operator *A*, each operator *D* instance implements a sorting mechanism. Frames arriving out of order are stored in a HashMap awaiting for a previous frame to arrive.

When the objects of the first frame that contains objects arrive (i.e., some frames may not contain objects) each object starts a trajectory. The algorithm iterates over every object in the input list and matches every centroid to its closest trajectory. Objects in the frames that follow are added to existing trajectories if the matching criterion is satisfied. If a match cannot be found for an object, its centroid will start a new trajectory. A centroid matches a trajectory if the Euclidean distance with the last point of that trajectory is less than a threshold (i.e., 50). In general, the threshold can be a function of each object's speed and angle with a trajectory. Objects with high speeds may introduce trajectory splits. Trajectories with less than $2 \, \text{s} \cdot fps$ points are considered to have been generated by noise and are filtered out (i.e., the duration of a trajectory must be more than 2 s). Figure 9 illustrates trajectories detected in a video. The red centroids are added to their closest trajectories. If no new objects are added to a trajectory for a predefined time duration $t$ (e.g., 10 s) that trajectory is considered stopped. This duration is set to $t * fps$, which corresponds to $t$ seconds regardless of the video's fps.

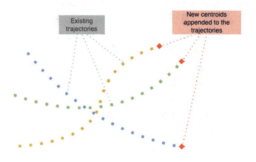

**Fig. 9.** Example object trajectories.

**Sink Operator:** It outputs video trajectories to a Kafka topic. Messages arriving to Kafka need to be serialized. The serialization dictates how a trajectory string will be transformed to a Kafka record. Every Kafka record representing a trajectory includes a key that identifies its video source.

## 4    Experimental Results

The purpose of the following experiments is to evaluate the runtime performance of MotionInsights based on response time and speed-up. The response time of the Flink deployment is compared to the responses of its non-parallel implementation. All measurements are taken on the Google Cloud platform. MotionInsights (i.e., Apache Kafka and Apache Flink) is deployed as a Kubernetes (K8s) cluster with two node pools. The background subtraction algorithm is GMM of OpenCV with its default parameters (e.g., $\alpha, T$ are the OpenCV's defaults). All videos are in raw RGB888 format. There are two clients producing videos concurrently at multiple resolutions and frame rates.

The first node pool is reserved to execute Apache Kafka, the clients (i.e., video producers), and the Flink Kubernetes Operator. The Operator provides easier installation of all the necessary components, simplified network configurations, and enhanced portability across different cloud providers. It is deployed on a single Virtual Machine (VM) node with 8 Virtual CPUs (vCPUs, 3.1 GHz base frequency, 3.8 GHz turbo frequency), 16 GB of RAM, 200 GB persistent SSD disk (1,200 MB/s read and write throughput), running the Container-Optimized OS of the platform. The second node pool is reserved by the Flink cluster that runs MotionInsights. It comprises a cluster with 7 *compute optimized* VMs with 2 vCPUs (3.1 base frequency, 3.9 GHz turbo frequency), 16 GB of RAM and 30 GB SSD storage.

## 4.1    Comparison with Sequential Implementation

In the first experiment, MotionInsights on Flink is compared against its monolith (i.e. non-parallel) implementation. The monolith application reads a video file from the disk, runs background subtraction, connected component labeling, and trajectory calculation in a row. It is deployed as a Docker container, on a single VM instance with 8 CPUs and 32 GB of RAM. The resources of this VM instance are equal to the total resources of the Flink cluster. The parallel application reads video from Kafka. The Flink cluster has a parallelism of 7. Each video is sourced 5x and average response times are calculated. Table 2 summarizes the results of this experiment.

For high resolution (i.e., 2K) video, the parallel implementation is up to 5.5x faster. For resolutions below 720 p, the monolith system is faster. For low input throughput, the communication overhead within the Flink's processing environment has a significant impact on the performance. Network-intensive operations such as *keyBy( )*, serialization and deserialization of the data at each operator, introduce an overhead. Also, the number of blocks per frame is much less for low input throughput [2]. There may be parallel instances of operators that receive few or no keys at all. These operators might remain underutilized or idle. Higher-resolution videos yield a greater number of keys and well populated groups, resulting in improved resource utilization of each operator instance.

**Table 2.** Response times of sequential implementations and parallel implementation (parallelism =7) with compute optimized machines.

| Resolution | Width× Height | Input Throughput (MB/sec) | Monolith (sec) | Parallel (sec) |
|---|---|---|---|---|
| 576p | 768× 576 | 39.8 | 9.93 | 138.12 |
| 720p | 1280× 720 | 82.9 | 161.78 | 193.00 |
| 1080p | 1920 × 1080 | 186.6 | 475.10 | 216.85 |
| 1440p | 2560 × 1440 | 331.5 | 1.413.13 | 252.76 |

## 4.2    Speed-Up of Parallel System

In the following, we study the improvement of the response time as a function of the number of machines in the Flink cluster (i.e., parallelism). For each parallelism, there is an equal number of Kafka partitions receiving video in the round-robin. There are two clients producing video concurrently so that, the input throughput is twice the throughput of the previous experiment (with one client). Table 3 summarizes the results of this experiment. The performance of the Flink deployment achieves its best results for high-resolution video and parallelism of 7. This is acceptable for processing video in batch since the parallel implementation is much faster than its sequential counterpart. For streaming (i.e., real-time) video, the results reveal a critical concern: The output throughput does not match the input throughput for high resolution and high transmission rate and the system is not operationally sustainable. The video data will accumulate in the Kafka (disk) partitions waiting to be processed, and the system will fail as soon as the disk capacity is exceeded.

**Table 3.** Performance of parallel implementation on compute optimized machines.

| Resolution | Input through-put (MB/sec) | Parallelism | Total GBs | Response Time (sec) | Output Throughput (MB/sec) |
|---|---|---|---|---|---|
| 576 p | 66.0 | 3 | 7.68 | 91.91 | 83.56 |
| | | 5 | 7.68 | 64.72 | 118.66 |
| | | 7 | 7.68 | 55.60 | 138.12 |
| 720 p | 165.8 | 3 | 16.56 | 148.27 | 111.68 |
| | | 5 | 16.56 | 125.78 | 131.65 |
| | | 7 | 16.56 | 85.80 | 193.00 |
| 1080 + 576 p | 226.4 | 7 | 22.50 | 99.15 | 226.9 |
| 1080 p | 373.2 | 3 | 37.32 | 373.68 | 99.87 |
| | | 5 | 37.32 | 267.62 | 139.37 |
| | | 7 | 37.32 | 172.10 | 216.85 |
| 1440 p | 663.0 | 3 | 66.32 | 633.36 | 104.71 |
| | | 5 | 66.32 | 460.30 | 144.07 |
| | | 7 | 66.32 | 262.38 | 252.76 |

Exploring the limit that allows the system to remain sustainable an extra measurement (in gray span) is added to Table 3. This extra measurement is taken by sourcing two videos (one of 576 p and one of 1080 p) by two concurrent clients both yielding 226.4 MB/sec input throughput. This result shows that Flink (with parallelism of 7) marginally matches the input throughput of 226.43 MB/sec. For higher input throughput, the machines struggle to complete their tasks, implying that the solution to the real-time performance problem would be to engage even more powerful machines [2].

## 5   Conclusions and Future Work

MotionInsights facilitates multiple object tracking and trajectory extraction on real-time and batch video. The system is deployed as a Flink cluster on Kubernetes. and achieves up to 6x faster processing on video compared to its sequential implementation. An interesting improvement would be to take advantage of Flink's advanced fault-tolerance guarantees. Automatic scaling [15] can also be incorporated into MotionInsights to enable automatic adjustment of parallelism by allocating (or de-allocating) computational resources automatically. Queries are an important feature for long-running streams. Queries on the trajectories might reveal the motion of objects in different locations in given time intervals. Integrating YOLO model would have a substantial impact on the accuracy of the resulting trajectories. It would also provide additional support for object recognition.

**Acknowledgment.** We thank Prof. Nik Giatrakos of TUC for insightful comments on Apache Flink performance and to Dimitris Kastrinakis who provided us with the sources of the Video2Flink system. We are also grateful to Google for the Google Cloud Platform Education Grants program. The work has received funding from the European Union's Horizon 2020 - Research and Innovation Framework Programme H2020-SU-SEC-2019, under Grant Agreement No 883272- BorderUAS.

# References

1. Antonakakis, M., et al.: Real-time object detection using an ultra-high-resolution camera on embedded systems. In: IEEE International Conference on Imaging Systems and Techniques (IST 2022), pp. 1–6. Kaohsiung, Taiwan (Jun 2022). https://ieeexplore.ieee.org/document/9827742

2. Banelas, D.: Motion and object detection from streamingvideo on Apache Flink. Diploma thesis, Technical University of Crete (TUC), Chania, Crete, Greece (Nov 2023). https://dias.library.tuc.gr/view/97992

3. Carbone, P., Katsifodimos, A., Ewen, S., Markl, V., Haridi, S., Tzoumas, K.: Apache Flink: stream and Batch Processing in a Single Engine. IEEE Data Eng. Bull. 38(4), 28–38 (2015). https://api.semanticscholar.org/CorpusID:3519738

4. Huang, Q., et al.: SVE: distributed Video Processing at Facebook Scale. In: ACM Symposium on Operating Systems Principles (SOSP 2017), pp. 87–103. Shanghai, China (Oct 2017). https://doi.org/10.1145/3132747.3132775

5. Huang, Q., et al.: SVE: distributed video processing at Facebook scale (SOSP 2017). In: Proceedings of 26th Symposium on Operating Systems Principles, pp. 87–103. New York, NY, USA (Oct 2017). https://doi.org/10.1145/3132747.3132775

6. Kastrinakis, D., Petrakis, E.G.M.: Video2Flink: Real-Time Video Partitioning in Apache Flink and the Cloud. Machine Vision and Applications 34(3) (Apr 2023). https://doi.org/10.1007/s00138-023-01391-5

7. Kim, Y.K., Kim, Y., Jeong, C.S.: RIDE: real-time massive image processing platform on distributed environment. EURASIP J. Image Video Process. 39(1) (2018). https://doi.org/10.1186/s13640-018-0279-5

8. Kubernetes: Kubernetes - Production-Grade Container Orchestration (Jun 2022). https://kubernetes.io

9. Narkhede, N., Shapira, G., Palino, T.: Kafka The Definitive Guide, Real Time Data and Stream Processing at Scale. O'Reilly Media (2017). https://spark.apache.org/

10. Pawar, D.: GPU based background subtraction using CUDA: state of the art. In: International Conference on Wireless Communications, Signal Processing and Networking (WiSPNET 2017), pp. 1201–1204. Chennai, India (Mar 2017). https://ieeexplore.ieee.org/document/8299953

11. Seinstra, F.J., Geusebroek, J.M., Koelma, D., Snoek, C.G., Worring, M., Smeulders, A.W.: High-performance distributed video content analysis with parallel-horus. IEEE MultiMedia 14(4), 64–75 (2007). https://ieeexplore.ieee.org/document/4354159

12. Spark: Apache Spark - Unified Engine for Large-Scale Sata Analytics (2022). https://flink.apache.org/, the Apache Software Foundation

13. Stauffer, C., Grimson, W.: Adaptive background mixture models for real-time tracking. In: IEEE Computer Society Conference on Computer Vision and Pattern Recognition (CVPR 1999), pp. 246–252. Fort Collins, Colorado, USA (Jun 1999). https://ieeexplore.ieee.org/document/784637

14. Uddin, M.A., Alam, A., Tu, N.A., Islam, M.S., Lee, Y.K.: SIAT: a distributed video analytics framework for intelligent video surveillance. Symmetry **11**(7) (2019). https://doi.org/10.3390/sym11070911

15. Zafeirakopoulos, A.N., Petrakis, E.G.M.: HYAS: hybrid autoscaler agent for apache Flink. In: Web Engineering: 23rd International Conference (ICWE 2023), pp. 34–48. Alicante, Spain (Jun 2023). https://doi.org/10.1007/978-3-031-34444-2_3

# An Ontology-Based Approach to Improve the Lead Time for Industrial Services

Luiza Bartels de Oliveira[(⊠)], Marco Antônio Pereira Araújo,
and Mário Antônio Ribeiro Dantas

Graduate Course in Computer Science (PPGCC), Juiz de Fora Federal University, Juiz de Fora
MG 36036-900, Brazil
luiza.bartels@estudante.ufjf.br,
{marco.araujo,marco.dantas}@ufjf.br

**Abstract.** Inadequate data organization within a company can lead to decreased efficiency, increased costs, and extended delivery times. This is particularly evident in the internal electronic maintenance laboratory of a mining company, where disorganized data obstructs the conversion of information into actionable knowledge, negatively impacting delivery efficiency. This work focuses on the challenge of organizing dispersed, often tacit knowledge among a few employees and their neglected reports from the maintenance laboratory. Following the development of a specific domain ontology, the subsequent step involved creating a prototype system called FastMain, with the aim of significantly reducing the execution time for maintenance services. Preliminary results indicate that FastMain achieved an average 17% reduction in maintenance lead time. The implementation anticipates substantial benefits, not only in reducing component maintenance time but also in alleviating the costly process of hiring and training new employees. With FastMain, new employees can easily access maintenance information through troubleshooting searches, allowing experienced workers to intervene only when necessary. This approach expedites the learning process for new employees and diminishes the workload of experienced workers, enabling them to concentrate more on core responsibilities.

## 1 Introduction

All motor vehicles utilize Electronic Control Units (ECUs) for optimal performance in modern automotive engineering. These electronic components optimize engine efficiency, manage emissions, control transmission, and ensure smooth operation. Similarly, heavy machinery relies on interconnected electronic modules for efficient and safe operation in challenging environments such as extreme temperatures, intense vibrations, dust presence, and electrical discharge risks.

In heavy construction and mineral extraction environments, both human operators and machines face challenges due to extreme conditions. ECUs play a crucial role in optimizing engine efficiency, managing emissions, and ensuring smooth performance in motor vehicles and heavy machinery. The interconnected electronic modules in heavy machinery contribute to efficient and safe operation in adverse environments, requiring ongoing maintenance to address failures and prevent deterioration.

L. Barolli (Ed.): AINA 2024, LNDECT 199, pp. 415–425, 2024.
https://doi.org/10.1007/978-3-031-57840-3_38

Maintenance in heavy construction and mining varies from quick resolution of simple issues to dedicated days for highly complex tasks. The complexity depends on diverse machines and environmental challenges. A mining company with a large fleet of heavy machines established a dedicated electronic laboratory for maintaining central electronic modules and components.

Global scarcity of raw materials for fertilizers and high commodity prices have significantly increased the use of heavy machinery in mining sites. This surge in activity has created a rising demand for maintenance, including electronic components. However, the laboratory's limited workforce has resulted in a waiting queue for electronic components awaiting maintenance.

Another challenge is the lab's location at the company's headquarters, distant from mining sites. Consequently, components requiring maintenance must be transported from machines to the lab, including sending a failure report with each component.

Manually filled labels on components often lead to confusion due to incomplete, illegible reports, or the use of slang, technical jargon, or regional terms in failure descriptions. This lack of clarity requires additional investigation, prolonging the maintenance time. However, recipients often store the attached printed report with detailed failure information, correction processes, and test results arbitrarily. This compromises traceability and the correlation between the failure and its resolution.

Recognizing the importance of knowledge as an intangible asset, the organization has been seeking ways to optimize data management. As mentioned by [14], knowledge plays a significant role in organizations worldwide, driving competition through a knowledge-oriented approach and enhancing processes.

In the context mentioned above, we proceed with the motivation that led to this research, the identification of the problem, the approach to the found solution, and finally, the defined objectives.

### 1.1 Motivation, Problem and Objective

Considering the highly specialized nature of maintenance services and the fact that only a select group of professionals possesses the necessary expertise, and knowing that effective knowledge management has been proven to generate considerable financial benefits and elevate the standard of quality in deliveries, there is an evident urgency to undertake a movement aimed at organizing the knowledge held by this restricted group of experts.

The emphasis given to this approach is noteworthy, to the extent that there is the Global Mining Guidelines (GMG), a worldwide group dedicated to promoting best practice guidelines. One branch of study in this group is focused on asset management, where guidelines and resources are developed to assist mining companies in optimizing the performance of their assets and maximizing value throughout the life cycle of their equipment. This collective effort aims to significantly enhance asset management in the mining industry. As well emphasized in their work, [4] concludes that companies in the mining industry poorly manage knowledge, relying on simple historical analyses and personal experiences of those involved in maintenance.

The challenge of consolidating and systematizing this knowledge acquired over years of experience is crucial not only to enhance the efficiency and effectiveness of maintenance operations but also to ensure that this valuable intellectual asset is preserved and shared in a way that benefits the organization as a whole.

Promoting a knowledge management culture within the company can contribute significantly to the ongoing success of operations, fostering innovation, more agile problem-solving, and ultimately, a lasting competitive advantage.

In this context, the problem addressed in this work relates to the search for effective models or methods to organize the currently dispersed, often tacit knowledge among the few employees and their neglected reports, issued by the electronic maintenance laboratory.

The main objective is to develop approaches that allow the organization and more effective use of the knowledge currently present in the reports and retained by employees. At the same time, there is an effort to create an accurate representation of the work environment of these professionals and the electronic maintenance activities they perform, with the purpose of optimizing the application of this valuable knowledge. Additionally, the goal is to make the domain available for applications that seek to enhance maintenance processes according to current needs.

This article is structured as follows. Section 2 provides an overview of a secondary study and related works that contributed ideas to the current work. Section 3 details the development of the FastMain approach, which is the central focus of this article. Sections 4 and 5 present the results obtained from implementing the approach and the conclusions drawn from them and also outline potential outcomes and future directions stemming from this study.

## 2  Related Works

One of the initial stages in the study involved conducting a comprehensive secondary study through a systematic mapping of existing literature. The main objective was to gain a deeper understanding and, above all, identify the methodologies employed in primary studies related to the acquisition and processing of semantic information by expert systems, with a focus on providing decision support.

To conduct the Systematic Literature Mapping, the methodology proposed by [20] was followed, which utilizes the PICOC strategy. This approach involves the identification of keywords that describe the Population, Intervention, Comparison, Outcomes, and Contexts related to the research object. The components of the PICOC acronym are detailed in Table 1.

The choice not to conduct a search for a specific context was intentional, as the goal was to obtain a comprehensive overview of the state of the art on the subject, exploring various perspectives and applications.

As a result, the following search string was obtained: ("knowledge-based system" OR "expert system") AND "knowledge engineering" AND "semantics" AND "knowledge acquisition" AND "quality of experience" AND "collaboration" AND "decision support."

The exclusion criteria were: book chapters, books, articles derived from literature reviews, articles that did not produce results related to the keywords "collaboration" and "decision support," and articles not published in journals. The inclusion criteria

**Table 1.** PICOC

| Criteria | Scope |
|---|---|
| Population | Knowledge-based systems or expert systems used for task management |
| Intervention | Understanding how knowledge engineering is used for data acquisition and the role of semantics within these systems |
| Comparison | None |
| Outcome | Understanding how these systems are used to promote collaboration in the workplace and how they provide decision support |
| Context | None |

were: articles published in journals and articles that yielded practical results related to the keywords "collaboration" and "decision support."

On the 6th June 2022, a total of 417 articles were retrieved from the search engines Association for Computing Machinery (ACM), ScienceDirect, and Institute of Electrical and Electronics Engineers (IEEE). After applying the selection criteria, a total of 85 articles were analyzed in order to gain a better understanding of how problems related to semantic data are being addressed by the academic community. Some of the top inspirational papers that brought inspiration and clarification to the development of the current work are described below.

In his study, [7] states that proper knowledge acquisition is one of the key problems in building knowledge management systems. The study helped to gain insights into how knowledge acquisition could be better conducted and how to involve context specialists, who were also the end users, in order to enrich the ontology construction.

Achieving successful acquisition, [10] argues that the knowledge acquired from previous problems should be reused as much as possible. The results of the study demonstrate that the knowledge-driven analysis approach can assist in identifying failures, bottlenecks, and opportunities for improvement in processes. This is exactly what is sought when introducing an ontological artifact into the laboratory context.

The research subject from [16,18,19] have observed that with technological advancement and globalization, managing the information generated by employees can bring a competitive advantage in the market and even promote cost reduction. This will be possible because industrial projects originate from activities that require many discussions and sharing of ideas to reach a decision, as observed by [1,9,17]. Thus, [12] believes that with effective knowledge management, companies will tend to create a teaching culture, where employees optimize their abilities by learning from mapped successes and errors.

Not only relevant in the industrial context, but the processing of semantic data in the medical context is relevant because the lack of specialized systems for automatic information extraction causes important patient clinical data to be underutilized (or still not reused) and poorly managed due to the high rate of manual data entry due to [2,3,5,6,11,21]. It is also expected that an expert system will provide physicians with more accurate patient history, aiding in making decisions about diagnosis and enabling more accurate medical treatment. Although they are studies in the medical context, analyzing them helped to realize how similar problems permeate all fields of knowledge and that there is a standardization in the search for their solution by using semantic processing supported by ontological artifacts.

In their study, [2] further complement that, after discussions and interviews with a group of physicians, it was concluded that the biggest source of errors in data entry is caused by those who perform such entries. They also observed that inconsistencies in clinical data lead to low performance in medical practice and hinder analysis of the patient history to make a diagnosis. This one, in particular, draws attention to addressing the failure of who and how to harm an expert system, a point that should be taken into consideration in the context of the electronics laboratory.

Studies produced by [8, 13] warn that the greater the number of medical decision support systems, the more heterogeneous the information becomes. Therefore, it is important to make efforts to create a platform for applications that need to unify clinical data from multiple databases.

The selected articles offered valuable insights for the development of the ontology itself, as well as for its future application in works stemming from this study. After reviewing the articles, the choice to use ontologies was motivated by the pursuit of an approach that inherently provides interoperability between systems. This is justified by the fact that corporate environments may receive varied data inputs. Furthermore, the decision to employ ontologies aims to explicitly represent the meanings of the data in a clear manner.

## 3   Methodology

Until now, we've covered the theoretical foundation, the company's context, and the identified failures. Moving forward, our focus shifts to developing an ontological artifact to address and resolve the identified problem.

To summarize the prior conducted study [15], where a specific ontology was developed to represent the domain of electronic maintenance services, entitled OMMEL (Ontology for Maintenance Management of an Electronics Laboratory), Fig. 1 illustrates the subject, verb, and predicate relationship established between the concepts and verbs listed after the knowledge acquisition step. In this scenario, a component named *CARTÃO_KOMATSU_STATEX_2* has a single *PartNumber* and *SAPCode*, but displays multiple *SerialNumber*, as well as different issues, descriptions, and diagnoses.

This illustration highlights the model's ability to handle complex and varied cases, showcasing the system's flexibility and robustness in capturing specific information and distinct relationships among various attributes and entities. This capability is crucial to ensure the effectiveness of the conceptual model in accurately and comprehensively representing real-world scenarios. Moreover, it underscores the utility of this ontology in a system, making it a valuable tool for decision-making. The ability to analyze information, draw conclusions, and support decision-making significantly contributes to the practical applicability and value of this ontology in dynamic and complex environments, as in the case of the electronics laboratory.

Following the conclusion of this study, the next step involved developing a prototype system to support a real-world case study. This system, nicknamed FastMain (an abbreviation for Fast Maintenance), aims to significantly reduce the execution time for maintenance services conducted by the electronics laboratory. Figure 2 depicts its architecture.

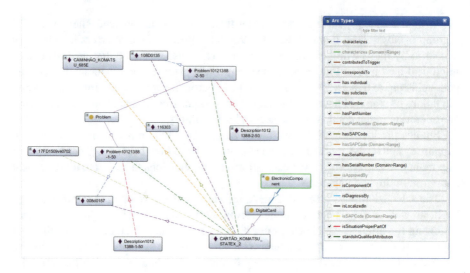

**Fig. 1.** Example of correlations between individuals in the OMMEL ontology

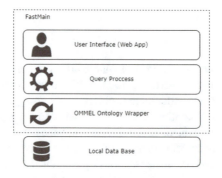

**Fig. 2.** Architecture of the FastMain approach

With the aim of accelerating development, available tools and libraries were employed. The ontology implementation was carried out using the Protégé tool, which provides a range of significant benefits throughout the entire life cycle of an ontology.

Protégé also supports various ontology languages, such as OWL, RDF, and Turtle, providing flexibility and suitability for different ontological needs and standards.

The local database consists of a spreadsheet, updated by the laboratory's own collaborators after each maintenance operation.

In the realm of ontology-based systems development, two languages stand out: Java and Python. It is known by the variety of available libraries to process RDF data. The latter was chosen due to its user-friendly nature, quick learning curve, and a wide array of libraries that facilitate the processing of ontological artifacts, promoting agile and efficient development.

Thus, the ontology wrapper layer utilized the pandas library to read and consume new entries from the database. Additionally, it employed the rdflib library to transform the data obtained from the database into a subject-verb-predicate structure, enabling effective integration with the ontological artifact.

Afterward, in the query processing layer, a SPARQL query engine enables users to have the capability to seek answers through the user interface, employing three specific methods: 1 - search by serial number; 2 - search by part number; 3 - search by a similar description. Algorithms 1 and 2 delineate the query processes in pseudocode, detailing each step for searching by part number or serial number and searching by a similar description, respectively.

The search for a similar description relies on the fuzzywuzzy library, which computes the Levenshtein Distance between two strings. A useful tool in cases where exact string matching may not be possible due to typos, spelling variations, or other minor discrepancies.

---

**Algorithm 1. Search for PartNumber e SerialNumber**

1: Get PartNumber user input
2: Get SerialNumber user input
3: Get SearchMode user input
4: Perform SPARQL Query and return results depending on SerialNumber or PartNumber
5: **for** Iterate over answers **do**
6:     $Component \leftarrow Component$
7:     $Description \leftarrow Description$
8:     $Diagnosis \leftarrow Diagnosis$
9:     $Solution \leftarrow Solution$
10:     $Date \leftarrow Date$
11: **end for**
12: Renders answers to the user interface

---

**Algorithm 2. Search for similar problem description**

1: Get Problem Description user input
2: Perform a general SPARQL Query to return al description instances
3: **if** User Description has more than 50% similarity compared to the Original Description **then**
4:     Create list of SerialNumber
5: **end if**
6: Perform SPARQL Query and return results depending on SerialNumber
7: **for** Iterate over answers **do**
8:     $Component \leftarrow Component$
9:     $Description \leftarrow Description$
10:     $Diagnosis \leftarrow Diagnosis$
11:     $Solution \leftarrow Solution$
12:     $Date \leftarrow Date$
13: **end for**
14: Renders answers to the user interface

Although the ontology has its model and implementation in the English language, its instances or instances values are mostly in Portuguese as it is the native language of the users. For the same reason, the entire user interface is also developed in Brazilian Portuguese. We believe that even the use of an internationally spoken language would influence the results due to the low proficiency of end users.

The Python-written backends render web applications running on a local server, on a dedicated machine (Intel® Core™ i5-1135G7, 16 GB DDR4-3200 MHz (SODIMM), 512 GB SSD M.2 2280 PCIe TLC Opal) provided in the laboratory, to avoid violating cybersecurity rules advocated by the company. The results will be presented in the Sect. 4.

## 4   Results

The tests were initiated in the first week of December 2023, so the results presented are the outcome of a research still in progress.

Every user search, whether by SerialNumber, PartNumber, or a similar description, yields the same information in a tabular format. The "#" represents a simple counter, "Componente" is the name of the electronic component, "Descrição" is the description of the problem, "DiagnÓstico" is the diagnosis of the problem, "Solução" is the solution to the problem, and "Dias" indicates how many days ago that situation occurred. Figure 3 depicts the explanation above.

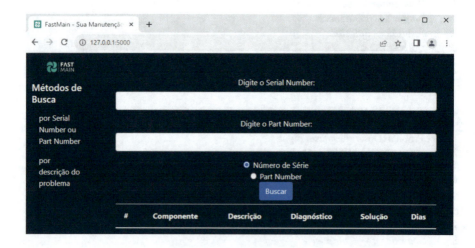

**Fig. 3.** FastMain user interface

The search by SerialNumber and PartNumber provides precise results. The search for a similar description returns a diffuse, but with meaningful content based on the description entered by the user. For instance, the user's input "isoamplificador não" may yield results like "isoamplificador não funciona", "não amplifica", "ISSO danificado",

"ISOAMPLIFICADOR com falha no sinal de saída". Clearly, it utilized the word or part of it to display the results.

During business days between December 4th and 15th, nine items were impacted by the FastMain system, resulting in an average of 17% reduction in maintenance lead time.

However, due to the small sample size, this average may not accurately reflect the situation of approximately 422 electronic components undergoing maintenance throughout the year. To obtain a sample with 95% confidence and a 5% margin of error, it would be necessary to consider a sample of 202 components.

## 5   Conclusion and Future Works

Although the study is in its early stages, there is a genuine expectation of improvement in lead time. Even a small reduction in the maintenance time for a single item throughout the year can lead to significant benefits.

Including in the table whether applying a solution to a problem led to component obsolescence or not can lead to a significant reduction in maintenance time, approaching approximately 100%, especially when there is confidence that the system provides reliable answers.

The laboratory workers anticipates substantial benefits from the implementation of this system, extending beyond the reduction in the time required for component maintenance. The hiring of a new employee is a costly process, involving not only salary expenses but also costs associated with training and integration. Depending on the level of experience, it may be necessary for a more experienced employee to dedicate up to three months to guide, convey the company culture, and teach specific service procedures. During this period, the experienced employee needs to allocate part of their working time to provide this guidance.

With the implementation of FastMain, new employees will have easy access to the information needed for maintenance through troubleshooting searches. This means that experienced employees will only need to intervene punctually since the new team member can independently find the necessary material. This approach not only accelerates the learning process for the new employee but also reduces the workload of more experienced staff, allowing them to focus more on their core responsibilities.

The FastMain approach is not yet showcasing its full potential. As it is a development focused on the industry, which is eager for results, the decision was made to make incremental developments after testing minimal viable products, following agile development methodologies. This development methodology also facilitated testing and increased laboratory staff participation in the search for an approach that truly brings practical value to their daily work.

The upcoming stages envisioned for the progression of this work include:

- establishing a connection between obsolescence and the problem's solution;
- associating the solution with the experience of the worker who provided it;
- clearly presenting to the user the percentage of reliability for each result.

**Acknowledgement.** This work was partially funded by CAPES/Brazil, CNPq/Brazil, INESC P&D Brasil.

# References

1. Afacan, Y., Demirkan, H.: An ontology-based universal design knowledge support system. J. Knowl.-Based Syst. **244**, 530–541 (2011)
2. Afzal, M., Hussain, M., Ali Khan, W., Ali, T., Lee, S., Huh, E., Farooq Ahmad, H., Jamshed, A., Iqbal, H., Irfan, M., Abbas Hydari, M.: Comprehensible knowledge model creation for cancer treatment decision making. J. Comput. Biol. Med. **82**, 119–129 (2017)
3. Chang, C., Tsai, M.: Knowledge-based navigation system for building health diagnosis. J. Adv. Eng. Inform. **272**, 246–260 (2013)
4. Curilem, M. et al.: Prediction of the criticality of a heavy duty mining equipment: Latin America congress on computational intelligence (LA-CCI). Curitiba, Brazil **2015**, 1–5 (2015). https://doi.org/10.1109/LA-CCI.2015.7435946
5. Doumbouya, M., Kamsu-Foguem, B., Kenfack, H., Foguem, C.: Argumentative reasoning and taxonomic analysis for the identification of medical errors. J. Eng. Appl. Artif. Intell. **46**, 166–179 (2015)
6. Fox, J., Gutenstein, M., Khan, O., South, M., Thomson, R.: OpenClinical.net: a platform for creating and sharing knowledge and promoting best practice in healthcare. J. Comput. Ind. **66**, 63–72 (2015)
7. Garcia, A., Vivacqua, A.: Grounding knowledge acquisition with ontology explanation: a case study. J. Web Seman. **57**, 100487 (2018)
8. German, E., Leibowitz, A., Shahar, Y.: An architecture for linking medical decision-support applications to clinical databases and its evaluation. J. Biomed. Inform. **2**(42), 203–218 (2009)
9. Guo, L., Yan, F., Li, T., Yang, T., Lu, Y.: An automatic method for constructing machining process knowledge base from knowledge graph. J. Robot. Comput.-Integr. Manuf. **73**, 102222 (2022)
10. Jabrouni, H., Kamsu-Foguem, B., Geneste, L., Vaysse, C.: Continuous improvement through knowledge-guided analysis in experience feedback. J. Eng. Appl. Artif. Intell. **245**, 1419–1431 (2011)
11. Kamsu-Foguem, B., Tchuenté-Foguem, G., Allart, L., Zennir, Y., Vilhelm, C., Mehdaoui, H., Zitouni, D., Hubert, H., Lemdani, M., Ravaux, P.: User-centered visual analysis using a hybrid reasoning architecture for intensive care units. J. Decis. Support Syst. **511**, 496–509 (2012)
12. Monticolo, D., Mihaita, S., Darwich, H., Hilaire, V.: An agent-based system to build project memories during engineering projects. J. Knowl. Based Syst. **68**, 88–102 (2014)
13. Mykkänen, J., Porrasmaa, J., Rannanheimo, J., Korpela, M.: A process for specifying integration for multi-tier applications in healthcare. Int. J. Med. Inform. **702**, 173–182 (2003)
14. North, K., Kumta, G.: Knowledge Management. STBE, Springer, Cham (2014). https://doi.org/10.1007/978-3-319-03698-4
15. Oliveira, L., Araujo, M., Dantas, M.: A case study on the development of an ontology for maintenance services of heavy machinery electronic components. In: Proceedings of the 12th Latin-American Symposium on Dependable and Secure Computing (LADC 2023). Association for Computing Machinery, New York, NY, USA, pp. 188–191 (2023)
16. Ouertani, M., Baïna, S., Gzara, L., Morel, G.: Traceability and management of dispersed product knowledge during design and manufacturing. J. Comput. Aided Des. **435**, 546–562 (2011)
17. Peng, G., Wang, H., Zhang, H., Zhao, Y., Johnson, A.: A collaborative system for capturing and reusing in-context design knowledge with an integrated representation model. J. Adv. Eng. Inform. **33**, 314–329 (2017)

18. Ruiz, M., Costal, D., España, S., Franch, X., Pastor, Ó.: GoBIS: An integrated framework to analyse the goal and business process perspectives in information systems. J. Inf. Syst. **53**, 330–345 (2015)
19. Spoladore, D., Pessot, E.: An evaluation of agile Ontology Engineering Methodologies for the digital transformation of companies. J. Comput. Ind. **140**, 103690 (2022)
20. Wohlin, C., Runeson, P., Höst, M., Ohlsson, M., Regnell, B.: Experimentation in Software Engineering. Springer, Heidelberg (2012). https://doi.org/10.1007/978-3-642-29044-2
21. Zhang, Y., Gou, L., Zhou, T., Lin, D., Zheng, J., Li, Y., Li, J.: An ontology-based approach to patient follow-up assessment for continuous and personalized chronic disease management. J. Biomed. Inform. **72**, 45–59 (2017)

# Using Scrum to Build Tourism Information Mobile Application

Edyta Wieslawa Zmudczynska[1] and Hsing-Chung Chen[1,2(✉)]

[1] Department of Computer Science and Information Engineering, Asia University, Taichung, Taiwan, R.O.C.
112121014@live.asia.edu.tw, cdma2000@asia.edu.tw, shin8409@ms6.hinet.net

[2] Research Consultant with Department of Medical Research, China Medical University Hospital, China Medical University, Taichung, Taiwan

**Abstract.** This research aimed to discern the requisites of nature-based tourism enthusiasts in Taiwan concerning service attributes and communication. The goal was to devise a mobile application for service promotion using the Scrum method. Conducting primary research in November 2023 via interviews with a cohort of 10 individuals, the investigation delved into the behaviors and challenges faced by foreign nature-based tourism enthusiasts in Taiwan. It also explored their communication preferences and expectations regarding mobile application features. The research outcomes facilitated the delineation of the profile characterizing a foreign tourist. This, in turn, led to strategic recommendations for enhancing promotional endeavors related to the country's nature-based tourism. Leveraging insights from primary research and applying the Scrum methodology, a mobile application project was conceptualized to amplify the visibility of this tourism offering.

## 1 Introduction

Tourism, a potent catalyst for global economic advancement, presents a dynamic landscape for both emerging and established economies. The sector's escalating revenues underscore its pivotal role in economic progress. Forecasts indicate that the confluence of digitization, innovative offerings, enhanced accessibility, and evolving societal dynamics will propel further expansion in this industry. To remain competitive in this swiftly transforming global scenario, tourism managers must adeptly navigate change while upholding the principles of sustainable development.

Taiwan, an open-minded and culturally affluent nation, boasts a plethora of natural wonders. Its tropical and temperate climate zones, coupled with diverse topography, foster an environment teeming with biodiversity, creating an exclusive habitat for nature tourism. The island stands out globally for its high butterfly density, plant diversity, and an impressive array of species—123 mammals, 788 birds, 133 reptiles, 42 amphibians, 400 butterflies, and 3100 fish. Preservation efforts encompass national parks, nature reserves, wildlife sanctuaries, and natural habitats, safeguarding Taiwan's ecosystems (Lee, 2011; Ministry of Foreign Affairs, Taiwan, 2018). Positioned to be both an "origin" and a "destination" for nature tourism in the global landscape, Taiwan underscores the pivotal role of this market in the broader tourism context (Frost *et al.*, 2014).

Despite Taiwan's abundant natural offerings, conversations with nature tourism business owners revealed difficulties in attracting foreign tourists due to inadequate information availability. Current promotional efforts on platforms like Facebook, Instagram, and Line yield limited effectiveness. While the government has official tourism websites, they fall short of meeting evolving needs. This identified gap prompted an analysis, revealing the absence of a multilingual mobile application promoting nature tourism in Taiwan.

This paper introduces a mobile application project designed to serve as a conduit for promotional information for tourists, concurrently fostering increased demand for service providers. Moreover, the study addresses a void in academic research, exploring the nexus between mobile app usage, customer engagement, and potential implications for organizational-customer relationships (Tarute *et al.*, 2017).

The research aimed at two primary objectives: firstly, to discern the requirements of tourists involved in nature-based tourism in Taiwan, specifically concerning service attributes and communication; and secondly, to create a mobile application utilizing Scrum method. The extensive research scope involved secondary exploration into the characteristics of the Scrum method in mobile app development and its tool application, along with examining the features of mobile applications and their role in promotional activities. The primary research phase comprised interviews with nature-based tourism participants in Taiwan to profile their preferences and challenges. Additionally, the study involved formulating promotional strategies and developing a mobile application for nature-based tourism in Taiwan, all within the framework of Scrum method.

Scrum's structured approach and adaptability allowed for efficient collaboration and continuous improvement. The research contributes by filling an information gap for foreign tourists and innovatively enhancing the tourism experience. The application incorporates features aligned with mobile app success principles identified in the literature review. Additionally, the study pioneers the application of Scrum method in academic research, demonstrating its effectiveness.

The paper examines challenges in Taiwan's nature-based tourism, reviews mobile app literature, and applies the Scrum framework to create the "Natural TaiWander" app. Results emphasize Taiwan's nature appeal and the app's potential impact. The conclusion underscores Scrum's research significance and proposes future study areas.

## 2   Literature Review

Contemporary mobile technologies, especially mobile applications, play a crucial role in today's promotional activities. Their effective utilization translates into customer engagement and a positive impact on the image of an organization, product, or brand. The concept of mobile app marketing is gaining increasing significance, focusing on promotional campaigns from the moment of app download to transforming customers into regular users and brand advocates. This phenomenon is not merely theoretical, as suggested by Hutton and Rodnick (2009) or Tarute *et al.* (2017), but becomes a reality, positively influencing customer interaction.

This strategy is particularly notable for its efficacy in establishing direct contact with consumers through the medium of mobile applications. The effectiveness of mobile app marketing campaigns is underscored by their ability to evoke positive sentiments regarding app usability, functionality, and user enjoyment. This is manifested through the integration of augmented visual elements and supplementary information, fostering emotional engagement that proves instrumental in forging robust relationships between organizations and consumers (Tarute *et al.*, 2017; Hutton and Rodnick, 2009; Oragui, 2018). Noteworthy is the transition observed by scholars from conventional mobile and internet marketing paradigms towards the ascendancy of mobile app marketing.

Several authors extol the virtues of mobile applications as exceptional mediums for consumer communication, transcending the conventional realms of websites and multimedia banners. This is achieved by fostering interactive relationships and engagement with customers (Calder *et al.*, 2009; Inokullu *et al.*, 2014; Bellman *et al.*, 2011). Kang (2014) and Onkonkwo (2016) posit that innovative applications of information technology can fundamentally shape user relationships by introducing novel approaches concerning place, time, and context. Recognizing the pivotal role of engagement in customer relationships is paramount for establishing enduring business relations (Dovaliene *et al.*, 2015).

Consumer motivation for interacting with mobile apps is multifaceted, encapsulating utilitarian aspects focused on functionality, as well as hedonic elements geared towards providing pleasure. The inherent convenience of app usage facilitates the seamless attainment of user goals while concurrently offering entertainment and enjoyment, thereby incentivizing regular usage (Bridges and Florsheim, 2008; Kim *et al.*, 2013; Hamka *et al.*, 2014; Tarute *et al.*, 2017).

Critical principles underpinning mobile app success encompass mobility, usability, novelty, simplicity, smooth navigation, good organization, user orientation, and personalization. Augmenting these, considerations of data security and customer safety are recognized as indispensable components contributing to the overall comfort of app usage (Singhal *et al.*, 2011; Mahmood *et al.*, 2016; Kim *et al.*, 2013; Zheng and Jin, 2016).

In the realm of tourism, a variety of experience-filtering techniques are already employed to facilitate selections and recommendations. The proliferation of richer and more comprehensive personal data is expected to further disseminate travel recommendations widely. Particularly on mobile devices, recommendations—whether socially generated or automatic—can be delivered in real-time alongside mapping services (Missaoui *et al.*, 2019).

The utilization of digital tourism introduces three pivotal elements aimed at enhancing consumer or tourist experiences, ultimately fostering satisfaction and cultivating tourist loyalty. These elements encompass Smart Tourism Technologies, Memorable Tourism Experiences, and Tourist Satisfaction. According to Azis (2020), the integration of smart tourism technology enables tourists to attain unforgettable travel experiences, thereby contributing to overall satisfaction and the subsequent development of tourist loyalty.

Singhal et al. (2011) posits seven principles that determine the success of mobile applications. These principles include mobility, allowing transactions from any location at any time (Mallat *et al.*, 2009); utility, providing expected features and solutions to users; novelty or uniqueness of ideas, distinguishing offerings from competitive proposals; simplicity and user-friendliness, evident in seamless transactions (Dewan and Chen, 2005; Timalsina *et al.*, 2012; El-Ebiary *et al.*, 2018); navigational fluidity, ensuring efficient application usage; effective organization, facilitating intuitive exploration of necessary functions and information; and user orientation and personalization, exemplified by the provision of language choices (Gugnani et al., 2006; Teixeira *et al.*, 2014).

Furthermore, the pervasive integration of mobile applications in the tourism service market is indicative of their escalating prevalence. Their utilization not only augments perceived brand value but also exerts a discernible influence on consumer purchase intentions. The advent of information and communication technologies has spurred the inception of novel business models for tour operators, underscoring the transformative impact of mobile applications on contemporary marketing strategies (Bellman et al., 2011; Fedeli, 2017).

## 3   Method

The section outlines the methodology employed in developing a multilingual mobile application for foreign tourists in Taiwan, utilizing the Scrum approach. The authors detail the application and adaptation of the Scrum framework to their specific research context, providing insights into the roles (Product Owner, Scrum Master, Development Team), key events (Sprint Planning, Daily Scrum, Sprint Review, Sprint Retrospective), and artifacts (Product Backlog, Sprint Backlog, Product Increment) within the Scrum process. The passage also introduces the tools and techniques used to support the Scrum implementation, with Fig. 1 visualizing the Scrum process (Schwaber amd Sutherland, 2017).

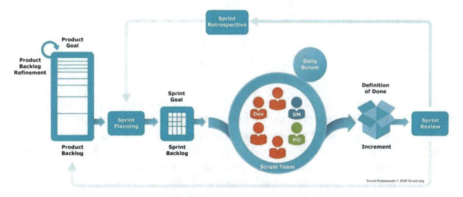

**Fig. 1.** Scrum Framework (Schwaber and Sutherland, 2017).

This framework provides a structured approach to product development, with clear roles, events, and artifacts that guide the process. It promotes transparency, inspection, and adaptation, which are key principles of Scrum (Schwaber and Sutherland, 2017).

The methodology of the work is as shown in Fig. 2 below:

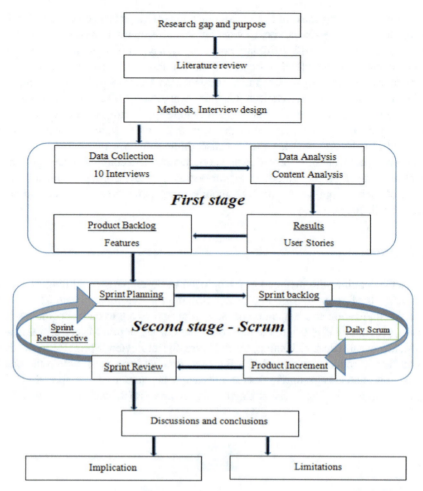

**Fig. 2.** Research Framework.

In pursuit of the designated research objective, a dual-pronged approach utilizing both secondary and primary sources of information was adopted. Secondary data encompassed insights derived from pertinent literature, scientific articles, market research findings, and statistical analyses. Primary data acquisition transpired through semi-structured interviews conducted in November 2023, involving a purposefully selected sample of 10 individuals actively engaged in Taiwan's nature-based tourism offerings. The interviews, carefully designed by the authors, were stratified into inquiries addressing:

The significance of specific elements within nature tourism.

Challenges encountered by tourists in availing nature tourism services.

The significance of specific elements within a mobile application.

After outlining the profile of tourists engaging in Taiwan's nature-based tourism and formulating strategic recommendations for promotion, the research seamlessly incorporated user stories into the mobile application development using the Scrum method.

The agile project management process, including Sprint Planning, Daily Scrum, Sprint Review, and Sprint Retrospective, played a crucial role in guiding and enhancing the research and application development. The structured execution of each Scrum event contributed significantly to the overall success of the project. The structure and objectives of each Scrum event, as applied within our methodology, are presented in Table 1.

**Table 1.** Scrum Method Events in Mobile Application Development.

| Events | Sprint Planning | Daily Scrum | Sprint Review | Sprint Retrospective |
|---|---|---|---|---|
| Purpose | Agreement on sprint scope and objectives | Daily progress update and planning | Demonstration and feedback gathering | Reflect on team performance and plan improvements |
| Participants | Product owner and development team | Development team members | Development team, product owner, stakeholders | Development team |
| Activities | Define sprint goals, select product backlog items for implementation | Briefly share progress, challenges, and plans for the day | Showcase product increment, collect feedback and suggestions | Identify strengths, weaknesses, propose actions for enhancement |
| Outcome | Agreed-upon scope and objectives for the upcoming sprint | Team synchronization and awareness of individual tasks | Stakeholder input for improvement and validation of progress | Insights for continuous improvement in team processes |

To implement Scrum in this research, a team was formed with a product owner (main researcher), a Scrum master (senior researcher), and a developer (junior researcher). The product vision focused on creating a mobile app for non-Chinese tourists in Taiwan, offering information on attractions, restaurants, hotels, transportation, weather, and culture. The product backlog was created through user feedback gathered from interviews with 10 foreign tourists who had visited Taiwan. Further analysis identified key features and requirements, including featuring a map, search engine, personalization, social media, rating system, and language translation. Each feature and requirement were articulated as user stories, concise descriptions of user needs and motivations (Table 2).

We also added acceptance criteria, which are the conditions that the user story must satisfy to be considered done. We estimated the value and the complexity of each user story, using the MoSCoW method:

*Must Have*

*A map with navigation and location-based services:* This is essential for any travel application to help users navigate their surroundings.

*Categories:* This allows users to easily find the information they're looking for.

*Information in English:* As the app is targeted towards English-speaking tourists in Taiwan, this tool is crucial for overcoming language barriers.

**Table 2.** Backlogs list.

| |
|---|
| User Story 1: As a user, I want a map with navigation and location-based services so that I can navigate Taiwan easily |
| User Story 2: As a user, I want categories so that I can find relevant information quickly |
| User Story 3: As a user, I want a personalization and customization system so that the app can cater to my specific needs and preferences |
| User Story 4: As a user, I want a social media and communication platform so that I can share my experiences and connect with other travelers |
| User Story 5: As a user, I want a rating and review system so that I can make informed decisions based on other travelers' experiences |
| User Story 6: As a user, I want information in English so that I can overcome language barriers while traveling in Taiwan |

*Should Have*

*A personalization and customization system:* While not essential, this feature would significantly enhance user experience by tailoring information.

*A rating and review system:* This would provide users with valuable insights from other travelers, although the app could technically function without it.

*Could Have*

*A social media and communication platform:* This would be a nice feature for users to share experiences and connect with others. However, it does not be as critical to the app's main functions.

*Won't Have (at this time)*

Any other features that might be nice to have but aren't necessary for the initial launch of the app. These could be added in future updates.

We organized user stories into epics, large stories broken into smaller ones. We added time-boxed spikes for exploring unknown areas. Our one-week sprint aimed to deliver a minimum viable product (MVP) for the mobile app. We implemented the Scrum events and artifacts as described below:

In Sprint Planning, we selected high-value, low-complexity backlog items aligned with research questions. We also considered the dependencies, the risks of the product backlog item(s). We then tried to balance the workload and the skills of the development team members. We created our sprint backlog, which consisted of 10 user stories and 2 spikes:

Spike 1: Research on integrating a map service with the application.
Spike 2: Research on implementing English translation, social media, and personalization.

and we assigned each user story and spike to a development team member.

Daily Scrums kept us updated, and a burndown chart tracked progress. Sprint Review showcased our product to stakeholders, and a Zoom meeting included foreign tourists for feedback. We followed the feedback grid formats consisting of likes, wishes, questions, and concerns. Sprint Retrospective had the team reflect on strengths, weaknesses, and improvement actions using the starfish format, which consists of five categories: keep doing, less of, more of, stop doing, and start doing.

## 4   Results

The research findings highlight the attractiveness of the surveyed region for nature tourism but identify a significant barrier in the form of limited access to information, particularly in English. To overcome this, the text proposes the development of a mobile application to serve as a virtual intermediary between nature tourism businesses in Taiwan and foreign visitors. The application aims to provide comprehensive information, including articles, contacts, galleries, and online reservations. The proposed app is expected to address the information deficit, promote nature tourism, and contribute to the growth of businesses in the sector.

Considering typical associations with nature and natural environments for the proposed application, the name "Natural TaiWander" has been suggested. This is a play on words emphasizing the app's focus on exploring Taiwan's natural wonders, incorporating "wander" and "Taiwan." The key elements of the application are:

1. Explore the Enchanting Features - The main page of the Natural TaiWander mobile app is divided into five segments, namely Discover Taiwan, Going Out, Natural Nourishment, Nature's Gems, and Eventures (Fig. 3).

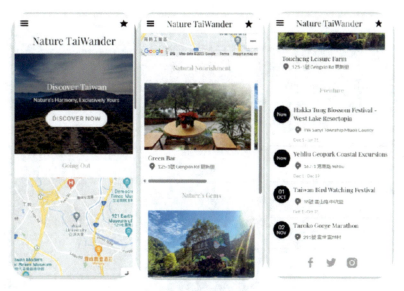

**Fig. 3.** Simplifying App Usage through Categorization

2. Dive into Visual Wonders - Immerse yourself in captivating visuals (photos, videos) showcasing the realm of nature tourism offerings (Fig. 4).

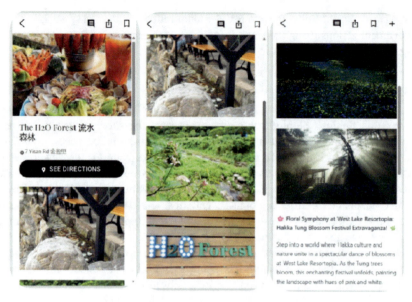

**Fig. 4.** Photo-sharing Gallery for Nature Tourism Offerings

3. Navigate, Connect, Explore - Obtain contact information, find locations, buy ticket, and plan your route seamlessly (Fig. 5).

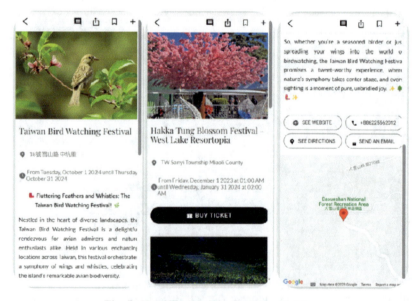

**Fig. 5.** Hotel/Tourism Site Contact with Map

4. Seamless Social Media Integration - Effortless sharing within the app to your account on social media platforms like Facebook/Instagram/Twitter.
5. Express Your Thoughts - Empower tourists to share their opinions about the app and the showcased offerings through user reviews (Fig. 6).

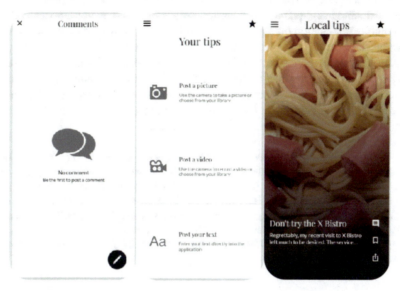

**Fig. 6.** Review Feature, enabling tourists to share their experiences

## 5   Conclusions

This scientific study aimed to address the challenges faced by nature-based tourism services in Taiwan, particularly in attracting foreign tourists. The research identified a gap in information availability and communication channels between service providers and potential guests. The study not only outlined the characteristics and preferences of tourists engaging in nature-based tourism but also proposed a novel solution through the development of a multilingual mobile application using the Scrum method. The research highlighted the significance of mobile applications in contemporary promotional activities, emphasizing their role in customer engagement and the positive impact on organizational image. The effectiveness of mobile applications in establishing direct contact with consumers, fostering interactive relationships. The Scrum methodology provided a structured and iterative approach to the development process, allowing for continuous improvement and adaptation. The application of Scrum principles to academic research contributes to the growing body of knowledge on agile methodologies and their applicability beyond traditional software development contexts. The proposed application, with its innovative features, is poised to contribute to the enhancement of the tourism experience for foreign visitors in Taiwan, potentially serving as a model for similar initiatives in other regions. While this research contributes valuable insights into the challenges and potential solutions for promoting nature-based tourism in Taiwan through a Scrum-based mobile application, it is essential to acknowledge several limitations. Future studies should consider larger and more diverse samples to enhance the robustness of the results. Future research could focus on conducting more extensive user experience testing with a larger and more diverse group of tourists. This would provide valuable insights into the usability, satisfaction, and overall impact of the proposed mobile application on the tourism experience. Moreover, to assess the sustained impact of the mobile application, longitudinal studies could be conducted to track user engagement and the application's influence on tourists' decision-making processes over an extended period. Investigating the specific needs and preferences of tourists from various cultural backgrounds can enhance the application's effectiveness. Comparative studies could be conducted to tailor the application to the diverse expectations of global tourists.

**Acknowledgments.** This work was supported by the Chelpis Quantum Tech Co., Ltd., Taiwan, under the Grant number of Asia University: I112IB120. This work was supported by the National Science and Technology Council (NSTC), Taiwan, under NSTC Grant numbers: 111-2218-E-468-001-MBK, 110-2218-E-468-001-MBK, 110-2221-E-468-007, 111-2218-E-002-037 and 110-2218-E-002-044. This work was also supported in part by Asia University, Taiwan, and China Medical University Hospital, China Medical University, Taiwan, under grant numbers below. ASIA-111-CMUH-16, ASIA-110-CMUH-22, ASIA108-CMUH-05, ASIA-106-CMUH-04, and ASIA-105-CMUH-04.

# References

Bellman, S., Potter, R.F., Hassard, S.T., Robinson, J.A., Varan, D.: The effectiveness of branded mobile phone apps. J. Interact. Mark. **25**, 191–200 (2011)

Bridges, E., Florsheim, R.: Hedonic and utilitarian shopping goals: the online experience. J. Bus. Res. **61**, 309–314 (2008)

Calder, B.J., Malthouse, E.C., Schaedel, U.: An experimental study of the relationship between online engagement and advertising effectiveness. J. Interact. Mark. **23**, 321–331 (2009)

CEIC (2020). https://www.ceicdata.com/en

Dewan, S.G., Chen, L.D.: Mobile payment adoption in the USA: a cross-industry, cross-platform solution. J. Inf. Privacy Secur. **1**, 4–28 (2005)

Dovaliene, A., Masiulyte, A., Piligrimiene, Z.: The relations between customer engagement, perceived value, and satisfaction: the case of mobile applications. J. Procedia – Soc. Behav. Sci. **213**, 659–664 (2015)

El-Ebiary, Y.Y.A., Waheeb, A.U., Alaesa, L.Y.A., Hilles, S.M.S.: Mobile commerce in Malaysia—opportunities and challenges. Adv. Sci. Lett. **24**, 4126–4128 (2018)

Fedeli, G.: The role and potential of ICTs in the tourism industry: the visitor attractions sector. In: Kalbaska, N., Ge, J., Murphy, J., Sigala, M. (eds.) ENTER2017 eTourism Conference PhD Workshop Research Proposals, pp. 14–20. Universita' della Svizzera italiana, Lugano (2017)

Frost, W., Laing, J., Beeton, S.: The future of nature-based tourism in the Asia-Pacific region. J. Travel Res. **53**, 721–732 (2014)

Gugnani, V., Arora, K., Shukla, V.N.: Issues and challenges in developing multilingual applications for mobile: Indic Languages Perspective (2006). http://www.w3.org/2006/07/MWI-EC/PC/cdac_Mobilepaper.pdf

Hamka, F., Bouwman, H., de Reuver, M., Kroesen, M.: Mobile customer segmentation based on smartphone measurement. Telematics Inform. **31**, 220–227 (2014)

Hutton, G., Rodnick, S.: Smartphone opens up new opportunities for smart marketing. Admap **44**, 22–24 (2009)

Inukollu, V.N., Keshamoni, D.D., Kang, T., Inukollu, M.: Factors influencing quality of mobile apps: role of mobile app development life cycle. Int. J. Softw. Eng. Appl. **5**, 15–34 (2014)

Kang, S.: Factors influencing intention of mobile application use. Int. J. Mob. Commun. **12**, 360–379 (2014)

Kim, Y.H., Kim, D.J., Wachter, K.: A study of mobile user engagement (MoEN): engagement motivations, perceived value, satisfaction, and continued engagement intention. Decis. Support Syst. **56**, 361–370 (2013)

Lee, S.H.: Diversity and innovation: the trends and changes of Taiwan tourism. J. Ritsumeikan Soc. Sci. Human. **3**, 37 (2011)

Mahmood, S., Amen, B., Nabi, R.: Mobile application security platforms survey. Int. J. Comput. Appl. **133**, 40–46 (2016)

Mallat, N., Rossi, M., Tuunainen, V.K., Öörni, A.: The impact of use context on mobile services acceptance: the case of mobile ticketing. Inf. Manag. **46**, 190–195 (2009)

Ministry of Foreign Affairs (Taiwan, R.O.C.) (2018). https://www.taiwan.gov.tw

Missaoui, S., Kassem, F., Viviani, M., Agostini, A., Faiz, R., Pasi, G.: LOOKER: a mobile, personalized recommender system in the tourism domain based on social media user-generated content. Pers. Ubiquit. Comput. **23**(2), 181–197 (2019)

Azis, N., Amin, M., Chan, S., Aprilia, C.: How smart tourism technologies affect tourist destination loyalty. J. Hosp. Tour. Technol. **11**(4), 603–625 (2020). https://doi.org/10.1108/JHTT-01-2020-0005

Okonkwo, C.: Critical success factors of mobile application development. (2016). https://pdfs.semanticscholar.org/4296/e88ece457328a6e98ce39e4fba51fb06c750.pdf

Oragui, D.: What is Mobile App Marketing? (2018). https://themanifest.com/app-development/what-is-mobile-app-marketing

Schwaber, K., Sutherland, J.: The Scrum Guide (2017). https://www.scrumguides.org/docs/scrumguide/v2017/2017-Scrum-Guide-US.pdf

Singhal, S., et al.: The Wireless Application Protocol: Writing Applications for the Mobile Internet. Addison-Wesley, Boston (2011)

Tarute, A., Nikou, S., Gataute, R.: Mobile application-driven consumer engagement. Telematics Inform. J. **34**, 145–156 (2017)

Teixeira, A., Francisco, P., Almeida, N., Pereira, C., Silva, S.: Services to support use and development of speech input for multilingual multimodal applications for mobile scenarios. In: The Ninth International Conference on Internet and Web Applications and Services Paris, pp. 41–46 (2014)

Timalsina, S.K., Bhusal, R., Moh, S.: NFC and its application to mobile payment: overview and comparison. In: 8th International Conference on Information Science and Digital Content Technology Jeju, pp. 203–206 (2012)

Zheng, W., Jin, L.: A consumer decision-making model in m-commerce: the role of reputation systems in mobile app purchases. Inf. Resour. Manag. J. **29**, 37–58 (2016)

# Video Compression Method Using Vector Quantization

Yusuke Gotoh[1(✉)], Toranosuke Ohashi[2], and Kiki Adhinugraha[3]

[1] Faculty of Environmental, Life, Natural Science and Technology, Institute of Academic and Research, Okayama University, Okayama, Japan
y-gotoh@okayama-u.ac.jp
[2] Graduate School of Natural Science and Technology, Okayama University, Okayama, Japan
[3] Department of Computer Science and Information Technology, La Trobe University, Melbourne, Australia

**Abstract.** As the number of users viewing video content on the Internet increases, video traffic is growing rapidly. When viewing video over the Internet, poor communication conditions and degraded viewing quality cause video and audio to be halting during playback due to the increased time needed for processing the playback of video content. In order to suppress such degradation in viewing quality, many researchers have proposed video compression models that leverage machine learning as a method to transmit high-quality video using fewer bits. These studies use convolutional neural networks to reduce the dimensionality of the image data. They also use scalar quantization and vector quantization to convert continuous values into discrete values. Previous research works have proposed video compression models using only scalar quantization, but not vector quantization. In this paper, we propose a video compression method using vector quantization and evaluate the video compression model. The proposed method reduces the processing time of video compression by fine-tuning the next frame using the weights of the fine-tuned frame. Therefore, the proposed method can fine-tune more frames within a given time. From the evaluation of the video compression model using the proposed method, the image similarity is higher than that of a video compression model using a method without fine-tuning.

## 1 Introduction

The popularity of video delivery services has grown rapidly in recent years, and video content is expected to become a growing proportion of Internet traffic [1]. When users watch video over the Internet, the viewing quality is affected by various communication conditions. Accordingly, deteriorated quality leads to video and audio interruptions due to the increased playback processing time of the video content.

To improve user viewing quality, many convolutional neural network-based image compression models [2–5] and video compression models [6–8] have been proposed to transmit high-quality video images using fewer bits. These models consist of an encoder, a quantizer, and a decoder. The encoder reduces the dimensionality of the data. The quantizer uses vector or scalar quantization to convert continuous data values into discrete values that can be encoded. The decoder reconverts the low-dimensional data discretized by the encoder and quantizer back to the original image data.

L. Barolli (Ed.): AINA 2024, LNDECT 199, pp. 438–449, 2024.
https://doi.org/10.1007/978-3-031-57840-3_40

When quantizing image data, image compression models [2–4] using scalar quantization improve image compression by using generalized divisive normalization (GDN) [9] to reduce the amount of mutual information between variables. Xiaotong et al. proposed an image compression model [5] using vector quantization. By optimizing the codec parameters for each input image, the developed model achieved better compression than conventional models at low bit rates [10, 11].

In general, when quantizing data having high correlation between variables, such as video and images, vector quantization is superior to scalar quantization because it can process multiple dimensions simultaneously [12]. However, in conventional video compression models, only methods using scalar quantization have been considered, and no method using vector quantization has been proposed so far.

In this paper, we propose a video compression method using vector quantization and evaluate a video compression model based on the proposed method. The contributions of this paper are as follows:

- The proposed method increases the compression effect in coding by improving the image compression model [5] using vector quantization and then fine-tuning this model to the frames composing the input video.
- By using the weights of the fine-tuned frames to fine-tune the next frames, the proposed method can reduce the processing time of video compression and fine-tune more frames in a given time.

The remainder of the paper is organized as follows. In Sect. 2, we explain the video compression model. Scalar and vector quantization methods are introduced in Sect. 3. In Sect. 4, we propose a video compression method using vector quantization. We evaluate our proposed method in Sect. 5. Finally, we offer our conclusions in Sect. 6.

## 2   Video Compression

### 2.1   Conventional Video Compression Model

Conventional video compression models require manual feature extraction and selection of classification models for videos. Design techniques for video compression models include intra-prediction, which eliminates spatial redundancy using similarities between different regions within the same frame, and inter-prediction, which eliminates temporal redundancy using correspondence between pixels in different frames. Intra-prediction typically uses convolutional neural networks. These prediction techniques are also important for machine learning-based video compression models.

Many methods using intra-prediction have been proposed. Deep Video Compression (DVC) [6] is a method that predicts the correspondence of pixels between consecutive frames in a video using a model based on optical flow prediction. DVC eliminates temporal redundancy by using a warping operation that reconstructs the next predicted frame based on the results of the optical flow prediction and encodes the difference from the current frame. However, since DVC uses pre-trained models based on optic flow prediction, it cannot perform end-to-end learning for all parameters between the pre-processing step of feature extraction and the final selection of a classification model.

Scale-Space Flow (SSF ) [7] is a method for creating a model that deals with uncertainty by adding a scale parameter to the flow prediction. SSF does not require a learned model for flow prediction but does require warping operations to reconstruct the next prediction frame.

Video Compression Transformer (VCT) [8] uses a transformer to create a model based on frame-to-frame dependencies. VCT also predicts the distribution of any image representation in latent space with low resolution and high accuracy. Moreover, it can easily implement models without flow prediction or warping operations. Therefore, in this paper, we design a video compression model based on VCT.

### 2.2 Optimization of Video Compression Model with Quantization

When optimizing a video compression model using the gradient descent method, which computes the parameter that minimizes the difference between the measured and predicted values, the gradient disappears due to quantization. In this case, learning does not progress because the gradient is not propagated within the model and the encoder parameters are not updated.

To solve the gradient loss problem, a straight-through estimator (STE) [13] method has been proposed that simply copies the gradient to the quantization function during back propagation. This method does not lose any gradient information during quantization. However, since this does not account for the error introduced by quantization, the actual gradient may be lost and the learning may be inaccurate. Therefore, many scalar quantization compression models [3] approximate the quantization error by adding noise from a uniform distribution to the encoder output. On the other hand, the method used in the vector quantization compression model is Noise Substitution in Vector Quantization (NSVQ) [14], which uses noise to approximate the quantization error. The evaluation of compressed models using NSVQ shows that it compresses data more accurately and enables faster learning than STE, exponential moving average, and Mini-batch K-means [15]. Consequently, in this paper, we use NSVQ to learn vector quantization.

## 3 Quantization Methods

### 3.1 Scalar Quantization

In the scalar quantization method, each dimension of the data is quantized independently, compressing the data without regard to the correlations among the data. As shown in Fig. 1, let the following parameters be finite: input variable $x$, quantization threshold $x_i$ $(x_0 < x_1 < \cdots < x_{N-1})$, representative quantization value $y_i$, and number of quantization levels $N$. The scalar quantization $\mathcal{Q}(\cdots)$ yields a quantization index $i(i = \mathcal{Q}(x))$ satisfying $\min_i |x - y_i|$. On the other hand, the inverse quantization $\mathcal{Q}^{-1}(\cdot)$ yields $y_i(y_i = \mathcal{Q}^{-1}(i))$ with $i$ as input.

### 3.2 Vector Quantization

Vector quantization is a method of quantizing an entire vector at once. By processing multiple dimensions simultaneously, it effectively compresses highly correlated data.

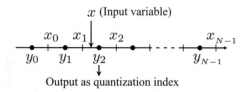

Fig. 1. Input and output in scalar quantization

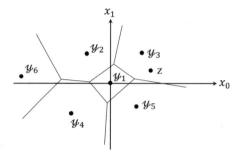

Fig. 2. Example of vector quantization

Let $d$-dimensional input vector $z = (x_1, x_2, \cdots, x_d)^T$ and quantized representative vector $\{y_1, y_2, \cdots, y_N\}$, $y_i \in \mathbb{R}^d$. The vector quantization $\mathcal{Q}(\cdot)$ yields the quantization index $i(i = \mathcal{Q}(z))$ satisfying $\min_i |z - y_i|$. On the other hand, the inverse vector quantization $\mathcal{Q}^{-1}(\cdot)$ takes $i$ as input and outputs $y_i(y_i = \mathcal{Q}^{-1}(i))$. The set of quantized representative vectors $\{y_1, y_2, \cdots, y_N\}$ is called the codebook.

An example of vector quantization is shown in Fig. 2. In the case of 2-dimensional 6-level quantization, the 2-dimensional input vector $z = (x_1, x_2)^T$ and the quantized representative vector is $\{y_1, y_2, \cdots, y_6\}$, where the codebook is $\{(0.0, 0.0)^T, (-0.8, 1.0)^T, (1.0, 1.0)^T, (-1.2, -1.0)^T, (0.8, -0.8)^T,$ and $(-2.5, 0.3)^T\}$, respectively.

In this paper, we perform vector quantization on the encoder's output latent space representation. The latent space representation is divided into blocks of size $r \times r \times s$ with no overlap. For example, if the dimension of the latent space representation is $w \times h \times m$, $z_i \in \mathbb{R}^d$, $d = r^2 \times s$, $i = 1, 2, \cdots, I$, and $I = \frac{w \times h \times m}{r^2 \times s}$. In this study, as in the previous research [5], $r = 1$ and $s = 8$.

### 3.3  Examples of Quantization Using Images

In vector quantization, the vector representation of data consisting of audio and images occurs, in many cases, only in a part of the vector space. Therefore, vector quantization has proven to be very effective in compressing such data [12].

For the grayscale image [16] shown in Fig. 3, the values $(x, y)$ of adjacent pixels are represented as pairs in a 2-dimensional vector space. As shown in Fig. 4, in the vector space each side is represented as a square whose value ranges from 0 to 255, and pixel values are dense on the $x = y$ diagonal. In the case of a typical image, the difference between adjacent pixel values is small. Therefore, the point in the 2-dimensional vector space created by these pixel values is close to the $x = y$ line.

**Fig. 3.** Grayscale image

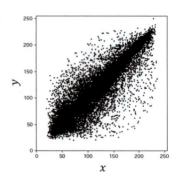

**Fig. 4.** Correlation of neighboring pixels

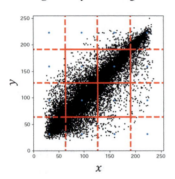

**Fig. 5.** Vector space partitioning in scalar quantization

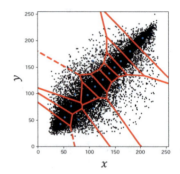

**Fig. 6.** Arrangement of cells with the smallest average quantization error in vector quantization

The case of scalar quantization at 2 bits per pixel is shown in Fig. 5. Each pixel is rounded to 4 discrete values and divided into $4 \times 4 = 16$ cells in vector space. The top-left and bottom-right cells have nearly empty vectors, while the four adjoining cells to them have sparse vectors. Therefore, the average quantization error in this method is high, and scalar quantization is inefficient.

Next, we consider vector quantization, which allocates 4 bits per 2 pixels. In vector quantization, a vector-quantized histogram is generated using a codebook that serves as a dictionary based on the local features of the image to be trained. The 16 vectors in the codebook can be freely arranged in the 2-dimensional vector space shown in Fig. 4. For the region shown in Fig. 6, the cells are arranged so that as many vectors as possible are densely packed on the $x = y$ diagonal in order to minimize the average quantization error calculated in Fig. 4. In the case of vector quantization, the cell size becomes smaller where the vectors are dense. Therefore, codebooks should be used for as many vectors as possible.

Many image and video compression models can efficiently compress video with scalar quantization by de-correlating the latent space representation using a generalized divisive normalization (GDN) layer [9]. On the other hand, Xiaotong et al. showed that an image compression model [5] using vector quantization outperforms conventional models at low bit rates.

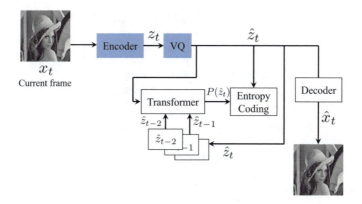

**Fig. 7.** Video compression model using vector quantization

**Table 1.** Structure of encoder and decoder

| layer |
| --- |
| $(3 \times 3 \times 32$, pad=1, stride=1) |
| Res. block $\times 3$, filter=8 |
| $(3 \times 3 \times 64$, pad=1, stride=2) |
| Res. block $\times 3$, filter=16 |
| $(3 \times 3 \times 64$, pad=1, stride=2) |
| Res. block $\times 3$, filter=32 |
| $(3 \times 3 \times 64$, pad=1, stride=2) |
| Res. block $\times 3$, filter=64 |
| $(3 \times 3 \times m$, pad=1, stride=1) |
| VQ and de-quantization |
| $(1 \times 1 \times m$, pad=0, stride=1) |
| $(3 \times 3 \times 256$, pad=1, stride=1) |
| Res. block $\times 3$, filter=64 |
| Pixel Shuffle |
| Res. block $\times 3$, filter=32 |
| Pixel Shuffle |
| Res. block $\times 3$, filter=16 |
| Pixel Shuffle |
| Res. block $\times 3$, filter=8 |
| $(9 \times 9 \times 16$, pad=4, stride=1) |
| $(3 \times 3 \times 3$, pad=1, stride=1) + Tanh |
| output |

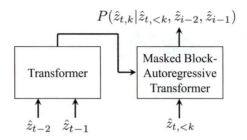

$$P(\hat{z}_{t,k}|\hat{z}_{t,<k}, \hat{z}_{i-2}, \hat{z}_{i-1})$$

**Fig. 8.** Transformer

## 4 Proposed Method

### 4.1 Proposed Model for Video Compression

In this paper, we propose a video compression method using vector quantization. First, we show the video compression model to which the proposed method is applied in Fig. 7. The model consists of an encoder that compresses frames, a decoder that restores compressed frames to their original data size, and a transformer that predicts the probability distribution of vector values as latent variables from past frames.

The parameters of the encoder and decoder were set based on an image compression model [5] using vector quantization. The structure of the encoder and decoder is shown in Table 1. The encoder uses four convolutional layers to reduce the number of dimensions by a factor of 16. The output of the encoder is divided into non-overlapping blocks of $r \times r \times s$. Each block is replaced by the pattern of codeword vectors closest to the source word (hereinafter, the centroid) by the quantizer. The decoder reconstructs the image based on the quantized latent variable vector values.

The transformer used in this paper is shown in Fig. 8. Its input includes the tokens of the two previous frames and the current frame. The transformer models the dependencies between frames based on the VCT [8] and predicts the coding vector to which the latent vector belongs. For the two immediately preceding frames, the latent space representation of size $w \times h \times m$ is used to divide the latent space into tokens of size $p \times p \times m$ ($p < w$ and $p < h$), allowing for overlap in space. The current frame is also partitioned into tokens of size $c \times c \times m$ ($c < p$) that use the same latent space representation but do not allow space overlap. In this paper, $c = 4$, $p = 8$, and $m = 192$, as used in VCT.

The minimum average code length at the encoder and decoder is set by the Shannon cross entropy given by Eq. (1):

$$R = \mathbb{E}_{\hat{z}}[-\log_2 P_\theta(\hat{z})], \tag{1}$$

where $\hat{z}$ is the quantized vector and $P_\theta$ is the distribution of coding vectors to which $\hat{z}$ is predicted to belong by the transform.

### 4.2 Video Compression with Fine-Tuning

Many image compression models are trained only once on a large dataset. Therefore, it is difficult to apply such a model to each input image, thus preventing efforts to improve the performance of image compression. On the other hand, methods that optimize codec parameters for each input image [5, 10, 11] perform well in image compression.

Video compression differs from image compression in that fine-tuning is performed on all of the frames that make up the video, which increases the processing time. To solve this problem, we take advantage of the fact that adjacent frames in a video are often similar in structure. The weights optimized for the previous frame are also valid for the next frame. Therefore, the proposed method reduces the number of fine-tuning operations per frame and inherits the weights of fine-tuned frames so that fine-tuning can be performed using these weights in the next frame.

The objective function is shown in Eq. (2):

$$\mathcal{L} = MSE(x, D(Q(E(x)))), \tag{2}$$

where the encoder is $E$, the decoder is $D$, and the quantization function is $Q$. In this case, the parameters of the decoder $D$ and the codebook of the quantizer $Q$ are fixed, and only the encoder $E$ is updated.

## 5 Evaluation

### 5.1 Dataset

We used the Vimeo-90k dataset [17] to train a video compression model by the proposed method. This dataset was constructed to evaluate optical flow prediction and video super-resolution tasks. The dataset consists of 89,800 pairs of three consecutive video frames, with different content for each frame. In this paper, we also used a method of rescaling images [8] at a random magnification to remove distortions caused by video compression.

We used MCL-JCV [18] and UVG [19] as the evaluation datasets. MCL-JCV consists of 30 movies at 1080p captured at 25 or 30 frames per second. UVG consists of 16 movies at 1080p or 2160p and 50 or 120 fps. In addition, the videos in the UVG were downscaled from 2160p to 1080p using a bilinear method due to the memory size of the computer used in this paper.

### 5.2 Comparative Methods

In this paper, we compare the performances of two different methods. The first is Vector Quantization VCT (VQVCT), which is based on VCT [8] and introduces vector quantization. The second is Fine-Tuning VCT (FT-VQVCT), which fine-tunes the encoder to the input frame.

### 5.3 Evaluation Methods

We used PSNR and MS-SSIM [20] in the RGB domain as evaluation metrics. PSNR is the signal-to-noise ratio of the original image to the compressed image. A high PSNR value indicates that the original and restored images are similar along with high quality. MS-SSIM is a measure that considers structural information, contrast, brightness, and

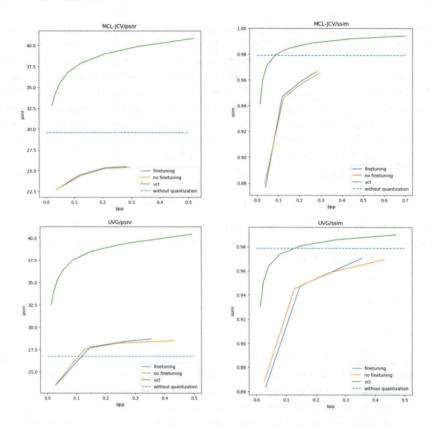

**Fig. 9.** PSNR and MS-SSIM as a function of bitrate change

hue in an image. Compared to PSNR, MS-SSIM often shows results based on human visual perception. The value of MS-SSIM ranges from 0 to 1, with a value closer to 1 indicating a higher-quality image.

### 5.4   Image Similarity with Bitrate

The three rate distortion curves for FT-VQVCT, VQVCT, and VCT using MCL-JCV and UVG are shown in Fig. 9. The image similarity, when trained as a pure autoencoder without quantization, is shown by the dotted line. The horizontal axis is the bitrate, and the vertical axis is the PSNR or MS-SSIM.

Figure 9 shows that PSNR and MS-SSIM are higher with FT-VQVCT than with VQVCT but lower than with VCT. PSNR and MS-SSIM with a pure auto-encoder are lower than with VCT at higher bit rates.

### 5.5   Image Similarity with Elapsed Time

We evaluated the image similarity of the fine-tuning method with respect to the weights that were fine-tuned in the previous frame by inheriting these weights, and we evaluated

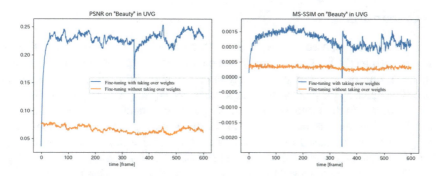

**Fig. 10.** Image similarity based on elapsed time

that of the fine-tuning method without inheriting these weights. Figure 10 shows the results of measuring the difference in image similarity between these two methods and the method without fine-tuning for all frames at a given time. The vertical axis is the PSNR or MS-SSIM, and the horizontal axis is the elapsed time. We used "Beauty"[21] from the UVG dataset as the video for evaluation. The learning rate was set to $5e-4$.

From Fig. 10, the number of updates per frame was 4 for the fine-tuning method with inherited weights and 10 for the fine-tuning method without inherited weights. The image similarity of the former method was lower than that of the latter method in the first part of the video but higher in the following time frames.

## 5.6 Discussion

First, we evaluated image similarity as a function of bit rate. The image similarity of FT-VQVCT with the proposed method was higher than that of VQVCT without fine-tuning, regardless of the change in bit rate. Therefore, we confirmed that the method of fine-tuning models to input images is effective for both video compression and image compression by vector quantization.

Next, we evaluated the image similarity as a function of elapsed time. The image similarity obtained by continuing the fine-tuning with the weights inherited from the previous frame was higher than that obtained by fine-tuning without these weights, since the number of updates per frame was reduced. Therefore, we confirmed that the use of frame-to-frame similarity can suppress the increase in processing time that occurs when fine-tuning is performed for all frames in video compression.

On the other hand, the PSNR and MS-SSIM in the rate distortion curve of FT-VQVCT were significantly lower than those of VCT. PSNR and MS-SSIM were also lower when trained as a pure autoencoder compared to VCT. Therefore, we confirmed that the performance of the proposed method can be further improved by optimizing the structure of the autoencoder.

## 6   Conclusion

In this paper, we proposed a compression method for video using vector quantization. The proposed method is based on an image compression model using vector quantization, and it improves the compression effect in coding by fine-tuning the model to the frames that make up the input video. To reduce the processing time required to fine-tune all of the frames in the video, the number of updates per frame is reduced by inheriting the weights fine-tuned in the previous frame. By evaluating the rate-distortion curves, we compared the proposed method, a method without fine-tuning, and conventional VCT. The results show that the PSNR and MS-SSIM of the rate-distortion curves for FT-VQVCT using the proposed method were higher than those using the method without fine-tuning. In the evaluation of image similarity by elapsed time, we compared the method that inherits the weights from the fine-tuning in the previous frame with the method that does not inherit these weights. We confirmed that when the number of updates is small, the image similarity obtained by the method that inherits the weights is higher than that obtained by the method that does not inherit the weights.

In the future, we plan to optimize the autoencoder structure and improve the codebook based on the input frames.

**Acknowledgement.** This work was supported by JSPS KAKENHI Grant Numbers JP21H03429 and JP22H03587, the JGC-S Scholarship Foundation, and the JSPS Bilateral Joint Research Project (JPJSBP120229932).

## References

1. Cisco: Cicso Annual Internet Report (2018-2023) White Paper - Cisco (online) (2022). https://www.cisco.com/c/en/us/solutions/collateral/executive-perspectives/annual-internet-report/white-paper-c11-741490.html
2. Agustsson, E.: Soft-to-hard vector quantization for end-to-end learning compressible representations. In: Proceedings of 31st Conference on Neural Information Processing Systems (NIPS 2017), pp. 1–11 (online) (2017). https://api.semanticscholar.org/CorpusID:850237
3. Ballé, J., Laparra, V., Simoncelli, E.P.: End-to-end optimized image compression. In: Proceedings of 5th International Conference on Learning Representations, (ICLR 2017), pp.1–27 (online) (2016). http://arxiv.org/abs/1611.01704
4. Ballé, J., Minnen, D., Singh, S., Hwang, S.J., Johnston, N.: Variational image compression with a scale hyperprior. In: Proceedings of 6th International Conference on Learning Representations (ICLR 2018), pp. 1–23 (online) (2018). https://openreview.net/forum?id=rkcQFMZRb
5. Lu, X., Wang, H., Dong, W., Wu, F., Zheng, Z., Shi, G.: Learning a deep vector quantization network for image compression. IEEE Access **7**, 118815–118825 (online) (2019). https://doi.org/10.1109/ACCESS.2019.2934731
6. Lu, G., Ouyang, W., Xu, D., Zhang, X., Cai, C., Gao, Z.: DVC: an end-to-end deep video compression framework. In: Proceedings of 2019 IEEE/CVF Conference on Computer Vision and Pattern Recognition (CVPR), pp. 10998–11007 (online) (2019). https://doi.org/10.1109/CVPR.2019.01126
7. Agustsson, E., Minnen, D., Johnston, N., Ballé, J., Hwang, S.J., Toderici, G.: Scale-space flow for end-to-end optimized video compression. In: Proceedings of 2020 IEEE/CVF Conference on Computer Vision and Pattern Recognition (CVPR), pp. 8500–8509 (online) (2020). https://doi.org/10.1109/CVPR42600.2020.00853

8. Mentzer, F., et al.: VCT: a video compression transformer. In: Proceedings of 36th Conference on Neural Information Processing Systems (NeurIPS 2022), pp.1–13 (online) (2022). https://openreview.net/pdf?id=lme1MKnSMb

9. Ballé, J., Laparra, V., Simoncelli, E.P.: Density modeling of images using a generalized normalization transformation. In: Proceedings of International Conference on Learning Representations (ICLR 2016), pp.1–14 (online) (2016). https://arxiv.org/abs/1511.06281

10. Dong, W., Shi, G., and Xu, J.: Adaptive nonseparable interpolation for image compression with directional wavelet transform. IEEE Signal Process. Lett. 15, 233–236 (online) (2008). https://doi.org/10.1109/LSP.2007.914929

11. Sullivan, G.J., Ohm, J.-R., Han, W.-J., Wiegand, T.: Overview of the high efficiency video coding (HEVC) standard. IEEE Trans. Circ. Syst. Video Technol. **22**(12), 1649–1668 (online) (2012). https://doi.org/10.1109/TCSVT.2012.2221191

12. Cover, T.M., Thomas, J.A.: Elements of Information Theory. Wiley Series in Telecommunications and Signal Processing, Wiley-Interscience, USA (2006)

13. Bengio, Y., Léonard, N., and Courville, A.C.: Estimating or propagating gradients through stochastic neurons for conditional computation. Comput. Res. Repository (CoRR), 1–14 (2016). http://arxiv.org/abs/1305.2982

14. Vali, M.H., Böckström, T.: NSVQ: noise substitution in vector quantization for machine learning. IEEE Access **10**, 13598–13610 (online) (2022). https://doi.org/10.1109/ACCESS. 2022.3147670

15. Sculley, D.: Web scale K-Means clustering. In: Proceedings of the 19th International Conference on World Wide Web (WWW), pp. 1177–1178 (2010)

16. Chen, T., Wang, J., Yonglei, Z.: Combined Digital Signature and Digital Watermark Scheme for Image Authentication. In: Proceedings of the 2001 International Conferences on Info-Tech and Info-Net, vol. 5, pp. 78–82 (online) (2001). https://doi.org/10.1109/ICII.2001. 983498

17. Xue, T., Chen, B., Wu, J., Wei, D., Freeman, W.T.: Video enhancement with task-oriented flow. Inter. J. Comput. Vis. **127**(8), 1106–1125 (online) (2019). https://doi.org/10.1007/ s11263-018-01144-2

18. Wang, H., et al.: MCL-JCV: A JND-based H.264/AVC video quality assessment dataset, In: Proceedings of the 2016 IEEE International Conference on Image Processing (ICIP), pp. 1509–1513 (online) (2016). https://doi.org/10.1109/ICIP.2016.7532610

19. Mercat, A., Viitanen, M., and Vanne, J.: UVG Dataset: 50/120fps 4K Sequences for Video Codec Analysis and Development, pp. 297–302 (online) (2020). https://doi.org/10.1145/ 3339825.3394937

20. Wang, Z., Simoncelli, E., Bovik, A.: multiscale structural similarity for image quality assessment. In: Proceedings of the 37th Asilomar Conference on Signals, Systems & Computers, vol. 2, pp. 1398–1402 (online) (2003). https://doi.org/10.1109/ACSSC.2003.1292216

21. Mercat, A., Viitanen, M., Vanne, J.: UVG Dataset: 50/120fps 4K sequences for video codec analysis and development. In: Proceedings of the 11th ACM Multimedia Systems Conference (MMSys 2020), pp. 297–302 (online) (2020). https://doi.org/10.1145/3339825. 3394937

# Author Index

L. Barolli (Ed.): AINA 2024, LNDECT 199, pp. 451–452, 2024.
https://doi.org/10.1007/978-3-031-57840-3

Printed in the United States
by Baker & Taylor Publisher Services